Information Systems and Security

Edited by Audrey Coon

CLANRYE
INTERNATIONAL
www.clanryeinternational.com

Clanrye International,
750 Third Avenue, 9th Floor,
New York, NY 10017, USA

ISBN: 978-1-63240-592-0

Cataloging-in-Publication Data

Information systems and security / edited by Audrey Coon.
   p. cm.
Includes bibliographical references and index.
ISBN 978-1-63240-592-0
1. Information storage and retrieval systems. 2. Information technology. 3. Information resources. 4. Computer systems.
5. Computer networks. 6. Wireless communication systems. 7. Electronic information resources. I. Coon, Audrey.
Z699 .I54 2017
025.06--dc23

For information on all Clanrye International publications
visit our website at www.clanryeinternational.com

Printed in the United States of America.

Contents

Preface..VII

Chapter 1   **Preface to special issue on miscellaneous emerging security problems**..................................1
              Kai Chen, Jun Shao

Chapter 2   **A framework for usable end-user privacy control in social software systems**............................5
              Maryam Najafian Razavi, Denis Gillet

Chapter 3   **FraTAct for Transforming a Nescient Process Activity into an Intelligent Process Activity**..24
              Rafiqul Haque, Nenad B. Krdzavac

Chapter 4   **Effective user selection algorithm for quantized precoding in massive MIMO**......................35
              Nayan fang, Jie Zeng, Xin Su, Yujun Kuang

Chapter 5   **Hierarchical Codebook Design for Massive MIMO**...41
              Xin Su, Shichao Yu, Jie Zeng, Yujun Kuang

Chapter 6   **Salus: Kernel Support for Secure Process Compartments**..46
              Raoul Strackx, Pieter Agten, Niels Avonds, Frank Piessens

Chapter 7   **Trust in social computing The case of peer-to-peer file sharing networks**............................62
              Heng Xu, Tamara Dinev, Han Li

Chapter 8   **Single and Multiple UAV Cyber-Attack Simulation and Performance Evaluation**.............75
              Ahmad Y. Javaid, Weiqing Sun, Mansoor Alam

Chapter 9   **Advancements of Outlier Detection**...86
              Ji Zhang

Chapter 10   **Non-stationary Parallel Multisplitting Two-Stage Iterative Methods with Self-Adaptive Weighting Schemes**..112
              Guo Yan Meng, ChuanLong Wang, XiHong Yan, QingShan Zhao

Chapter 11   **Towards Privacy-Preserving Web Metering Via User-Centric Hardware**........................120
              Fahad Alarfi, M aribel Fernández

Chapter 12   **Privacy Preserving Large-Scale Rating Data Publishing**...136
              Xiaoxun Sun, Lili Sun

Chapter 13   **Concurrent Operations of O_2-Tree on Shared Memory Multicore Architectures**.................148
              Daniel Ohene-Kwofie, E. J. Otoo, Gideon Nimako

Chapter 14   **End-to-End Key Exchange through Disjoint Paths in P2P Networks**.............................166
              Daouda Ahmat, Damien Magoni, Tegawendé F. Bissyandé

Chapter 15   **How did you know that about me? Protecting users against unwanted inferences**............................181
Sara Motahari, Julia Mayer, Quentin Jones

Chapter 16   **Exploring Relay Cooperation for Secure and Reliable Transmission in Two-Hop
Wireless Networks**..194
Yulong Shen, Yuanyu Zhang

Chapter 17   **Personalised Information Gathering and Recommender Systems: Techniques
and Trends**...213
Xiaohui Taoy, Xujuan Zhouz, Cher Han Lauz, Yuefeng Li

Permissions

List of Contributors

Index

Preface

Information systems are virtual systems that enable the organization, storage and creation of information which can be accessed and manipulated in an easy and effortless manner. This book on information systems discusses topics related to the processes and protocols involved in information storage and retrieval. Communication is the major field in which information systems are involved as these require rapid processing of large quantities of data as well as its storage, retrieval and manipulation. This book traces the progress of this field and highlights some of its key concepts and applications. A number of latest researches have been included to keep the readers up-to-date with the global concepts in this area of study. It aims to serve as a resource guide for students and experts alike and contribute to the growth of the discipline.

This book is the end result of constructive efforts and intensive research done by experts in this field. The aim of this book is to enlighten the readers with recent information in this area of research. The information provided in this profound book would serve as a valuable reference to students and researchers in this field.

At the end, I would like to thank all the authors for devoting their precious time and providing their valuable contributions to this book. I would also like to express my gratitude to my fellow colleagues who encouraged me throughout the process.

Editor

Preface to special issue on miscellaneous emerging security problems

Kai Chen[1,*], Jun Shao[2]

[1]Institute of Information Engineering, Chinese Academy of Sciences
[2]Zhejiang Gongshang University

Abstract

An introduction to the key topics and challenges in Botnet, buffer overflow vulnerability, privacy-preserving and others.

Keywords: security and safety, botnet, buffer overflow, privacy-preserving

1. Introduction

Security and safety become more and more important in cyber network. They impact not only the national critical departments (e.g., a state council) but also individuals' daily life. In the area of security and safety, botnet, buffer overflow vulnerability, privacy-preserving, and authentication are four hot topics impacting both those departments and individuals. With the billions users of smartphones, the threats become even severe.

In these areas, the threats from botnets are reaching an alarming level. A botnet is a collection of programs that communicate through Internet and perform specific tasks such as sending spam emails and performing DDoS attacks. The tasks are given by commands from an attacker via C&C (Command and Control) channels on Internet, which makes the programs as bot instances. The hosts containing the programs could be personal computers, workstations in enterprise networks or even sensitive servers in governments. In addition to those victims (e.g., websites in a DDoS attack performed by a botnet) directly attacked by bots, these hosts also become victims in an attack. Nowadays, with the rapid growth of smartphones, more and more botnets use smartphones as bots. Considering that there are billions of smartphones, botnets have much more targets than those in PC era.

A botnet is usually very good at concealing itself. The bots in hosts may not do anything harmful directly to the hosts. Sometimes, the only goal for these bots may be waiting for a command from a remote server to connect to some websites (e.g., for DDoS attacks). This connection can also be normally performed by an ordinary user, which is therefore difficult to detect. The bots could also reside in the host for a very long time waiting for a command. In this waiting period, nothing happens. In other words, no malicious behaviors can let users or anti-virus detectors aware the existence of the bots. Also, the waiting time can be too long to get noticed. Anti-virus software may also try to detect bots using their signatures. However, bots can also be highly customized (e.g., through obfuscation) to evade the detection. Moreover, before the awareness by those anti-virus software, enough time is given to the bots for their updates. Thus, to detect the malicious bots is generally very difficult from host sides.

Another direction for botnet detection is to detect botnet traffic. A basic idea is to use misuse detection. For example, some approaches generate signatures of known malicious traffic (e.g., using malicious hosts). But they cannot detect new C&C traffics. Some approaches classify traffic flows by protocols, length sequences of packages, and the encoding of URLs. However, bots' traffic can be disguised as normal ones especially when popular network protocols such as HTTP are used. Noises can also be injected to

---

*Corresponding author. Email: chenkai@iie.ac.cn

evade detection. Some approaches assume that C&C traffic connects to the same location within varying time windows. So they detect botnet through finding repeated combinations of traffic destinations. These approaches work when botnets like to use centralized architectures. Using these architectures, all the bots contact with one (or a few) C&C server(s) which distributes commands. This is easy for a botnet to implement, and is also easy to be detected. When the server is compromised, the botnet is destroyed.

However, botnets are becoming more resilient to detectors nowadays. They can structure their C&C channels in different ways. Recently, attackers start to build botnets on a pear-to-pear (P2P) architecture. Any node in the botnet could be a server to send commands, which makes them even more difficult to detect and take down. Anomaly detection can address the problem by observing deviations from normal traffic. However, this kind of detection usually needs a learning period. The malicious C&C traffics in this time period cannot be identified. Moreover, the "normal" traffic can be contaminated which makes future C&C traffics not easy to detect.

To solve this problem, in this special issue, Burghouwt, Spruit and Sips present a network-based anomaly detection approach. This approach estimates the trustworthiness of the traffic destinations. The flows from human input, prior traffic from a trusted destination, or a defined set of legitimate applications are considered as normal traffic. In this way, the approach can detect zero day malicious traffic destinations in real time. Even if the traffic is encrypted or uses normal popular protocols such as HTTP, the approach can still detect it without being noticed.

Vulnerabilities in software are highly related to botnet. To stealthily plant a bot into a system, lots of attackers will choose to exploit a vulnerability. In this process, nothing special happens when a bot is planted. Among different kinds of vulnerabilities, buffer overflow is one of the most dangerous. By overrunning the buffer's boundary and overwriting adjacent memory, attackers can leverage a carefully constructed input to overwrite important memory units (e.g., return address to a function). In this way, attackers can let a program run code in arbitrary memory addresses given by input data (constructed by attackers) to include arbitrary commands (e.g., downloading and executing the bot). When the bot spreads in the network, it can also exploit the vulnerabilities to infect more systems. In this process, those target systems look normal, which makes the detection quite difficult.

Buffer overflow vulnerabilities not only allow attackers to implant bots, but also give attackers opportunities to execute arbitrary commands. When the software with the vulnerabilities has root privileges, it becomes even more dangerous. All the data in the system may be

exposed to attackers. With the rapid growth of smartphones, a vulnerability in system will impact billions of users. Even for a single popular app with such vulnerabilities, it may impact millions of users.

To deal with such attacks, researchers can either identify and fix all buffer overflow vulnerabilities in software or do a defense against such attacks. Identifying all buffer overflow vulnerabilities (especially in commercial software without source code) is extremely difficult. In most cases, the vulnerabilities can only be triggered by specific inputs. To find such inputs, a tester may use white-box testing to identify those vulnerabilities. In this process, he can check whether each path contains an instruction which overflows a buffer. However, for an instruction in different execution paths, the instruction may or may not trigger a vulnerability. To traverse all the execution paths in software is very time-consuming and almost impossible. He can also leverage black-box testing, which feeds programs with different inputs and observes the abnormal running states of programs (e.g., crashes). The black-box testing avoids checking the running states of the program, which seems more efficient than white-box testing. However, whether a vulnerability can be triggered highly depends on the inputs. It is impossible to enumerate all the inputs for the software. Current approaches combine white-box testing with black-box testing for efficient detection. But they still cannot find all vulnerabilities. Once there is one vulnerability missing, it could be exploited by attackers.

Another direction is to do a defense against the attacks. Some approaches dynamically check operations on buffers to see whether the operations exceed the boundaries of those buffers. But the checking process will impose high overhead against normal execution. A random value, or canary, is added to the memory unit right adjacent to important memory units in stack such as return address or saved frame pointer. Based on the design, once the buffer is overflowed to overwrite the important memory units, the canary will also be overflowed, which will warn the program (usually through exceptions) that an attack is happening.

Data Execution Prevention (DEP) marks important memory regions (e.g., stack) as unexecutable. Once a program tries to execute instruction inside these regions (usually caused by injecting code through buffer overflow), an exception (e.g., on hardware level) will be triggered. However, attacks based on Return-Oriented Programming link and execute a sequence of small code snippets (called gadgets) which are not in stack. In other words, attackers can let the return address point to an instruction in code area (e.g., the program itself or dynamic linked libraries). The instruction is chosen by attackers in advance, which is followed by a return instruction. After the instruction is executed, the return instruction will divert the program to another instruction according to the return address in stack which is given by attackers (still through injection). In this way, attackers can inject the program with a sequence of addresses to let the program run any code without the detection by DEP.

Address Space Layout Randomization (ASLR) is another approach to prevent shellcode from being successful. It randomizes the location of executables and libraries in memory or even in-memory structures. When attackers try to chain the gadgets that they find before attacks, the gadgets may not be in the original places in memory as those in attackers' computers. In this way, attacks can be prevented. However, ASLR must maintain page alignment (4KB on x86) within limited memory address space (typically less than 2GB on 32-bit x86), which also limits the spaces for target code. The correct address could be guessed by attackers depending on how often they can try. Combining two or more defenses (e.g., DEP and ASLR) will make the attacks even harder. But new attacks such as Just-in Time (JIT) Spraying can still bypass them.

To make a better defense, in this special issue, Krugel and Müller propose a compiler-level protection. This protection separates a stack to two stacks (i.e., control and data stacks). It protects sensitive data (e.g., return addresses and saved frame pointers) on a separated stack (i.e., control stack). In this way, when a buffer is overflowed, only the data in data stacks will be overwritten. The sensitive data will still keep the original values (in data stack) in this overflow. The authors implemented this idea on LLVM compiler infrastructure and made detailed evaluations on it. While protecting the sensitive data, this approach imposes little overhead on the running performance.

Privacy is always a major issue nowadays. It does not mean that the information related to individuals cannot be revealed to others, but it does mean that people has the right (ability) to decide who could collect, store, or use the information. However, many of existing web applications are against users' privacy, i.e., to complete their functionalities', they have to record user's identity information, action histories, or behavior patterns. One representative is web metering.

The web metering is a web application where interested enquirers can obtain the evidence of the number of visits done by users to the website. This evidence is the crucial factor to decide the price of advertisements on the websites. As we know, most websites nowadays earn their money not from the users who visit them but from the advertisements on the websites. The data used as the evidence is required to be non-repudiated, and can be linked to user's identity information or other actions.

The existing web metering schemes can be classified into the following three categories: user centric, web server centric and third party centric. In the first category, several cryptographic primitives, including digital signature, hash chain, and secret sharing, are applied. Although the applied cryptographic primitives make the evidence non-repudiated, the user's identity will be revealed to verify the validity of the evidence.

In the second category, the web server makes use of several techniques to obtain the non-repudiated evidence, including e-coupon, solving computational complexity problems, and audited hardware box. However, the resultant schemes cannot guarantee that the obtained data are always non-repudiated if the web server is not always honest, let alone the user's privacy.

In the third category, the data from the user side always go through the third party; hence, the user's privacy is not protected from the third party.

It seems that designing a privacy-preserving web metering scheme is a quite difficult task. Inspired by the hardware-based security systems, a new hardware-based web metering scheme is proposed in this special issue. There are three entities in the proposed web metering scheme: user, web server, and audit agency. To complete the web metering, the web server should obtain the non-repudiated data from a new key of the user. The certificate of the new key is generated by the audit agency after validating the hardware on the user side. To protect the user's privacy, the certificate is based on some anonymous authentication methods. More details of the new web metering scheme can be found in the third paper named "Towards Privacy-Preserving Web Metering Via User-Centric Hardware" by Alarifi and Fernĺcndez. In the paper, they also analyze and compare the existing web metering schemes and show the gained privacy benefits of the proposed scheme. Furthermore, the proposed scheme supports different security countermeasures and users' privacy settings, such as security w/o privacy.

One of the basic techniques used in the paper by Alarifi and Fernĺcndez is the anonymous authentication. Till now, there existing many cryptographic primitives providing anonymous authentication, including blind signature, group signature, ring signature, direct anonymous attestation, anonymous credential, and so on.

In a blind signature scheme, the signer will generate a signature without knowing the corresponding message, while the signature requester can obtain the pair (signature, message). With this pair, the requester can be authenticated by others in an anonymous way. In the real execution, the signer should make sure that the one knowing the corresponding signed message is an authenticated user.

Group signature can provide the limited anonymity on the authentication. In particular, every group member can sign messages on behalf of the group, and nobody outside of the group can reveal the identity of the signer, while the group manager can trace back to the signer. In other words, the signature is anonymous except the group manager.

Ring signature is quite similar to group signature, but without the group manager. Everyone can build up a group without agreements from others, and can also sign message on behalf of the built group. Unlike group signature, the resultant signature in the ring signature scheme is fully anonymous.

Direct anonymous attestation has been adopted by the Trusted Computing Group to protect the privacy of the Trusted Platform Module (TPM) platform. The key of direct anonymous attestation is to use a zero-knowledge proof to show the validity of the credential provided by the TPM platform without violating the TPM platform's privacy.

Anonymous credential is an anonymous system where the user can obtain credentials from an organization, and then at some later pint, s/he can be authenticated by others using the credentials without revealing her/his identity.

The careful reader may find that all the above cryptographic techniques for anonymous authentication can be used in the privacy-preserving web metering, or other applications where privacy and non-repudiation are both desired.

References

[1] KOPKA H. and DALY P.W. (2003) *A Guide to LaTeX* (Addison-Wesley), 4th ed.

[2] LAMPORT L. (1994) *LaTeX: a Document Preparation System* (Addison-Wesley), 2nd ed.

[3] MITTELBACH F. and GOOSSENS M (2004) *The LaTeX Companion* (Addison-Wesley), 2nd ed.

A framework for usable end-user privacy control in social software systems

Maryam Najafian Razavi*, Denis Gillet

Ecole Polytechnique Fédérale de Lausanne (EPFL), 1015 Lausanne, Switzerland

Abstract

Recent studies have shown that many users struggle to properly manage selective sharing of the diverse information artefacts they deposit in social software tools. Most tools define privacy based on the 'network of friends' model, in which all 'friends' are created equal and all relationships are reciprocal. This model fails to support the privacy expectations that non-technical users bring from their real-life experiences, such as enabling different degrees of intimacy within one's network and providing flexible, natural means of managing the volatile social relationships that social software systems confront. Furthermore, the model suffers from lack of empirical grounding and systematic evaluation. This paper presents a framework for building privacy management mechanisms for social software systems that is intuitive and easy to use for the average, non-technical user population of these systems. The framework is based on a grounded theory study of users' information sharing behaviour in a social software tool. Results inform the design of OpnTag, a social software prototype that facilitates personal and social information management and sharing. Preliminary empirical data suggest that our proposed privacy framework is flexible enough to meet users' varying information sharing needs in different contexts while maintaining adequate support for usability.

Keywords: grounded theory, people tagging, privacy, social software, usability, Web 2.0

1. Introduction

Social software systems are a family of Web 2.0 applications, characterised by their primarily user-driven content and the ability to mediate personal and social information across collectivities such as teams, communities, and organisations. The advent of the 'Social Web' has made users producers as well as consumers of information, resulting in publishing and distributing huge amounts of user-created data. Examples of social software systems include social authoring tools (e.g. wikis), social bookmarking tools (e.g. del.icio.us), and social networking tools (e.g. LinkedIn and Facebook). Users are widely adopting these tools for personal and social information management because they provide significant enhancements in utility and cost over similar desktop tools, in the sense that not only they allow their users to create personal information spaces that are easily accessible from anywhere on the Web, but also give them the tools to share their various information artefacts with others and take advantage of others' shared artefacts. These two advantages are in many cases so strong that users are either explicitly willing to give up control of that information or do so without any real awareness of the degree to which they are doing so. Recent research in the area of knowledge management (KM), however, has recognised the need to improve people's ability to control who sees what from the information they deposit in their online personal spaces (Erickson, 2006). Nevertheless, the topic of personal privacy—how people manage privacy of their own information with respect to other individuals (as opposed to organisational privacy (Iachello and Hong, 2007)—remains largely unexplored in the research literature.

Current state of the art with privacy management in social software is that most tools either define access control as a private/public dichotomy, not accounting for the various other shades of privacy in between (e.g. del.icio.us), or they need users to self-administer fine-grained privacy control on their data through a privacy setting

*Corresponding author. Email: maryam.najafian-razavi@epfl.ch

page (e.g. Facebook), which is complicated and cumbersome for the average, non-technical users. These two approaches pretty much present the two ends of the spectrum for privacy management in social software systems, with most tools falling somewhere in between. While mechanisms at one end of the spectrum do not provide sufficient control over privacy of one's information, mechanisms at the other end are not usable enough for the average, non-technical user population of social software systems.

The purpose of this work is to present a usable framework for providing end-users with better control over selective sharing of information they deposit in a social software system. The objectives of the work fall into three general areas:

(i) Developing a better understanding of users' perspectives on personal privacy in social software domain, in terms of the extent of the problem, specific privacy needs and concerns, and strategies that users employ in order to achieve their desired levels of personal privacy.

(ii) Identifying factors that impact users' information sharing behaviour in this domain, in order to build a conceptual model of personal privacy that matches users' mental models.

(iii) Devising guidelines for building privacy management mechanisms in this domain that satisfy users' varying privacy needs, and yet, are usable for the average, non-technical user population of social software systems.

In order to meet these objectives, we first employed a grounded theory study to develop an understanding of the information sharing process in the context of social software systems. Based on the results of the study, we propose a set of design heuristics for privacy management in social software, and consolidate those heuristics into a framework for privacy in this domain. Our proposed framework supports per-artefact privacy management as opposed to per-category privacy management supported by most other tools, and enables definition of social contacts of non-equal weights through creation of egocentric groups.

In order to create a test bed where the suitability of the proposed principles can be tested, we next introduce OpnTag, a social tool we developed whose privacy management mechanism instantiates the proposed framework, and present the results of an empirical evaluation that provide initial validation that the framework is flexible enough to meet users' varying privacy needs and indicate areas for future enhancements.

2. Related work

In recent years, use of social software has moved from niche phenomenon to mass adoption (Gross *et al.*,

2005; Millen *et al.*, 2006). This increase in use has been accompanied by diversity of purposes and access patterns. As a result, researchers have studied several issues that pertain to these tools, including people's attitudes towards disclosing personal data. Gross *et al.* (2005) report on a study of patterns of information revelation in online social networks and their privacy implications. Their results are based on actual field data from more than 4000 users of Facebook. They report that patterns of information revelation depend on a number of factors, including pretence of identifiability, type of information revealed or elicited, and the degree of information visibility. Along the same line, Darrah *et al.* (2001) observes that people tend to devise strategies to restrict their own accessibility to others while simultaneously seeking to maximise their ability to reach people, and Westin (2003) argues that privacy management is the continuous act of balancing the desire for privacy with the desire for communication and disclosure.

Researchers have also studied users' attitude towards revealing information in several other contexts, including workplace and location-aware mobile services. Olson *et al.* (2005) took a quantitative approach in conducting an in-depth survey of people's willingness to share a range of everyday information (such as web sites they visit or their health status) with various others, including family members or co-workers. They pointed out that whether data are anonymised or can be tied directly to people play a major role in people's willingness to disclose. Other relevant factors reported include general attitude towards privacy based on Westin privacy indexes (privacy unconcerned, pragmatist, or fundamentalist) (Westin, 1991), and personal judgement regarding 'appropriateness' (i.e. relevance) of sharing certain information with certain groups.

In another work, Patil and Lai (2005) conducted a study on the privacy vs. awareness trade-off to identify the kinds of information that users of an awareness application are willing to share with various others (team-mates, family, friends, managers, etc.) for various purposes in the context of the workplace. They identified which clusters of awareness information are more likely to be shared with whom and in what context (i.e. 'team members' received comparable levels of awareness sharing with 'family' during work hours).

Whalen and Gates (2005) reported on a small-scale study of the type of personal information that users would be willing to disclose in open online environments, primarily focussing on uncontrolled spaces such as search engines. Their results, although limited in scope, point to the existence of consistencies in the way people treat certain classes of information, which suggests it might be possible to group related information into clusters that are treated similarly.

Recent work in KM has also recognised the need to improve people's ability to control who sees what in their personal information. Erickson (2006) explored the concept of personal information management in group context, by arguing that when personal information is to be shared with a group, the way it is used, and managed changes. In that article, he defined Group Information Management and identified many research questions that need to be explored, including how personal information is shared within a networked group, the norms of personal information sharing within groups, and the way those norms are negotiated in the group.

Palen and Dourish (2003) clarified the difference between the problem of personal information privacy and that of access control, by arguing that privacy is a continuous process of negotiating boundaries of disclosure, identity, and time, rather than a definitive entitlement. They observed that people in social software systems might act simultaneously in different spaces: as individuals, as members of a family, members of some occupational group, etc. In each of these affiliations, they may choose to disclose different information to different audiences. Palen and Dourish then exposed the unsuitability of existing access control models for privacy management since the conventional separation of one's network into 'roles' (as done by existing access control models) fails to capture the fluid nature of these various genres of disclosure in which one acts.

We therefore follow Palen and Dourish's lead and adopt the term 'privacy management' to mean the user-centred expression of personal (and organisational) constraints on information sharing. This is distinguished from 'access control', which is the means by which systems enable and enforce these choices. The main problem of privacy management in social software systems then is how to reconcile the co-presence of various groups that one identifies with, by providing users with flexible, non-overwhelming means to control what to share with whom when, which is the focus of this work.

3. A study of information sharing behaviour in social software

As a first step in this research, we performed a grounded theory study to understand end-users' information sharing behavior in social software systems and to identify specific privacy needs and concerns in this domain. This section describes the study and the theory that was derived from it.

3.1. Methodology

The research method adopted in the study was grounded theory (Glaser, 1978, 1998; Glaser and Strauss, 1967). It is a primarily inductive investigation process in which the researcher aims to formulate a small-scale, focussed theory that is derived from the continuous interplay between data analysis and data collection. Rather than starting with a preconceived theory that needs to be proven, the researcher begins with a general area of study and allows the theory to emerge from the data. Such theory has been claimed to have a better chance at resembling reality, compared to theory that is derived by putting together a series of concepts from solely speculation on how one 'thinks' things should work (Glaser, 1992).

3.2. Locating the study

Since a grounded theory method looks for emergence of theory from the data, grounded theory researchers are advised to choose samples in a way that maximises access to the phenomenon under study by selecting selecting most evident cases (Glaser, 1992, 1998). Informants chosen for study must be expert participants with rich and extensive prior experience with the phenomenon in order to be able to provide the researcher with a valid account of their experiences. For these reasons, we needed to adhere to three criteria in locating our study:

Finding the right tool. Firstly, we needed to choose a social software tool that provides some form of privacy management, preferably at an advanced level. After an extensive review of the existing tools, we chose Elgg (Tosh and Werdmuller, 2004), an open source social software system with integrated blog, wiki, social bookmarking, and social networking functionality. The two key features that motivated our choice of environment were its support for the creation of *ad hoc* groups and communities where privacy issues potentially arise, and its strong emphasis on its permission architecture, which had resulted in reasonable support for privacy control at a fairly granular level that other tools simply did not have.

Finding the right users. Secondly, we needed to find a situation where the tool was used extensively, preferably over a long period of time, so that users were properly familiar with it and were not novice users. While exploring various options to identify such user community among the general user population of Elgg, we came across a community of high school students enrolled in a special program for gifted kids called Trans, who were using Elgg for over a year as a requirement for their curriculum. This provided us with a user community for our selected tool who were using it on an ongoing basis for a reasonably long period of time.

Finding the right context of use. Finally, we needed to locate a context of use where both concepts of information sharing with various groups and privacy were paramount. Students in the Trans community were required to fill in their personal profile, write reflections in their weblogs on the topics covered in the classroom on a daily basis, and join and participate in a special community created for their group. For each of these artefacts (weblog posts, profile items, and personal reflections posted to the community blog), they had the option of regulating access (i.e. make it visible to only oneself, the instructor, a specific community, or everyone). Since active use of the environment was part of the their curriculum, these students had in fact a rich experience in using various features of the tools, which was an essential requirement for the emergence of the issue of privacy preferences and selective disclosure of information.

Confirmation of the suitability of the context of use was the final step in the process of locating this study. It must be noted, though, that even though the study is situated in the context of Elgg, constant effort has been made not to limit the discussions to the specifics of the application. Instead, we treated Elgg just as a focal point to ensure that the subjects had the experience with a system that allowed them to manage their privacy directly.

3.3. Data collection

Our initial set of participants included nine students from the Trans community. The participants' ages ranged from 15 to 17, and the gender balance was rather evenly split (five females and four males). All nine participants were quite confident with the tool and with the Web in general. We selected semi-structured, in-depth interviews as our data gathering strategy, suggested as one of the best fits with the grounded theory methodology (Glaser, 1998; Glaser and Strauss, 1967). Unlike structured interviews, semi-structured interviews have a flexible and dynamic style of questioning directed towards understanding the significance of experiences from the informants' perspectives. Our interview strategy involved asking open-ended questions to allow informants to discuss what would be important from their perspective. We then used both planned and unplanned probing to uncover details and specific descriptions of the informants' experiences. The interviews were structured around a list of topics based on the research questions that needed to be answered, including questions about sharing preferences with regard to the type of information, the person or group with whom the information was shared, and the purpose and incentive behind sharing or holding information. Although we started with the same set of questions with each informant, because of the open nature of semi-structured interviews, each interview took a different turn based on the specific ideas and experiences of the participant in question. When that happened, we followed the participant's lead, allowing for new and important issues to uncover. Each interview was between 30 and 40 min in length. All interviews were tape-recorded with the informants' permissions and later transcribed to provide accurate records for analysis. Standard procedures were followed to maintain the confidentiality of the interview data and the anonymity of the informants.

The analysis of the data gathered from our initial interviews with the nine participants resulted in identifying the basic social processes (BSPs), which are the core concepts around which the grounded theory is built. After identifying the BSPs, we used a procedure called theoretical sampling (Glaser, 1978) to develop new insights and refine the insights we had already gained. For this stage, we consciously selected three more participants from the same group (one female and two males) who had extensive experience with other social software applications in addition to Elgg (i.e. other ePortfolios, forums, and weblogs). We also redirected the interview questions in a way to reflect our new goal of verifying the emerging theoretical themes and their relationships. The experiences of these three participants particularly helped in identifying places where the current privacy mechanism was considered insufficient and users felt the need to switch to other platforms in order to achieve their goal.

After analysing the data from all 12 interviews, we realised we could identify interchangeable examples showing the same phenomenon in different instances, and there were not any new concepts and/or relationships being developed. This was an indicator of theoretical saturation (Glaser and Strauss, 1967), the point at which we ceased data collection.

3.4. Data analysis

Our grounded theory was formulated from data using a constant comparative method of analysis with three stages: the first stage of analysis, called open coding, involved breaking the interview transcripts down into discrete incidents (i.e. ideas, events, and actions) which were then closely examined and compared for significant concepts. These concepts were abstractions in the sense that they represented an aggregated account of many participants' story. We used the qualitative analysis software NVivo at this stage to label incidents in the data with code words and to write theoretical notes that captured momentary thoughts.

The second stage of analysis, called theoretical coding, involved taking the concepts that emerged during open coding and reassembling them with propositions about the relationships between those concepts. The relationships, like the concepts, emerged from the data through a process of constant comparison. Neither the concepts nor the relationships were preconceived or forced upon the data.

The third stage of analysis, called selective coding, involved delimiting coding to only those concepts and relationships that related to the core explanatory concept reflecting the main theme of the study. At the end of this stage, we were able to produce a more focussed theory with a smaller set of high-level concepts.

4. The grounded theory

The concept map in Figure 1 represents the theory that emerged from the data in this study. The concepts in

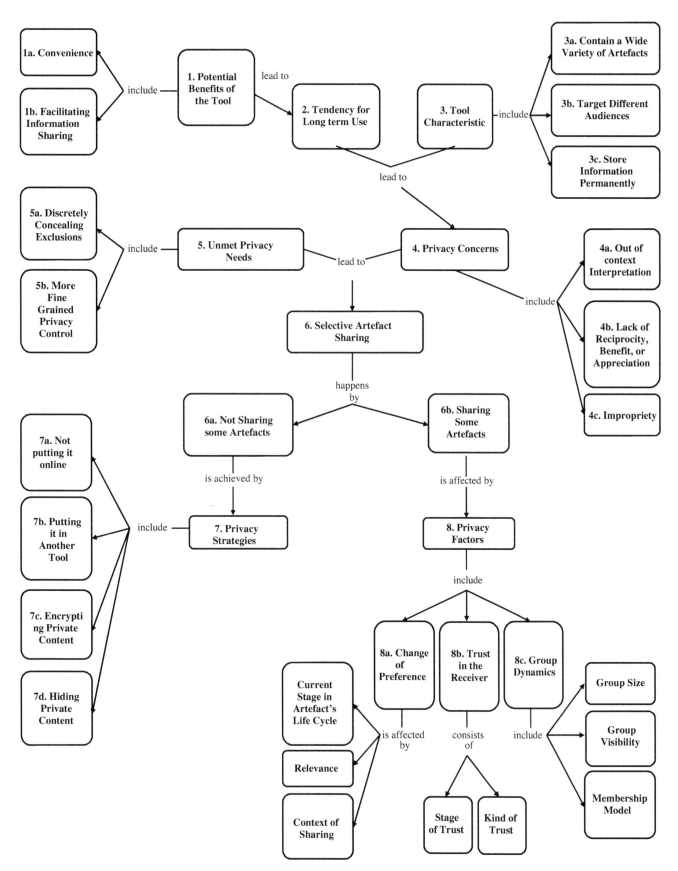

Figure 1. A concept map of information sharing behaviour in a social software system.

the map have been marked by numbers so that they can be referred to in the following description by their corresponding numbers (used in brackets), thus making navigation of the map easier. In summary, the theory suggested that because of the many potential benefits of the tool [1], such as convenience [1a] and ease of information sharing [1b], participants were willing to use it over a long period of time. However, there were certain inherent characteristics of the tool [3] that in combination with users' willingness for long-term use [2] gave rise to privacy concerns [4]. First, the artefacts contained in social tools cover a wide spectrum, ranging from personal to professional to social information [3a]. Second, various information artefacts may be targeted to different groups of audiences that are not necessarily static [3b], and finally, information disposed in social tools is persistent and permanently searchable [3c]. Some of the privacy concerns mentioned by our participants included out-of-context interpretation [4a]; lack of reciprocity, benefit, or appreciation [4b]; and impropriety (considering certain content inappropriate for some audiences) [4c]. We also observed that there were certain privacy needs of users that the tool failed to support [5], including the need to control privacy at a more fine-grained level [5b] and the need to discretely conceal exclusions [5a].

The combination of privacy needs and concerns gave rise to the need for selective information sharing [6], which happens by sharing certain artefacts with selected lead to individuals [6b] while holding back form sharing others [6a]. To withhold from sharing, users were employing certain strategies [7], including using other platforms with better privacy management mechanisms for their more private content [7b]; refraining from deploying certain content in an online environment because of lack of acceptable privacy levels [7a]; putting the more sensitive content somewhere (i.e. a web page or weblog), but not providing a link to it from places where their real identity was known [7d]; and finally, writing their more private content in some sort of a 'code language' so that it is meaningless to anyone other than the user him/herself [7c].

Deciding on sharing an artefact on the other hand was affected by a set of privacy factors [8]. One such factor was change of preference [8a]. The study showed that rather than a binary scale of public vs. private, users' judgements of privacy of resources often reflected a transition from private, to semi-private/restricted share, to public, depending on the state of the artefact, the relationship between the owner and the receiver of artefact, and the context of sharing. We also found out that users' assessment of the persons or groups who will be the receivers of information played a strong role in making decisions about information sharing [8b]: users tended to share less with people/groups with whom they were in the initial stages of trust, and as their trust moved towards a more mature level over time, they began to feel more comfortable and share more. Furthermore, their decision was affected by both the kind of trust they had in the receiver (e.g. cognitive or emotional) and the stage of their trust with the receiver (e.g. initial, intermediate, or mature).

And finally, our results indicated that users' willingness to share something they have vested interest in also depends on their perception of how it will be used, with the dynamics of the groups or communities where the information is going to be shared being the most influential factor in deciding about information sharing [8c]. Our study revealed that users often hold back from sharing information in anticipation of lack of reciprocity, benefit, or appreciation, and loss of credit for their work. The theory suggested that when group/community dynamics are clear enough to convey to the users how their information will be used within the group, users are better equipped to make informed decisions regarding how much they want to share within the group. Moreover, this predictability may be critical to making the decision to share information in the given context at all. Tables 1–8 provide examples of users' comments that led to various

Table 1. Sample user quotes on the benefits of Elgg and tendency for long-term use.

P#	Example comment
4	It is very useful to have everything in my Elgg, because then other people can see what work samples I have. Like say, if I am applying for scholarships, then I want them to have samples of my work without me having to send them things, so they can access the site whenever they want to. It's like I have an online resume.
1	There are two benefits to using Elgg; first, it has good integration potential, meaning, I can access my course materials all in one house. I can even present from where my data is. Second, it makes it easy to connect with classmates because they are all on Elgg, too. So, if I want to, say, organise an event or something, I know they will all see it.
3	For me, the main use of Elgg would be after Trans: it is my ePortfolio; it will be part of the application package I send out to schools. That's the reason I am using it in the first place. Besides, all my Trans friends are using Elgg now; so we can stay in touch if we continue using it after Trans.

Table 2. Sample user quotes on privacy concerns.

P#	Example comment
1	...I don't always share information on scholarships; because they are highly competitive, by sharing I would just put myself in a less advantaged position. I know other people don't share such stuff, either; so...
12	I usually prefer people not to know that I am coming to this program because that sort of affects the way that people think about me. By keeping my educational and social information from certain people who really don't know a lot about me, I am treated more like an equal.
5	...even though that is an important part of my identity [referring to a certain interest], I just decided to take that off my ePortfolio, because although I don't mind my fellow Trans students know that, I don't want to find people who don't know me think I am weird...

Table 3. Sample user quotes on privacy needs.

P#	Concept	Example comment
8	Life cycle	Sometimes, I would put it [referring to samples of creative writing] on private because it has too much information about me that I don't want sharing over the Internet, or sometimes it has more private things 'to me' to go public. But then sometimes they become 'outdated' or I need to put them up as samples for assignments, or examples for a question.
3	Life cycle	My reflections are usually private, but for example, for [a particular course], we need to write down our reflections so that [the instructor] could see what we took out of the sessions. That's when I need to move something from private to public: it's because I need [the instructor]'s comments on it.
2	Relevance	My critical reflections are public for now; but I will change them to private when I want to provide my Elgg for scholarships. They don't need to see all my critical reflections. Not that they are self put-down or anything; in fact, criticism is the way we make progress, right? it will just be irrelevant for the purpose.
12	context	We have created a group for our [a course] group project in the past. There was this [...] assignment that we had and everyone needed to contribute by writing in the journal. So we uploaded the file into Elgg file repository and initially, gave access to it to only the group. Then when it was done, we also let [the instructor] see it, like we added her to the friends in the group. She was quite happy with the work, so she suggested we make it public so that others can see it, too.
1	Context	I move things between private and public in my ePortfolio, which is mostly schoolwork. For example, say we have a lab assignment due on Thursday; I would post it up for me to look at in the private one, just to check that everything is completed before I submit. Only after the due date I post it in the public one, because of copying.

Table 4. Sample user quotes on privacy factors—change of preference.

P#	Example comment
9	The problem I have with that [current privacy mechanism in Elgg] is that when I let some people see something, other people can see that there is something, but they don't have access to that. So they are like: oh, can I look at it? and then sometimes, you just don't know whether you want to share with them or not, and it's kind of weird to say no right away. So, then sometimes, I just rather keep it all private or all public so not to have to make that decision.
8	What would have been nice to have, is for people who don't have access to it to see a blank page instead of a message like, sorry, you don't have access to this.
11	I would rather keep my reflections private, but for example, for [a particular course], we need to write down our reflections so that [the instructor] could see what we took out of the sessions. Then I need to move my reflections from private to public.

theory constructs. A more detailed description of the grounded theory process can be found in Razavi (2009).

5. Study limitations

The unified demographics of our particular sample might raise a question as to whether the results were affected by the specific characteristics of this group. Particularly, an argument can be made as to whether the fact that this group of participants was under age 18 had any effect on the results. While there were certain categories of artefacts that our participants would not share because of their age (e.g. none of them were allowed to post a real photograph of themselves on their profile), we did not

Table 5. Sample user quotes on privacy factors—trust.

P#	Concept	Example comment
11	Kind of trust	Right now I am on a forum and I remember in the beginning I was really careful about exposing personal information, such as where I go to school or posting a picture. I would just ignore and leave myself out of it. After a while, you sort of trust them a bit more. I haven't been as far as putting a picture on, but I would say oh, I would get my license in a couple of years or something like that. But I won't make a reference to the fact that I am not old enough—I would just say I will get it in a couple of years. So, I am still pretty cautious about it; because after all, my trust just comes from interacting with these people over time. I mean, I just 'feel' more comfortable after being in the group for a while.
2	Stage of trust	After interacting in a group for a while I would feel more comfortable sharing with the group but its not always very comfortable, just more comfortable than before; Like, from 'not very comfortable' to 'sort of comfortable'. I am not the kind of person who gets too comfortable over the net.
6	Stage of trust	If you participate in an online community and you talk to people and they begin to give their opinions about something, you feel you begin to know who that person is by what they say are their ideas and what they like, and you develop a sense of knowing who they are, and they are no longer unknown; because we fear what we don't know and so if we get to know what that person stands for, maybe we can trust them some more.
10	Kind of trust	I am not the kind of person who makes friends over the Internet easily and I don't really connect with forums well; but once that happened, though, I actually had my friend who had visited the forum for a long time. So, it was easier to connect because I had a really strong connection there.

Table 6. Sample user quotes on privacy factors—group dynamics.

P#	Concept	Example comment
11	Size	[What I share in a community] also depends on the size of the community. Because some communities are really popular; there are lots of people; so you can't really get to know everyone. I am usually more comfortable when it is small, like say ten people. That's a bit more personal, and I get better credit for my contributions.
4	Size	I once created a community for [...], which was a closed community. My experience with that community was actually very positive: everyone would contribute actively and give others feedback on their work. But then, we all sort of knew each other, so it was more like chatting with friends... It was a small community, though.
2	Membership model	The problem with anonymous communities [where providing real information is not a requirement for membership] is that you have no way of knowing who the comment is coming from... you can't trust them with their judgment: it could be a grade one kid or it could be a Ph.D. so it's not worth anything.
3	Membership model	[What I share in a particular community] would really depend on who else is in there. In Trans [the particular community they have for all Transition students] I know the students [who the community consists of], so I would share my opinion on certain things that I wouldn't mind sharing with them in person; but for some stuff, I would definitely not share.
5	Visibility	To me there is a strong distinction between private and public groups. Private groups are invitation only, so I would appear with my real name and share practically everything. The public ones are open to everyone though; so I usually use a pseudo name and I am cautious not to reveal any personal information.
12	Visibility	[When sharing stuff in a community] I'd like to know what they are doing with it, but they don't have to tell me. I mean I am offering it, so they can use it if they want to. If they want to tell me what they are doing with it, I would like to know that, too. I don't mind as long as they give me credit for it.

focus on this group of artefacts. Rather, we tried to stay away from these obvious cases, and focus on the more general area of sharing information they had vested interest in. Also, although this study is situated in the context of an educational environment, our participants used it for managing and sharing many more varieties of personal and social information in different contexts. As such, their experiences reflected diverse information sharing habits in various contexts, as evidenced by the fact that none of the privacy factors that emerged in the study (the changing nature of users' privacy preferences, the effect of trust on information sharing decision, and sharing differently in group of different dynamics) were specific to the educational context. As such, we believe that even though our study has well-defined boundaries in terms of the user population, types of information artefacts, the intended

Table 7. Sample user quotes on privacy strategies.

P#	Example comment
11	[What I share in a community] also depends on the size of the community. Because some communities are really popular; there are lots of people; so you can't really get to know everyone. I am usually more comfortable when it is small, like say ten people. That's a bit more personal, and I get better credit for my contributions.
11	Besides Elgg, I have two other ePortfolios, and a couple of weblogs. One is private and one is public. On the private ePortfolio, I have things that are actually more private, like it has information about me, that sort of stuff. The purpose of that is that I just want to write some stuff down, so that it is sort of 'said' somewhere. Sometimes I don't want to keep stuff in my mind, like for example, a journal or something, I would put it on the private one.
10	I use [another platform] for more private stuff because there are settings for public or friends-only or you get to choose who gets to see it. If it is something you want the teacher to see but not anyone else, you can just set it that way.
9	[my private blog] is open, but it's sort of hidden, it's not obvious how to find the page. I have not provided a link to it from anywhere. So, it's open, but it's sort of hard to find.
12	I use LiveJournal for stuff that I want to share only with my closest friends; for things that I consider really private, however, I wouldn't write them down anywhere online; because the easiest secret to keep is the one that is never told.

audiences, and the context of use, it does provide meaningful insights into users' privacy needs and strategies in the social software system domain.

6. From grounded theory to design Heuristics

The second phase of the work involves translating the social requirements (which were identified through the study) into technical requirements (which are the actual technical structure of a privacy system to support those requirements). The main objective of the grounded theory study was to improve understanding of information sharing phenomenon, in order to identify guidelines that can inform design. To achieve such goal, we next developed a set of heuristics for the design of privacy management mechanism based on the results of the grounded theory, and then consolidated these heuristics into a privacy framework for social software. Here is a description of these heuristics.

Heuristic 1: Privacy control must be available on a fine-grained basis. The first heuristic suggests that control of the privacy of information must be defined in terms of individual artefacts as well as their collections, and is based on the observation that many of our participants in the grounded theory study expressed the need for fine-grained privacy control. Although this is a confirmation of the long-standing model that access rights should be associated with individual objects (e.g. files) and collections (e.g. folders), the higher granularity and incremental object creation model in social software suggests that the way in which these rights are managed to protect privacy and facilitate sharing needs to be different in some essential ways: the diverse nature of content and audiences in the social software domain implies that different artefacts in the same category might have different privacy requirements (i.e. landscape photographs made visible to public, but family photographs restricted to friends). Moreover, often times users need to grant or

deny access rights other than just the read action to their artefacts (i.e. colleagues may view, but not modify), which suggests that social software systems need to support fine-grained privacy management not only for resources, but also for target audiences and actions.

Heuristic 2: Privacy preferences must be defined in context. Research shows that while non-technical users seem to have a good idea of what their personal privacy preferences are, often times they have difficulty articulating them in terms of a set of rules (Egelman and Kumaraguru, 2005). Personal preferences are also context-sensitive, which makes it even harder to enumerate specific privacy rules. Enabling privacy preferences at a fine granular level makes this problem even bigger. This is supported by the fact that while the need for managing privacy at a fine-grained level has been recognised by other social software systems as well, it has often found to pose a trade-off with usability. In Facebook, for example, users have to go to a separate privacy setting page and set privacy preferences for each of their various profile and public search visibility items individually. Privacy-related options for individual applications are found with the application and users have to be aware of the features to find the options and visit separate privacy pages for each. Although fine-grained, the result is a completely unintuitive system where non-technical users are highly unlikely to be able to set sensible privacy preferences or understand the ramifications of their choices. Interestingly, all this effort is needed for just regulating visibility of one's various artefacts (i.e. the read action). Facebook currently does not provide any mechanism for regulating other types of action; for example, who can edit an artefact or leave a comment, etc. To the best of our knowledge, neither do any of other existing heavily used social software systems.

Thus, our second heuristic is a direct follow-up to the previous one and suggests that a privacy management approach that requires users to indicate their privacy preferences to the system *a priori* (i.e. through a privacy setting page) may not work; rather, we propose that any attempt to support fine-grained privacy management must be paired with enabling users to express their privacy preferences in context (e.g. at the time an artefact is created or modified) when they have a better idea of whom they want to share the artefact with.

Heuristic 3: Privacy mechanisms must provide control over ownership. Another deduction that followed from our grounded theory study was that users have a fundamental assumption that when they put something in the tool, they should have control over its ownership as well as its visibility. Our study suggested that one reason behind users' reluctance to share information was the tool's inflexibility in providing them with the ability to control the transactional aspects of knowledge sharing activities (e.g. getting proper credit for their contribution or ensuring reciprocity). This heuristic suggests that in order for social software systems to properly support information sharing needs, they must provide a complete, persistent sense of the degree to which information that an individual creates or consumes is his/her own, the amount of control s/he has over the use of that information, and the ability to properly assess or exploit its value. In other words, in addition to providing the means for users to control access rights at different degrees between the extremes of private and public, tools need to also allow users to maintain personal ownership control over their shared information.

Heuristic 4: Privacy mechanisms must support various group models. From a user's point of view, the primary concern in managing information sharing is in the ability to define and/or understand the audience that will have access to a particular information artefact. Generally, the choice of audience for a particular artefact or personal attribute is expressed in terms of a group of others who one trusts with that particular piece of information, so we suggest that a privacy mechanism must enable users to understandably model their trust into groups in a flexible and dynamic way. We therefore propose that group management in social software must support various group models rather than a generic unified form, and that groups must be defined and controlled by users, rather than the system.

Heuristic 5: Privacy mechanisms must provide control and/or awareness over group dynamics. Privacy in social software is also affected by the semantics of social network relations. For example, membership in a group with public membership visibility may thereby disclose interests, preferences, or other personal information regarding group members. This means that if a group member discloses information about him or groups including himself, he (whether willingly or inadvertently) might also

be disclosing information about someone else. In other words, one member's treatment of his/her privacy has a direct effect on another member's privacy. This suggests that awareness of group dynamics is an essential need for a privacy management system; meaning, such dynamics must be both controlled by and clearly articulated to users.

Heuristic 6: Privacy mechanisms must allow definition of groups that reflect interpersonal relationships. This is a follow-up to the previous two heuristics, and suggests that one group model that must be supported in social software is the egocentric group that is defined based on users' interpersonal relationships. This follows from the observation that in social software domain, one's personal and social information are not always shared with identifiable, accountable individuals or groups, and sharing may happen in a variety of contexts, for example, competitive as well as collaborative. Moreover, people may act simultaneously in several contexts, holding multiple potentially conflicting relationships simultaneously. As such, a lot of users' information sharing needs is better described in terms of the relationship that exists between the owner of the artefact and the person or group with whom the information may be shared, specially since new intricacies have blurred the boundaries between public and private. Boyd (2006), for example, points out that US teenagers feel strongly about preserving a certain form of privacy: they want to be visible and searchable for their friends but not their parents. In terms of rights management, these observations strongly imply that the potential audience for some artefacts or attributes is likely defined in user's own terms, based on a variety of kinds of relationships that more closely resemble real-life privacy boundaries (e.g. one-sided and short-term relationships) and not in terms of any organisational 'roles' or groups. This will enable users to control the release of their personal information in the same manner they would control it in the real world, based on their relationship with the data receiver, rather than some externally imposed constraint such as the receiver's organisational role.

Heuristic 7: Privacy mechanisms must easily accommodate changes in preferences. The next heuristic is based on the dynamic nature of users' privacy preferences in the social software domain. While in general, any act of information sharing can be defined as 'a user sharing an artefact with a receiver based on their relationship', in the social software domain the information sharing act is often about establishing and maintaining a dynamic sharing relationship: over time, the nature and state of personal artefacts might change (i.e. research results getting published, patented ideas getting approval, personal opinions reconsidered), the receiving group with whom the information is shared might change (i.e. competitors joining a group or collaborators leaving), and the relationship between the owner and the receivers of information might change (i.e. people switching to a different project groups or changing affiliations). We thus propose that a privacy model that statically assigns access rights based on these factors at the time of an artefact's creation or modification will be insufficient. Rather, privacy mechanisms need to be flexible enough to accommodate frequent changes in users' privacy preferences in a non-labour-intensive way.

Heuristic 8: Action possibilities and their consequences must be clearly presented to users. Our last design heuristic emphasises the importance of interface clarity. Our study confirmed the intuition that users can be reluctant to share personal information when they are not sure how exactly to do things, or when the consequences of a sharing decision are unclear. A counterintuitive consequence of this is that some users might be more willing to share personal information in a space that affords virtually no privacy control (e.g. blogs or Myspace pages) than one which offers them an unclear set of privacy management tools. In our study, users were made aware that they could have some control of privacy and should manage the audience for their personal information by the promise of an access control system. However, many found it inadequate because either they could not perceived how to do something they wanted to do (i.e. users did not know they could make something visible only to one person, even though such functionality was supported by the tool), or they were not sure what the consequences of a sharing decision were (i.e. even though the tool provided different information sharing models through supporting both groups and communities, users were not clear on how they differed). As a result, they were not able to take advantage of certain aspects of the privacy management mechanism, because of the inability of the tool to convey their existence or consequences.

7. A framework for privacy in social software systems

The overall goal of the proposed design heuristics was to identify a minimal set of requirements (both technical and social) for privacy management in social software systems. While heuristics Heuristic 1 to Heuristic 7 describe what kinds of privacy control are necessary for managing and sharing personal artefacts, Heuristic 8 pertains to how these controls must be built and incorporated in order to be usable. We now consolidate these requirements into a framework for user-centred privacy in social software domain that describes privacy in terms of required controls over artefacts, audiences, relationships, and change; with an emphasis on the clear presentation of those controls to users.

Artefact control. The principle of artefact control is a consolidation of Heuristic 1, Heuristic 2, and Heuristic 3, and essentially reflects the finding that privacy management in social software must be defined on a per-artefact level as opposed to per-category; and that the access rights need to be applied in context (meaning, at the time of artefact creation or modification) as opposed to *a priori* (through a privacy setting page). Furthermore, in addition to control over visibility (the read action), users also need the ability to control other rights over their artefacts, for example, modification or deletion (the write action), and over further delegation of such rights.

Audience control. The principle of audience control is a consolidation of Heuristic 3, Heuristic 4, and Heuristic 5, and reflects the need to restrict both the visibility and ownership of artefacts to certain user-defined groups. Although most existing social software systems support some group functionality, we suggest that social software systems must provide the means not only for creation of these user-defined groups, but also for definition and control over various aspects of these groups (sizes, membership models, and visibility), and for controlling changes in those aspects. Furthermore, these controls need to be in the hands of users, rather than pre-defined by the system.

Relationship control. The principle of relationship control is a reiteration of Heuristic 6, and reflects the need for the ability to define information sharing based on a user's self-defined relationships with others. In essence, this emphasises that users need the ability to define groups of friends or collaborators in their own terms, and to use this model of their relationships with others as the basis for audience control.

Change control. The principle of change control is a reiteration of Heuristic 7 and is something of a cross-cutting concern within the other three controls. This principle reflects the observation that in the social software domain, the artefacts, the audiences, and the relationships used to define privacy and sharing patterns are all dynamic. A privacy and user interaction model must thus take into account that artefact life cycle and categorisations will change, that a user's requirements to share classes of artefacts with certain audiences will change, and that a user's relationships and trust patterns within those relationships will change, and that users come to expect their tools to provide flexible support for these changes in their privacy preferences when the social parameters that define the sharing model change.

Clarity. While the other heuristics focus on what kinds of control of privacy are needed, the last one focusses on how those controls must be presented to users in order to be usable. As such, it must be considered in parallel with the other four controls as presented in Figure 2. We use the term clarity to represent this heuristic; meaning any functionality to incorporate artefact, audience, relationship, or change control must be designed in a clear and understandable way; to ensure that in practice, the average, non-technical users would be able to take

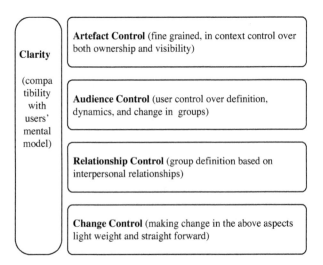

Figure 2. A framework for usable privacy control in social software.

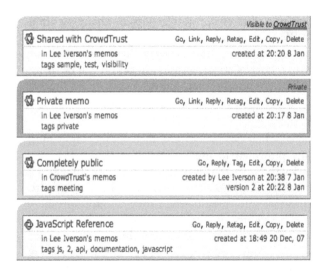

Figure 3. Memos in OpnTag: public, private, and selectively shared in a group.

advantage of the extra control over privacy that these user-centred controls are supposed to provide.

In order to illustrate the suitability of our proposed privacy framework for design, our next step was to design a system that instantiates the framework. As an example, and to provide an environment in which we can test these principles, we have developed an experimental system called OpnTag. We next present the technical structure of OpnTag, along with a discussion of how our framework as embodied in this system supports each of the five user-oriented privacy controls.

8. OpnTag

OpnTag (Iverson *et al.*, 2008) is an open source web application for note taking and bookmarking that we developed to address the information management needs of an individual performing in various social contexts. The fundamental unit of information storage in OpnTag is the memo, a tagged textual annotation that may optionally link to a web resource. Memos can function as bookmarks, notes, or wiki pages and are organised based on their intrinsic metadata (e.g. who owns or created them and when) and the tags applied to them by various users (Figure 3). Each memo has an owner, which presents who creates the memo and thus can edit and delete it, and a potentially restricted audience, which controls who can see that the memo exists and read it. Both the owner and the audience can either be an individual or defined as a group (Figure 4). Also, groups can be defined either by inviting other individuals or by applying people tags to individuals in one's network, thus categorising them into a group. Two types of groups exist in OpnTag: classic groups (Figure 5), with collectively-controlled visibility, size, and membership model; and egocentric groups (Figure 6), created through applying people tags to other

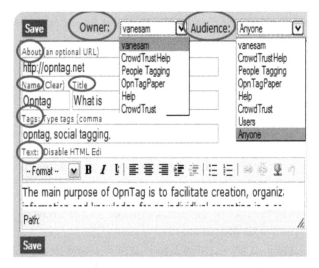

Figure 4. An individual memo in OpnTag with owner and audience list including both the individual and her groups.

individuals in the system, thus resulting groups with completely owner-controlled dynamics. For a more detailed description of the group functionality in OpnTag, see Razavi and Iverson (2008, 2009).

With individuals, classic groups, and egocentric groups, OpnTag's privacy control centres on the joint concepts of ownership and audience management, which ensure individual users retain control and credit over the artefacts they dispose in the system. Although a group can be specified as the designated owner of a memo, each memo is also visibly attributed to its individual creator, thus ensuring that each group member gets proper credit for the contributions s/he makes to the group's shared information space. Only a memo's creator can modify ownership, but any member of the owning group can change a memo's audience. This design choice allows the creator

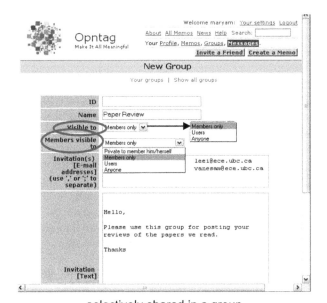

selectively shared in a group

Figure 5. Classic group definition page with group and member list visibility menus.

Figure 6. Egocentric group definition in OpnTag through applying tags to users.

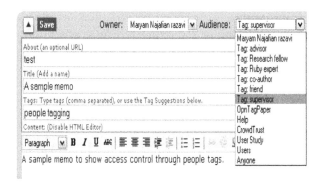

Figure 7. Selective sharing of a memo in OpnTag, with both classic and egocentric groups.

of a memo to decide whether access restrictions of a memo should be controlled collectively or individually. The latter case supports situations when there is a need to ensure that the access restriction policies stay with the shared data; i.e. it is not possible to make a memo visible beyond the intended audience set by its creator.

Audience restriction is the fundamental mechanism for selective sharing in OpnTag: at the time of creating or editing a memo, the creator has access to both his/her classic and egocentric groups and can thus adjust the audience of the memo to be either the owner himself (whether an individual or a group), a classic group that

the owner is a member of (with its collective membership dynamics), or one of the owner's egocentric groups (over which s/he has complete control). Figure 7 presents a screenshot of how this choice is made.

9. Satisfying privacy requirements

OpnTag's privacy management mechanism has been designed based on the five user-centred privacy controls. Artefact control in OpnTag is supported by providing ownership and audience management at the level of individual memos. For each memo, the user specifies the memo's owner (who can edit it), with the default owner of a memo set to the creator (either the individual or a group). By enabling context-sensitive ownership management at the memo level, OpnTag's privacy system supports definition of fine-grained privacy policies any time an artefact is created or modified, as opposed to providing a separate privacy page for defining general privacy policies on collections as most current tools do. In addition to fulfilling the fine-grained privacy management requirement, this also allows users to define their own privacy policies when they have a clear idea of their preferences. This is quite important in terms of adhering to the principle of artefact control, as the principle emphasises pairing fine-grained privacy control with in-context policy definition to alleviate difficulties that users often have with administrating such fine-grained privacy control. Moreover, with OpnTag's ownership management, privacy policies stay with the data, meaning, no one other than the memo's owner can change its audience; i.e. it is not possible for someone who is not an owner of a memo to make it visible beyond the intended audience set by its creator.

Audience control in OpnTag is supported through deep visibility management via user-defined groups and relationships. Every memo in Opntag has restricted visibility; meaning, it can only be seen by members of a designated group. OpnTag enables users to control various aspects of the classic groups they define, including size, visibility, and membership. The relationship dynamics of such groups is then determined by variations of these parameters; i.e. open/limited number of members, public/private visibility, and open/moderated/closed membership. This clear model of group characteristics highlights the potential trade-offs between risks and benefits of information sharing in a particular group and gives users a high-level overview of the effects of their sharing decisions. Knowing how a particular resource could potentially be used in a particular group, users can then tune their sharing decisions accordingly (e.g. one would be less likely to share sensitive information with an 'open' group that anyone could join than in a 'closed' group where new members must be invited). In addition, through people tagging, users can take complete control of the audiences for their artefacts.

Relationship control in OpnTag is supported through people tagging, which enables categorising one's network into user-defined, egocentric groups that may represent various kinds of relationships. Since egocentric groups are controlled entirely by the creator (the acts of creating or deleting a tag on a user, and controlling the visibility of the tag are solely controlled by the tagger), the act of tagging a person via their profile page is equivalent to asserting their membership in a group whose membership is entirely under tagger's control. As such, people tagging provides a lightweight and flexible mechanism for handling volatile relationships that frequently show up and fade out in natural social environments, but are often hard to manage in online world.

The combination of artefact, audience, and relationship control in OpnTag allows fine categorisation of resources, audiences, and actions, provides advanced group functionality, and enables users to define privacy preferences based on their often changing relationships. The fourth principle, change control, is supported by ensuring that ownership and visibility management (modifying owner and/or audience of memos and tags), group management (creating, managing, joining, and leaving user groups), and people tagging are all handled in a flexible and straightforward way and that all the settings are modifiable at any time, making it easy for users to make frequent changes to their information space.

Finally, we have followed some of the known principles of usable design to achieve clarity. One such principle is to make privacy features (as secondary functionalities) highly visible and seamlessly available to users in the context of their primary actions (De Paula *et al.*, 2005). This is derived from the fact that privacy features often act as barriers to action, while usability principles aim to remove such barriers (Dourish *et al.*, 2004). Supporting proper visibility can thus help achieve the right balance between the two seemingly conflicting goals and ensure that privacy management features complement existing actions rather than inhibiting it. In OpnTag, relevant action possibilities for each privacy control are graphically clustered and presented together, while irrelevant or rarely used

information are omitted in order to reduce clutter: owner and audience selection for a memo (OpnTag's privacy controls options for artefacts) are visibly presented at the top of the memo edition page, and are the only functionalities presented at this level, separated from other functionalities that deal with the content of the memo (Figure 3). Likewise, group visibility, member list visibility, and people tag visibility (OpnTag's privacy control options regarding audience and relationship) are presented in the same context that classic or egocentric groups are created or modified. While this makes these features visibly and seamlessly available to users in the context of defining information artefacts, audiences, or relationships, it still gives them the option to skip such configuration and accept the defaults, if they would rather focus on their primary task.

Another usability principle is consistency (Nielsen, 1992). In OpnTag, various choices for each privacy control feature (owner and audience for memos, group and member list visibility for classic groups, and visibility of people tags) are consistently presented by drop-down menus, as a learned convention that is successfully and frequently used by people at all skill and experience levels. Designers of privacy systems are also advised to use feedback mechanisms to help users understand the implications of their privacy decisions (Bellotti and Sellen, 1993; Lederer *et al.*, 2003). OpnTag supports this through the combination of providing visualisation and pairing configuration with action. For both memos and people tags, the choice of audience causes the background colour to change accordingly. Such immediate visualisation of visibility and the choice that triggered it will help users understand how to generate rules that reflect their preferences, or to notice when the results of their actions do not correspond with their intended goal.

10. Usability evaluation

The two main features that distinguish our privacy framework from existing models of privacy management in current social software systems are the introduction of per-artefact control (as opposed to per-category control supported by most current tools), and use of people tagging which enables creation of nuanced relationship groups (as opposed to equal-weight, reciprocal relationships supported by the network-of-friends model). The result of these two innovative features is the promise of a more fine-grained, flexible, and dynamic privacy management that provides users with more control over selective sharing of their artefacts. Having developed OpnTag as a test bed, we next administered a small-scale usability evaluation to provide empirical evidence that the overall framework as embedded in OpnTag yields a usable privacy management mechanism. In other words, while as the first iteration in the design/evaluation process the focus of the evaluation study was mainly on finding out what aspects of the framework work and what aspects do not, an equally important objective was to ensure that the extra control over privacy is not to achieve at the price of sacrificing usability.

10.1. Participants

OpnTag target users are individuals operating in various personal, professional, and social group environments, meaning, our sample pool was practically general public and there were no salient characteristics to look for in participants to assert their suitability for participation in the study. As such, a participation request was distributed via email to mailing lists and social connections of existing OpnTag users. Ten people (six male, four female) who responded to the invitation were recruited for participation in the study. Our participants came from diverse backgrounds, including both technical and non-technical users, which was appropriate since OpnTag is designed for personal and social information management across a wide variety of usage contexts. All participants had some university education, with seven participants having a graduate degree and three an undergraduate degree.

10.2. Procedure

The procedure we employed for the purpose of evaluation was a laboratory study consisted of three stages: first, participants were asked to fill out a survey questionnaire consisting of 10 questions covering general areas of users' demographics (sex, age, and education level), profession, expertise with computers and the Internet, plus specific questions on their familiarity with popular social software systems and the privacy management features in them. The survey also included questions aimed at identifying users' privacy attitudes, i.e. what information they feel comfortable to reveal about themselves in social applications they use and whether they have ever experienced privacy violation problems in the past. All of our participants had some familiarity with social systems (i.e. at least using one other social system for a prolonged period in the past) and some concern for privacy (i.e. at least considering some artefacts private and some sharing uncomfortable). These two criteria ensured that they had an idea of the subject of the study and can relate to it, and that their answers were not affected by lack of knowledge or concern about the topic of investigation.

None of our participants had used OpnTag before. On average, participants reported using various social systems for 6 hours per week. None had any special expertise related to privacy; however, when reflecting on their experiences with privacy management in applications they used, seven participants said they find the feature useful and have used it at some point. Six said although they consider privacy management mechanisms necessary, they often find them too difficult and/or time consuming to use. Five said they do not trust social systems to put their private information there. Only two participants said they think social systems provide adequate privacy management. Seven out of 10 participants reported experiencing privacy violation at some point while using social systems, either as a result of their own action or others.

In the second stage, participants were asked to login to OpnTag as an imaginary persona and perform a set of pre-defined tasks, each involving creating memos of various degrees of sensitivity and sharing them with various people in the imaginary persona's social and/or professional network. For each task, participants needed to decide on the appropriate owner and audience for the memo they created, and on whether to assign their target audience to a classic or egocentric group. Participants were allowed to make any changes to the persona's information space that they felt necessary for the purpose of carrying out the tasks; including creating, modifying, or deleting groups and/or people tags, or changing the visibility of their groups/group members/people tags. The tasks were designed in a way to cover information sharing situations across a variety of privacy-sensitive contexts, ranging from inherently private, to semi-private, to public; and to require a mixture of visibility and ownership control.

In the third stage and upon completion of the tasks, we engaged each participant in a semi-structured interview. We used participants' actions during tasks as a starting point and tried to gather feedback on the reasons behind their actions and why they did things a certain way. We particularly looked for errors and signs of confusion and/or frustration, as well as comments on the strengths and weaknesses of the privacy management mechanism. This combination of methodologies allowed us to make detailed first-hand observations of how first-time users interacted with OpnTag's privacy management scheme and how they reflected on its utility and usability.

10.3. Results

Ease of use and effectiveness. We asked users to rate OpnTag's privacy management mechanism in terms of both ease of use and sense of privacy compared to the social software systems they were familiar with (Orkut, LinkedIn, and Facebook). We used a scale of 1 to 5 for rating, with 1 indicating the worst performance and 5 indicating the best. Ease of use was rated 4.2, with min. = 3 and max. = 5 (users thought there were just too many steps involved to navigate to a user profile for tagging). Users gave their perceived sense of information privacy an average rank of 4.0, with min. = 3 and max. = 5. Although not all of our participants took the optimum path for doing all scenarios, they were all able to navigate their way through the privacy management system to get the tasks done. From the total of 50 tasks that our participants performed unassisted (10 participants each doing five tasks), we only witnessed five errors. Furthermore, all five errors were results of improper understanding of the task in question; for example, making a memo visible to team members rather than colleagues, as mentioned in task description.

Overall, the concept of setting the owner and the audience for access management seemed to be fairly understandable to users: none of the users showed any signs of confusion or frustration. Also, the majority of our participants seemed to grasp the difference between granting 'write' vs. 'read' access: nine out of 10 users correctly created a group memo for the tasks that involved some form of collective contribution. Since this distinction is not supported by most of the existing tools, it was encouraging to see users quickly picking up and using a fairly new feature.

Usability.   Traditionally, usability of a software application is measured based on four quality components: speed, accuracy, success rate, and overall user satisfaction (Nielsen, 2001). However, researchers have often considered different criteria for measuring usability of a security or privacy mechanism. The reason is that privacy management is often a secondary goal in most systems, and therefore does not get the same consideration that many other aspects do (Egelman and Kumaraguru, 2005), which makes it difficult to set particular metrics for usability of privacy aspects (i.e. what exactly should be measured?). Whitten and Tygar (1998) were the first to propose a working definition of usability for security software based on the special characteristics of the usability problem for security, and to suggest several criteria for evaluating usability of a security system. A number of other researchers have also proposed similar and/or complementing guidelines for evaluating usability of security or privacy mechanisms (Chiasson *et al.*, 2006; Clark *et al.*, 2007; Cranor, 2005; Karat *et al.*, 2005). We found these criteria suitable for the purpose of our study. Here we reflect on the usability of OpnTag's privacy framework based on Whitten and Tygar's (1998) four usability criteria, plus the two complementary criteria suggested by Chiasson (2006).

(1) Users must be reliably made aware of the steps they have to take to perform a task
This is a restatement of the first guideline of (Whitten and Tygar, 1998) and suggests that the application must provide user with enough cues as to how to start the process for each task, and to identify the intermediate steps that are required to complete the task. In OpnTag, the acts of setting a memo's owner and audience are fairly straightforward, because those action possibilities are associated and presented with memo creation/modification functionality. Also, the fact that privacy management in OpnTag happens on a per-artefact basis made it easy for participants to figure out that the owner and audience are the two attributes of a memo that they need to set in order to share something with a certain audience.

(2) Users must be able to determine how to successfully perform the steps
Whitten and Tygar's second usability guideline suggests that once the user is made aware of what intermediary steps are necessary for each task, s/he must be able to figure out how to perform these steps. (Wharton *et al.*, 1994) suggest that users develop a mental model of how a system works, and that in order for users to be successful in performing the necessary steps required to complete a task, the model behind the system must match user's mental model.
In our study participants employed different privacy management strategies based on their privacy attitude and concerns (e.g. one participant would consider a memo private, while another one would make it public). Regardless of their privacy attitudes, our subjects were successful in achieving their desired level of privacy, properly disposing the created memo to the right audience. Moreover, OpnTags' privacy management system seemed to have a fast learning curve: after a short, initial training session, our subjects all seemed at ease with creating memos, defining or modifying groups and/or group members, tagging other users, and choosing the appropriate owner and audience for their memos.

(3) Users should not make dangerous errors from which they cannot recover
Since OpnTag is an information management system, the most dangerous error that can happen is exposing a memo to the wrong audience. However, in OpnTag memos are created in the current work space, meaning the default value for both a memo owner and audience is the user himself (if in user space) or the group (if in group's shared space). As such, even if users miss to set the right owner/audience, having rather conservative values as defaults helps decrease the chance of accidentally making a memo visible to a too large audience (e.g. public). Also, the different background colour-codes that reflect various levels of visibility for a memo provide a powerful visual cue to the user as to whether the memo is set to have the right visibility.

(4) User should know when they have completed a task
This is the first complementary criterion to Whitten and Tygar's proposed by (Chiasson *et al.*,

2006) (also mentioned by Cranor (2005)). This criterion suggests that one of the essential usability requirements is enabling users to tell when their task is completed, which implies that the feedback provided by the system to users during a task should be adequate to ensure they are aware of its successful completion.

We asked our subjects several questions in an attempt to gauge their perception of the appropriateness and adequacy of the feedback provided by the system (the owner name, colour-codes, and group name in the resulting memo). After each task, we asked the participant if they believed they performed the task correctly, and if yes, how could they tell if they have been successful in setting the appropriate privacy level for the artefact they created. Eight of our subjects mentioned the use of at least one of the feedback mechanisms to check their results, while the other two participants just assumed they did it right and had not paid attention to any of the feedback information. Overall, the consensus was that the combination of colour-codes and owner/audience in the memo list conveyed enough information to users to form an idea of the current privacy level of their various artefacts at a glance. Six participants mentioned the colour-code as the most useful feedback mechanism, probably because of its visualisation power. Interestingly though, our choice of colour-codes was not popular with the participants. One participant mentioned that blue (OpnTag's colour-code for public) and green (OpnTag's colour-code for private) are quite easy to be mistaken, and suggested we use other colours for the two cases that show the distinction more clearly. Another participant thought green is not the right colour-code for private, since it implies 'green light' in a way. This participant thought a more strong colour like red would be a better choice for the case.

(5) Users must be able to determine the current state of the system at all times
This is the second guideline form (Chiasson *et al.*, 2006), and suggests that it should be visible at a glance what is visible to whom. (Cranor, 2005) advocates the use of 'persistent indicators' that allow the user to see privacy information at a glance. OpnTag does this through its use of colour-codes.

(6) Users must be sufficiently comfortable with the interface to continue using it
This is the fourth principle of usable security of (Whitten and Tygar, 1998), and is an essential part of the principal of psychological acceptability quoted by (Bishop, 2005). After completing the

Table 8. Participants' comments on willingness to adopt the tool.

P#	Example comment
3	It would be nice to have something like this that limits the contribution to a certain group, even though it may be exposed beyond that group. For example, I would like the contributors to a discussion on board level design to be limited to: X [who is a board designer], Y [who is his boss], Z [my boss], and me. That's all people who are knowledgeable enough to contribute to this discussion, although the whole group may read it.
5	I strongly feel that something like this is needed in our workplace, even though personally I am against using something like this because of privacy reasons: I don't think one should put his ideas in a system on the Internet; unless he can control who would have access to it.
6	One thing I like about this [OpnTag] is the control; I like that it can act as an integrated environment; because you can separate your personal and work-related stuff. I think that's very important. The idea of having my desktop online sounds cool.

tasks, we asked our subjects if they can relate to the scenarios and whether a tool like OpnTag with advanced privacy management features would be more likely, less likely, or just as likely, to be incorporated in their daily personal and social information management activities. Most participants (9 out of 10) said that although the scenarios do not resemble their information sharing practices exactly, they could think of similar scenarios in their day-to-day activities where similar selective information sharing activities would be useful or necessary. Eight out of 10 participants said that they would try using such a tool for information management and sharing, and six participants said they feel a strong need for a tool with these sort of privacy management in their work place. Table 8 summarises some of participants' comments regarding willingness to adopt the tool.

11. Study limitations

There are a few limitations of the study that must be taken into account when interpreting the results. Ideally, we would have liked to perform a field evaluation with real users using OpnTag and its privacy system in real-life information sharing situations. However, while field studies report on users in their natural environment doing real tasks and as such, can demonstrate feasibility and in-context usefulness, they are time consuming to conduct and require mass adoption. A comprehensive review of literature on privacy evaluation methodologies (e.g. Chiasson *et al.*, 2006; Cranor *et al.*, 2006; DeWitt and Kuljis, 2006; Hawkey and Inkpen, 2007; Iachello and Hong,

3

FraTAct for Transforming A Nescient Process Activity Into an Intelligent Process Activity

Rafiqul Haque[1*], Nenad B. Krdzavac[1]

[1]Department of Accounting, Finance, and Information System, College of Business and Law, University College Cork, Cork City, Cork, Ireland.

Abstract

Existing business process technologies support defining only nescient activities. Currently there is no solution that underpins transforming a nescient activity into intelligent activity. In this paper, we address this shortcoming of the state of the art. We offer a framework 'FraTAct' for transforming regulation intensive nescient activities of a financial service business process into intelligent activities.

Financial service industries has been experiencing enormous challenges since the last decade. A recent financial crisis has unearthed various weaknesses in terms of administering the financial service industries. In order to prevent the future crisis, the regulators are constantly formulating new rules and also forcing the financial service industry to enact financial regulations in their financial service based application which automates financial operations.

A financial service application underpins the financial service business process that contains activities. A nescient activity within a financial service process is prone to the risk of producing an inconsistent outcome that results in severe legal consequences for a financial institute e.g., a bank. In order to avoid these legal consequences, a financial institute should develop their financial service processes by composing activities that should be intelligent to understand and comply with financial regulations. Intelligent activities will produce outcomes that are consistent to financial regulations. It will reduce the possibility of financial regulation noncompliance in financial service process based application.

Keywords: Business Process, Nescient Activity, Intelligent Activity, Ontology, Description Graph

1. Introduction

A process simply put, is a set of ordered activities [15]. Activities carry out *operations* while a process is running and producing outcomes. The notion of process is used in different domains such as software process, chemical process and so forth. In this paper, process refers to financial service process and activities are financial service activities. In a financial service process (FSP), the operations must be performed in compliance with *financial regulations* that are essentially financial rules. An operation produces an inconsistent outcome if it does not comply with the relevant regulations. Inconsistent compliance outcomes result in severe consequences for an organization regardless of whether it is a financial or business organization. The *Enron-Scandal* [12] is a practical example. Therefore, compliance is a highly significant issue in FSP.

Activities within FSPs are *regulation-intensive* which means that they are based upon financial regulations (e.g., Basel III accords [3]). However, some activities within an FSP may not be regulation-intensive. A regulation-intensive activity within an FSP is called *nescient* if it lacks explicit knowledge about the underlying regulations, it lacks correlation with financial regulations, and the activity cannot produce consistent outcomes. A nescient activity is not able

*Corresponding author. Email: arhaque@ucc.ie
†This research paper is an extension of our research published in [10].

to perform operations that comply with financial regulations. This promotes the need for activities that are able to perform operations in compliance with regulations.

Existing business process technologies (e.g., BPEL [21]) lack the constructs for defining (modeling) intelligent activities in a financial service process. Some approaches (e.g., Business Process Model and Notation (BPMN) [22])facilitate annotating meta-data of a process activity to give the semantics of that activity. However, these approaches cannot be used for FSPs due to the complexity and ambiguity in financial regulations. More specifically, an FSP is slightly different from the classical business process activity as it must be aware of one or more financial regulations which can be a conjunctive statements. It may not be easy to annotate an FSP with a conjunctive statement which expresses explicitly the semantics of different parts of it. In addition, transforming annotated financial regulations into machine readable ones by preserving their semantics is still an unsolved issue. A solution that supports the straightforward transformation of a nescient activity into an intelligent one is strongly required for *smart* financial service processes.

In this research, we offer a framework named *FraTAct* that underpins transforming nescient activities into intelligent activities within the FSP space. It is important to mention that the transformation happens *indirectly* i.e., the framework supports the nescient activities by providing the exact meaning of regulations and also assists in performing operations complying with financial regulation. In order to do so, the FraTAct framework uses the notion of the *knowledge base*. The knowledge base is developed using an ontology approach.

The remainder of this paper is organized as follows. A motivating example is presented in section 2. The preliminaries are described in section 3. In section 4, we describe the knowledge base for financial services. Section 5 presents the FraTAct framework. We discuss implementation of the proposed solution in section 6. The related works is discussed in the subsequent section. Finally, section 8 outlines the conclusion and future work.

2. Motivating Example

This sections sets out the *Common Equity Tier 1 Capital Ratio Calculation Process* model. The process model is developed based on the description provided in [3]. In order to make the example simple, a limited number of activities have been incorporated in the process. Fig. 1 shows the process model.

The process starts with performing the 'List Common Equity Tier 1 Financial Instruments' activity that

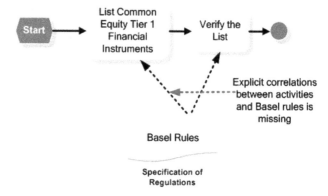

Figure 1. Common Equity Tier 1 Capital Ratio Calculation Process

produces a list of financial instruments. Then, the list is verified by performing *Verify the List* activity. After executing these activities successfully, the *end* event signals the *completion* of the process.

The 'Common Equity Tier 1 Capital Ratio Calculation' process is driven by regulations particularly, the Basel Capital Accords. The activities contained in this process are regulation-intensive. The operations performed by these activities must comply with the Basel accords. These accords are ambiguous. Therefore, understanding the semantics of these accords is a critical importance for the activities.

Activities contained within 'Common Equity Tier 1 Capital Ratio Calculation' are nescient as these activities lack the support of an approach that enables performing operations intelligently. The FraTAct framework supports transforming these activities into intelligent activities that carry out operations by realizing the semantics of the financial regulations derived from the Basel accords. The framework leads financial service process execution in compliance with the corresponding regulations and prevents unwanted consequences for financial institutes.

3. Preliminary

In this section, we provide a brief description of intelligent activity. In addition, this section explains ontologies and description graph.

3.1. Intelligent Activity

Intelligent activity is defined in [10] as follows:

Definition(Intelligent Activity): *Each regulation-intensive activity is an intelligent activity if and only if*

(i) it has explicit correlation with the regulations,

(ii) it relies on regulations which are encoded as axioms in a decidable description logic [13],

(iii) it cannot produce outcomes that are non-compliant to regulations,

(iv) Any activity that does not satisfy (i), (ii) and (iii) conditions will not be considered as intelligent activity within FSP.

For example, list common equity Tier 1 capital ratio calculation is a intelligent activity if it is able to interpret the semantics of the corresponding regulations (derived from the Basel Accords), complies with those regulation while executing the activities, and the activity produces the list that was desired by the users.

3.2. Ontologies and Description Graph

The Ontology Web Language (OWL2) [4] is used for modeling ontologies. The $\mathcal{SROIQ}$ DL [13] provides reasoning services for ontologies based on the tableau algorithm. The basic OWL2 concepts are classes, object properties, data-type properties, and individuals. Additionally, an ontology contains class axioms, and property assertions as well as individual assertions. Semantic Web Rule Language (SWRL) [14] is used to model rules in an ontology.

A Description Graph (DG) [18] is a directed labeled graph that contains a set of vertices, edges and a labeling function that assigns each node to an atomic concept, and each vertex to an atomic role. According to [18], a graph-extended knowledge base is a 4-tuple $\mathcal{K} = (\mathcal{T}; \mathcal{P}; \mathcal{G}; \mathcal{A})$ where $\mathcal{T}$ is a TBox, $\mathcal{P}$ is a program, $\mathcal{G}$ is a GBox, and $\mathcal{A}$ is an ABox. The $\mathcal{P}$ consists of a finite number of connected rules. The $\mathcal{G}$ contains graphs roles i.e. roles one can use in a DG [17]. Reasoning with ontology is, mostly, based on tableau [13] and hyper-tableau [18] algorithms.

4. Knowledge Base for Financial Services- A Graph-Extended Approach

As already mentioned, financial service components are built on a knowledge base (Fig. 2 shows the knowledge oriented service components). In this section, the core mechanism of knowledge base for financial services is described. Financial regulations are basic building blocks for financial services. Therefore, our aim is to develop knowledge bases for financial regulations. In this section, we explain how to encode regulation corresponding to common equity Tier 1 capital ratio calculation process (shown in Fig. 1) into a graph-extended knowledge base [18] and check satisfiability of the knowledge base using a hyper-tableau algorithm [17]. Modeling FSPs using graphs is not a new idea [28]. However, a problem arises when one needs to

apply automated reasoning on the processes which must satisfy certain regulations. The execution of each activity in the common equity Tier 1 capital ratio calculation process relies on regulations derived from e.g. Basel III accords.

Reasoning with the process (shown in Fig. 2) is provided indirectly over reasoning with regulations encoded as a graph-extended knowledge base. We do not specify regulations in the process itself. The reasoning with the given process is important to check whether the activity of financial service processes is compliant with the Basel regulations upon execution. It means that the activity has been executed successfully and is consistent with regulations from where it is taken [3]. To consume regulations encoded as graph-extended knowledge base, the reasoner must check satisfiability (see Fig. 2).

Figure 2. Encoding FSP to graph-extended knowledge base

To encode financial regulations underlying the activities within an FSP space, we use the graph-extended knowledge base, as follows:

(i) Entities related to each activity are encoded as nodes in DG,

(ii) The interconnection between elements is represented by graph roles,

(iii) DG rules and SWRL implement regulations related to an activity,

(iv) Restrictions on the graphs implements control flows between elements.

We encode the regulations into a graph-extended knowledge base for the following reasons:

(i) OWL2 is not expressive enough to encode financial regulations as non-tree structures [17],

(ii) Hyper-tableau algorithm is practically efficient in reasoning with more than one DG. Also it is possible to encode process regulations into more than one DG.

Execution of the *List of Financial Instruments of Core Capital* and the *List of Financial Instruments of Additional Capital* (see Fig. 1) must comply with the following regulations, taken from [3]:

(R1) If a financial institute has a financial instrument as common stock and the common stock can be converted to a currency, then the given currency is a principal amount for the financial institute.

(R2) If a financial institute has a currency as a principal

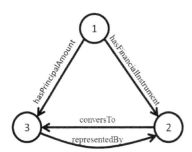

Figure 3. DGs for Common Equity Tier 1 Capital Ratio Calculation Process shown on Fig. 1

amount and the currency is represented by a common stock, then the financial institute has a financial instrument as given by common stock.

The regulations (R1), (R2) are encoded into the following first order logic formulas:

$$(\forall x, y, z)(FinancialInstitute(x) \wedge$$
$$hasFinancialInstrument(x,y) \wedge$$
$$convertsTo(y,z) \Rightarrow hasPrincipalAmount(x,z)) \quad (1)$$

$$(\forall x, y, z)(FinancialInstitute(x) \wedge$$
$$hasPrincipalAmount(x,y) \wedge$$
$$representedBy(y,z) \Rightarrow hasFinancialInstrument(x,z)) \quad (2)$$

$$(\forall x, y)(FinancialInstitute(x) \wedge$$
$$convertsTo(x,y) \Rightarrow representedBy(y,x)) \quad (3)$$

Formula (1) encodes regulation (R1) while formula (2) encodes regulation (R2). The formula (3) specifies the relationship between common stock and currency. Formulas (1), (2) can be expressed as property chains that are not allowed in OWL2. Instead, we use formalism based on DGs. Fig. (3) shows the DG which corresponds to regulations (R1) and (R2). The formal definition of the DG shown in Fig. 3 is: $G = (V, E, \lambda)$, where $V = \{1, 2, 3\}$, $E = \{(1, 2), (1, 3), (2, 3)\}$. Concept names are represented as nodes in the graph (see Fig. 3) as shown in formulation (4) [18]:

$$\lambda\langle 1\rangle = FinancialInstitute;$$
$$\lambda\langle 2\rangle = CommonStock; \lambda\langle 3\rangle = Currency; \quad (4)$$

The concept $FinancialInstitute$ is the main concept. Each node in the graph (see Fig. 3) is related to a concept name. The graph roles are defined as follows:

$$\lambda\langle 1, 2\rangle = hasFinancialInstrument;$$
$$\lambda\langle 1, 3\rangle = hasPrincipalAmount;$$
$$\lambda\langle 2, 3\rangle = convertsTo; \quad (5)$$

The extended signature of the graph (see Fig. 3) is:

$$N_C = \{FinancialInstitute, CommonStock, Currency\},$$
$$N_{Rt} = \emptyset, \quad (6)$$
$$N_{Rg} = \{hasFinancialInstrument, convertTo,$$
$$represnetedBy, hasPrincipalAmount\},$$
$$N_I = \{bankOfIrelan, stock1, euro\}$$

N_C represents the set of graph concepts, N_{Rg} is the set of graph roles, and N_I represents the set of individuals. The set of Tbox roles is empty and all roles are graph roles [18]. The formula $G(BankOfIreland, stock1, euro)$ represents an instantiation of the graph.

4.1. Reasoning with Financial Regulations

Reasoning with financial regulations involved with common equity Tier 1 capital ratio calculation process (see Fig. 1) includes:

- Preprocessing the DG which implements regulations consumed by common equity Tier 1 adequacy calculation process,

- Application of derivation rules to the graph-extended knowledge base.

The preprocessing step includes the process of encoding rules for the given DG. It includes rules in program $\mathcal{P}$, equality rules and disjointness rules. In our case, we encode the DG (shown on Fig. 3) into rules according to specification in [18]. Rules in program $\mathcal{P}$ can propagate constraints within DG. The following rules in program $\mathcal{P}$ corresponds to given DG (shown in Fig. 3):

$$hasFinancialInstrument(?x, ?y) \wedge conversTo(?y, ?z) \rightarrow$$
$$hasPrincipalAmount(?x, ?z) \quad (7)$$
$$hasPrincipalAmount(?x, ?y) \wedge representedBy(?y, ?z) \rightarrow$$
$$hasFinacialInstrument(?x, ?z)$$

The DG shown on Fig. 3, contains an inverse role i.e. $conversTo$ is inverse role of $representedBy$ role, so the inverse role can be formalized using the following DG rules:

$$conversTo(?x, ?y) \rightarrow representedBy(?y, ?x) \quad (8)$$

$$representedBy(?x, ?y) \rightarrow conversTo(?y, ?x) \quad (9)$$

Formula (10) specifies one equality rule, while the formula (11) specifies one disjointness rule.

$$G(x_1, y_1, z_1) \wedge G(x_1, y_2, z_2) \rightarrow y_1 = y_2 \quad (10)$$

$$G(x_1, y_1, z_1) \wedge G(x_2, x_1, z_2) \rightarrow \bot \quad (11)$$

Finally, the following semantic web rules must be defined:

$$hasFinancialInstrument(?x, ?y) \land$$
$$hasFinancialInstrument(?x, ?z) \rightarrow$$
$$SameAs(?y, ?z) \quad (12)$$
$$convertsTo(?x, ?y) \land convertsTo(?x, ?z) \rightarrow$$
$$SameAs(?y, ?z) \quad (13)$$
$$hasPrincipalAmount(?x, ?y) \land$$
$$hasPrincipalAmount(?x, ?z) \rightarrow$$
$$SameAs(?y, ?z) \quad (14)$$
$$representedBy(?x, ?y) \land$$
$$representedBy(?x, ?z) \rightarrow$$
$$SameAs(?y, ?z) \quad (15)$$

Rules of the form (12), (13), (14), and (15) are important for detecting inconsistency of graph-extend knowledge base shown in Fig. 3. For example, if individual *'BankOfIreland'* occurs in a new DG instance that does not contain individuals *'stock$_1$'* and *'euro'*, then the knowledge base inconsistency will not be detected. We fix this problem by defining all the roles, shown on Fig. 3, functional using rules of the form $R(x, y_1) \land R(x, y_2) \rightarrow y_1 = y_2$. After providing these rules, the HermiT reasoner would be able to detect inconsistency.

Applying derivation rules, the hyper-tableau algorithm checks the satisfiability of $(\mathcal{R}, \mathcal{A})$ where $\mathcal{R} = [\sim](\mathcal{G}) \cup \mathcal{P}$, and where $\mathcal{A}$ is defined Abox [17]. To prove satisfiability of given graph-extended knowledge base, the hyper-tableau algorithm tries to construct a model of a $(\mathcal{R}, \mathcal{A})$ by applying different derivation rules to $\mathcal{R}$, $\mathcal{A}$. Before applying the rules, it is important to define at least one Abox $\mathcal{A}$. If $\bot \in \mathcal{A}$ then there is a clash and the algorithm will detect inconsistency in the knowledge base.

5. FraTAct – Activity Transformation Framework

FraTAct (Framework for Transforming Activity) supports the transformation of a nescient activity of a financial service process into an intelligent activity. This framework combines two different but complementary technologies which include *ontology based technologies* and *service oriented technologies*. In this section, an overview of the framework is presented and the underlying technique of the framework is described.

The FraTAct framework is developed relying on Service Oriented Architecture (SOA) [27]. It adopts two characteristics of SOA including *loose coupling* and *ubiquity*. The FraTAct framework consists of a process design interface, a process engine, service component repository, and management components. FraTAct is a multi-layer framework depicted in Fig. 4.

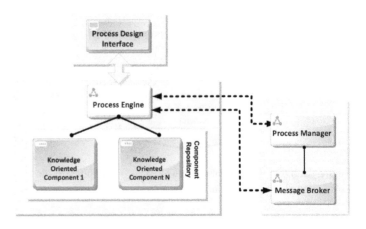

Figure 4. Activity Transaction Framework

In the following, the components of FraTAct framework are briefly described.

(i) *Process Design Interface*: The design interface facilitates the users performing design activities. It offers elements including *activities*, *decision gateways*, etc. for designing financial service business processes.

(ii) *Process Engine*: Process engine is a back-end component which resides at the process layer. A process after designed is deployed in the process engine that executes the process. The process engine executes processes by means of invoking the components for process activities (*aka* services). In other words, the process engine orchestrates the financial services business process activities.

(iii) *Component Repository*: Component repository resides at the component layer. It contains knowledge oriented financial service components that 'actualizes' the process activities. The term 'actualization' refers to *performing operations*. Precisely, the components perform operations of corresponding process activities. The process engine invokes one or more of these financial service components for performing a process activity.

Financial service components are the key constituents in a financial service system. These components perform operations of a corresponding process activity by relying on the financial regulation knowledge base. In effect, a process activity which used to be a simple abstraction that is not aware of financial regulations, turns into an intelligent one. It is worth noting that the intelligence of a financial service process activity is its ability to realize the exact semantics of financial regulations and carry out operations by complying them.

(iv) *Process Manager*: Process manager resides at management layer. It is responsible for performing management tasks such as substituting activities contained in business processes for adapting a change. In addition, it supports modification of business processes.

(v) *Message Broker*: It is a message management component residing at management layer. This component acts to manage messages exchanged between the components, components and process engine, and also between the components. The message broker is connected with the process manager. Any changes occur in processes may promote adjusting the messaging sequences, aggregating messages in message broker.

6. Implementation of FSBA using FraTAct Framework

Throughout this section, we will describe how we have implemented an FSBA using FraTAct framework. According to our understanding, an implementation of a Financial Service based Application(FSBA) will suffice in demonstrating how FraTAct assist in transforming a nescient activity into intelligent activity. Notably, we have implemented common Equity 'Tier 1 Capital Ratio Calculation Process' shown in Fig. 1. Sections include 6.2, 6.3, 6.4 will describe our implementation. In any case, we will provide an overview of the technological aspect of the framework in section 6.1.

6.1. Technological Overview of FraTAct Framework

FraTAct framework shown in Fig. 4 relies on OpenESB [26] that is a middleware comprises a list of components. We will not describe all of these components in this section rather we will briefly introduce the components that are used in FraTAct framework.

FraTAct is built by integrating a process design component. The component provides an interface for designing financial service business process. It offers elements used in designing financial service business processes. The framework integrates a scalable BPEL [21] service engine for executing financial service business processes.

Java Enterprise Edition (Java EE)is integrated into FraTAct. FraTAct Java EE facilitates building ontology oriented financial service components. Since FraTAct is built on the service oriented architecture paradigm, the components are offered as services using Web Service Description Language (WSDL) [34] which describes the service components specifically, the operations, and inbound and outbound messages. FraTAct relies on XML schemas for specifying inputs for the components that perform operations based on the given inputs.

FracTAct was deployed into Glassfish container [9] which is an application server. Furthermore, FraTAct utilizes Composite Application Service Assembly (CASA) to facilitate the users to assemble the financial service components in an FSBA. FraTAct relies on Java Business Integration module (JBI) that offers the binding components used to bind the components, messages, processes, and container. For exchanging messages, FraTAct depends on Simple Object Access Protocol (SOAP) [32].

FraTAct framework utilizes Glassfish administration module as a process manager (shown in Fig. 4) for managing the changes while financial service business processes are running. The module offers a user friendly interface for managing business processes. Moreover, FraTAct integrates the Hermes Java Messaging Service (Hermes JMS) [11] that is a message broker used in publishing and editing messages between service components that hosted at distributed locations.

6.2. Implementation of Service Component

At the first phase of our implementation, we develop service components. The development of service components rely on the mechanism described in section 4. We develop two components bsOntology.java and OntoConsistency.java that perform the *List Common Equity Tier 1 Financial Instruments* and *Verify the List* activities (see Fig. 1). These activities are regulation intensive, should be performed intelligently. Therefore, the components developed in this paper are ontology based. We describe the development of our ontology based service components in this section.

We implement the following financial instruments: Common shares, Stock surplus, Retained earnings, Comprehensive incomes, Other disclosed reserves, Subsidiaries common shares. Minority interests financial instrument is not part of the current implementation. The Implementation consists of three parts (see Fig. 5):

- Implementation financial regulations as local ontologies. For each financial instrument we have one local ontology. For example, common share local ontology.

- Implementation financial regulations as description graphs.

- Integration ontologies and description graphs.

We use OWL2 language to implement local ontologies. To check consistency of the ontologies we use Pellet reasoner [20]. When OWL2 is not expressive enough for implementing financial regulations, then we

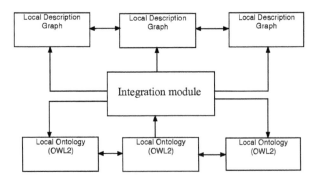

Figure 5. Federation of ontologies in graphs extended knowledge base

use description graphs. To do that, we use the hermit-reasoner API [31] as well as OWL API [25].

To create graph concepts and roles we use *AtomicConcept* and *AtomicRole* classes respectively imported from Hermit API [31]. For the creation of graph roles we allocate memory using the *AtomicRole* class. To check satisfiability of the graph-extended knowledge base, the Hermit reasoner takes ontology and DG as input. In order to do that, we use the *Reasoner* class imported from the Hermit API. For more details, how to implement some financial regulations using description graphs, please see [23]. Imported ontology can not contain built-in predicates because HermiT reasoner does not support reasoning with built-in predicates. For example, the HermiT reasoner is not useful in case of testing consistency of ontology during calculation equity capital in a balance sheet. For such cases, we use Pellet reasoner.

To implement regulations for all financial instruments, integration of ontologies and description graphs is employed. We use a hybrid approach (see Fig. 5) because of the following reasons:

- Local ontologies and graphs are easy to modify. The reason is that some regulations can be changed in the future and it requires modification of local ontologies.

- Integration module connects local ontologies and graphs. It will allow different type of reasoners to check consistency of different ontologies extended with description graphs or just checks the consistency of ontologies without extensions with description graphs.

- Local ontologies and description graphs can be interconnected independently of integration module. For example, one local ontology can be imported into another local ontology.

6.3. Implementation of Financial Service Process

The previous section describes the implementation of service components bsOntology.java and OntoConsistency.java. We wrapped these components using WSDL [34] as services bsOntologyService.wsdl and bsOntoConsistencyService.wsdl and published on Glassfish container. In addition, we developed a client side web service bsOntologyCChkWSDL.wsdl that provides interface for accepting the requests from the client.

Next, we implemented the *Common Equity Tier 1 Capital Ratio* process shown in Fig 1. In order to implement the process, we used BPEL (version 2.0) - an eXtensible Markup Language (XML) based *de facto* language for developing business processes. First, we defined a *Process Scope* that contains the process activities. The scope contains StartOntologyProcess activity that receives requests from client service bsOntologyCChkWSDL.wsdl. Fig. 6 shows the connection between StartOntologyProcess activity and bsOntologyCChkWSDL.wsdl service. The next activity contained in the scope AssignInput copies the input to the subsequent activity ProvideOntology within the scope. It is a *regulation intensive* activity and upon triggering this activity the process engine invokes the ontology oriented service component GetOntologyService.WSDL that actualizes List Common Equity Tier 1 Financial Instruments (see Fig. 1). As this component is an ontology based component, it performs activity by realizing the semantics of financial regulations and in effect, the activity List Common Equity Tier 1 Financial Instruments turns into an intelligent activity.

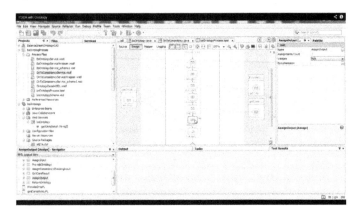

Figure 6. Implementation of Common Equity Tier 1 Capital Ratio Process

Then, we defined AssignConsistencyInput activity within the scope that copy the output produced by ProvideOntology activity into GetConsistencyChkResult which is another regulation intensive activity invokes the ChkConsistency.WSDL component actualizes the Verify the List activity contained in the

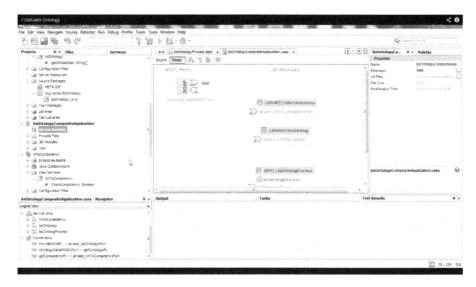

Figure 7. Graphical View of of Financial Service Application

process in Fig. 1. The output is copied by AssignOutput to ReturnOutput that delivers the output to client service bsOntologyCChkWSDL.wsdl. Notably, the start and end events are outside of this scope. Fig. 6 shows the bsOntologyProcess.

A BPEL script bsOntologyProcess.bpel is generated automatically which is then deployed on the processes engine. The BPEL script is added in Appendix.

6.4. Implementation of Common Equity Tier 1 FSBA

We describe how the financial service based application was developed. The application was developed by binding service components with 'Common Equity Tier 1 Capital Ratio Calculation' process that we implemented(described in the previous section). The 'Composite Application Service Assembly (CASA)' facilitates developing the common equity tier 1 capital ratio calculation service based application. We will describe how we assembled the service components into an application.

We generated the Jar files from process bsOntologyProcess.bpel and service components bsOntology.java and OntoConsistency.java. The jar files bsOntologyProcess.jar, bsOntology.jar, and OntoConsistency.jar are then imported on Java Business Integration module that binds the service components using SOAP protocol. The application was then built and deployed in the Glassfish application server. Fig. 7 depicts the development of common equity tier 1 FSBA.

The financial service based application of common equity calculation process is now ready to automate the common equity tier 1 capital ratio calculation process. The application performs the regulation intensive process activities intelligently.

Scalability and performance of the approach on a large scale of DGs depends on the HermiT reasoner

performances discussed in [19]. The reasoner shows significant advantages in reasoning with a large scale ontologies [19] comparing to other standard reasoners such as the Pellet [20]. The open issue is how to query financial ontologies extended with description graphs by using SPARQL.

7. Related Works

In this section, we discuss the related solutions revolving around the financial regulation knowledge base, reasoning, and process. To the best of our knowledge, financial regulation has been scoped in a very small amount of research. Thus, the actual need (e.g., machine readable definition of Basel-based regulations) of the financial service industry has been addressed very briefly in current solutions. Some very recent research including [29], [30], and [1] focus on investigating the issues and challenges, and defining a roadmap for governance, risk and compliance within business processes. Nonetheless, no concrete solution has been discussed in these works. Additionally, a language to capture the compliance requirements of business processes is proposed in [33] and [6]. The language is promising, but offers a very limited number of constructs that do not facilitate defining financial regulations in FSB. Solutions including [7] and [16] are offered to perform reasoning on business processes to verify the correctness of business processes. However, these solutions do not address the problem of this research.

8. Conclusion

In this paper, we have shown how to transform a nescient activity within an FSP into an intelligent activity. In order to support the transformation, we have shown how to encode financial regulations into a graph-extended knowledge base that assists process activities to perform operations intelligently. Additionally, we have shown how to perform reasoning with financial regulations. We have also described the current state of implementation.

We plan to extend the solution in order to enable a system to decide actions for the conditions that are evolved at runtime. In the future, we plan to cover more complex financial regulations. In addition, our intention is to cover financial regulations from various bodies. We also plan to enable these systems to support fuzzy reasoning. Keeping completeness for an incomplete ontology reasoner is a challenge [8] which we have experienced in this research. We will try to tackle this challenge in its future extension.

Currently, FraTAct is a specific framework which serves only financial domain but we plan to add more features that can cover other domains such as business to business(B2B). In B2B, businesses are governed by service-level agreement (SLA). The business processes activities must be accomplished by satisfying SLAs. We plan to develop knowledge-oriented component which provides explicit semantics of SLA clauses for corresponding process activities. We are considering to integrate technologies such as Semantic Markup for Web Services(OWL-S) [24].

Acknowledgement. This research is done within the scope of the Governance, Risk, and Compliance Technology Centre project funded by Enterprise Ireland. The authors thank to Patrick O'Sullivan for his contribution to improve the readability of the paper. The authors thank to the GRCTC Team for supplying inputs related to financial instruments and the Basel Accords. The authors is thankful to Boris Motik from University of Oxford, UK, for his advices in practical applications of DGs in modeling financial regulations.

References

[1] ABDULLAH S.A., SADIQ W.S., and MARTA I. (2010) Emerging Challenges in Information Systems Research for Regulatory Compliance Management. *International Conference on Advance Information System Engineering(CAiSE)*. pp. 251-265.

[2] Apache Jena, http://jena.apache.org/documentation/ontology/index.html

[3] BASEL III (2010) A Global Regulatory Framework. *Bank for International Settlement*.

[4] CUENCA GRAU B., HORROCKS I., MOTIK U., PARSIA P., PATEL-SCHNEIDER P.F., and SATTLER U. (2008) OWL 2: The next step for OWL. *Journal of Web Semantics*. Vol 6(4). pp. 309-322.

[5] IFRSs and XBRL, http://www.ifrs.org/XBRL/XBRL.htm,Lastvisited:May, 2012.

[6] ELGAMMAL E., Turetken O., and HEUVEL W.J.V.D. (2010) Using Patterns for the Analysis and Resolution of Compliance Violations. emphInternational Journal of Cooperative Information System. vol. 21(1), pp. 31-54.

[7] FRANCESCOMARINO C. D., GHIDINI C., ROSPOCHER M., SERAFINI L., and TONELLA P. (2008) Reasoning on Semantically Annotated Processes. *International Conference on Service Oriented Computing*. pp.132-146.

[8] GRAU B.C., MOTIK B., STOILOS G., HORROCKS I. (2012) Completeness Guarantees for Incomplete Ontology Reasoners: Theory and Practice. *Journal of Artificial Intelligence Research*. vol. 43. pp. 419-476.

[9] JAVA COMMUNITY PROCESS (2012). Glassfish Container. Available on: http://glassfish.java.net/

[10] HAQUE R., KRDZAVAC B.N., BUTLER T., (2012) Transforming Nescient Activity into Intelligent Activity. *In Proc. of the Data and Knowledge Engineering (ICDKE 2012)*. isbn:978-3-642-34678-1.

[11] HERMES. Hermes Java Messaging Services(Hermes JMS). Available on: http://www.hermesjms.com/confluence/display/HJMS/Home.

[12] HEALY M.P., and PALEPU G.K. (2003) The Fall of Enron. *Journal of Economic Perspectives*. vol. 17(2). pp. 3-26. Spring Publication.

[13] HORROCKS I., KUTZ O. , and SATTLER U.(2006) The Even More Irresistible SROIQ. *In Proc. of the 10th Int. Conf. on Principles of Knowledge Representation and Reasoning (KR 2006)*. pp. 57-67. (AAAI Press).

[14] HORROCKS I., PATEL-SCHNEIDER P.F., BOLEY H., TABET S., GROSOF B., and DEAN M.SWRL: A Semantic Web Rule Language Combining OWL and RuleML. *W3C Member Submission*.

[15] LEYMANN F. and ROLLER D. (1999) *Production Workflow: Concepts and Techniques*. Prentice Hall; Edition 1, ISBN-10: 0130217530.

[16] MISSIKOFF M., PROIETTI M.,and SMITH F. (2010) Reasoning on Business Processes and Ontologies in a Logic Programming Environment. *In the Proceedings of 3rd Interop-Vlab.It Workshop, CEUR-WS*. pp. 653.

[17] MOTIK B., SHEARER R., and HORROCKS I.(2009) Hypertableau Reasoning for Description Logics. *Journal of Artificial Intelligence Research*. pp.165–228.

[18] MOTIK B., CUENCA GRAU B., HORROCKS I., and SATTLER U. (2008) Representing Structured Objects using Description Graphs. In GERHARD B. and $Jer^o me$ L., editors. *In Proc. of the 11th Int. Joint Conf. on Principles of Knowledge Representation and Reasoning (KR)*. pp. 296–306. AAAI Press, Sydney, NSW, Australia.

[19] SHEARER R., MOTIK B.,and HORROCKS I.. (2008) HermiT: A Highly-Efficient OWL Reasoner. In RUTTENBERG A., SATTLER U., and DOLBEAR C., editors. *In Proc. of the 5th Int. Workshop on OWL: Experiences and Directions (OWLED 2008 EU)*. Karlsruhe, Germany.

[20] Pellet reasoner. http://clarkparsia.com/pellet/

[21] OASIS (2007). Business Process Execution Language. Available on: http://docs.oasis-open.org/wsbpel/2.0/wsbpel-v2.0.pdf

[22] OMG (2011). Business Process Modeling Notation. Available on:http://www.omg.org/spec/BPMN/2.0/

[23] N. Krdzavac, R. Haque, T. Butler, Web-based Reasoning With Balance Sheets, Workshop 'AI on the Web' at the 35th German Conference on Artificial Intelligence (KI2012, AIW 2012), 09/2012, Saarbrucken, Germany, (2012)

[24] OWL-S. Available: `http://www.w3.org/Submission/OWL-S/`. Last visited May, 2012.

[25] OWL API. Available: `http://owlapi.sourceforge.net/`. Last visited May, 2012.

[26] OPENESB COMMUNITY (2012). Open Enterprise Service Bus (OpenESB). Available on:`http://www.open-esb.net/`

[27] PAPAZOGLOU P.M., HEUVEL v.D.W., (2007) Service oriented architectures: approaches, technologies and research issues. *VLDB Journal*. vol. 16(3). pp. 389-415.

[28] POLYVYANYY A., and WESKE M. (2008) Hypergraph-Based Modeling of Ad-Hoc Business Processes. *Business Process Management Workshops*. pp. 278-289.

[29] SADIQ W.S., MUEHLEN Z.M., and INDULSKA M. (2012) Governance, risk and compliance: Applications in information systems. *Information Systems Frontiers*. vol.14(2). pp. 123-124.

[30] SADIQ W.S. (2011) A Roadmap for Research in Business Process Compliance. *In proceedings of BIS (Workshops)*. pp. 1-4

[31] SHEARER R., MOTIK B., and HORROCKS I.(2008) HermiT: A Highly-Efficient OWL Reasoner. *In Alan Ruttenberg, Ulrile Sattler, and Cathy Dolbear, editors, Proc. of the 5th Int. Workshop on OWL: Experiences and Directions (OWLED 2008 EU)*. pp. 26 - 27. `http://hermit-reasoner.com/`

[32] SOAP (2007). Simple Object Access Protocol(SOAP). Available on:`http://www.w3.org/TR/soap/`

[33] TURETKEN O., ELGAMMAL A., HEUVEL W.J.V.D., and PAPAZOGLOU P.M. Capturing Compliance Requirements: A Pattern-Based Approach. *IEEE Software*. vol.29(3). pp. 28- 36.

[34] WSDL (2001) Web Service Description Language. Available on: `http://www.w3.org/TR/wsdl`

9. Appendices

The BPEL script `bsOntologyProcess.bpel` is presented in the following.

BPEL Code:

```xml
<?xml version="1.0" encoding="UTF-8"?>
<process
    name="bsOntologyCChkProcess"
    targetNamespace="http://enterprise.netbeans.org/bpel/
    bsOntologyConsChkProcess/bsOntologyCChkProcess"
    xmlns:tns="http://enterprise.netbeans.org/bpel/
    bsOntologyConsChkProcess/bsOntologyCChkProcess"
    xmlns:xs="http://www.w3.org/2001/XMLSchema"
    xmlns:xsd="http://www.w3.org/2001/XMLSchema"
    xmlns="http://docs.oasis-open.org/wsbpel/2.0/process/
    executable"
    xmlns:sxt="http://www.sun.com/wsbpel/2.0/process/
    executable/SUNExtension/Trace"
    xmlns:sxed="http://www.sun.com/wsbpel/2.0/process/
    executable/SUNExtension/Editor"
    xmlns:sxeh="http://www.sun.com/wsbpel/2.0/process/
    executable/SUNExtension/ErrorHandling" xmlns:sxed2=
    "http://www.sun.com/wsbpel/2.0/process/executable/
    SUNExtension/Editor2"
    xmlns:ns0="http://xml.netbeans.org/schema/
    bsOntChkInputSchema">
```

```xml
<import namespace="http://j2ee.netbeans.org/wsdl/
bsOntologyConsChkProcess/bsOntologyCChkWSDL"
location="bsOntologyCChkWSDL.wsdl" importType=
"http://schemas.xmlsoap.org/wsdl/"/>
<import namespace="http://enterprise.netbeans.org/
bpel/OnToConsistencyServiceWrapper"
location="OnToConsistencyServiceWrapper.wsdl"
importType="http://schemas.xmlsoap.org/wsdl/"/>
<import namespace="http://OnToConsistency.nenad.com/"
location="OnToConsistencyService.wsdl"
importType="http://schemas.xmlsoap.org/wsdl/"/>
<partnerLinks>
    <partnerLink name="ProvideOntChkResultPL"
    xmlns:tns="http://enterprise.netbeans.org/bpel/
    OnToConsistencyServiceWrapper" partnerLinkType=
    "tns:OnToConsistencyLinkType" partnerRole=
    "OnToConsistencyRole"/>
    <partnerLink name="OntologyConsChkPL"
    xmlns:tns="http://j2ee.netbeans.org/wsdl/
    bsOntologyConsChkProcess/bsOntologyCChkWSDL"
    partnerLinkType="tns:bsOntologyCChkWSDL"
    myRole="bsOntologyCChkWSDLPortTypeRole"/>
</partnerLinks>
<variables>
    <variable name="CheckConsistencyOut" xmlns:tns=
    "http://OnToConsistency.nenad.com/" messageType=
    "tns:CheckConsistencyResponse"/>
    <variable name="CheckConsistencyIn"
    xmlns:tns="http://OnToConsistency.nenad.com/"
    messageType="tns:CheckConsistency"/>
    <variable name="BsOntologyCChkWSDLOperationOut"
    xmlns:tns="http://j2ee.netbeans.org/wsdl/
    bsOntologyConsChkProcess/
    bsOntologyCChkWSDL" messageType=
    "tns:bsOntologyCChkWSDLOperationResponse"/>
    <variable name="BsOntologyCChkWSDLOperationIn"
    xmlns:tns="http://j2ee.netbeans.org/wsdl/
    bsOntologyConsChkProcess/bsOntologyCChkWSDL"
    messageType="tns:bsOntologyCChkWSDLOperationRequest"/>
</variables>
<sequence>
    <receive name="StartOntoChkProcess"
    createInstance="yes"
    partnerLink="OntologyConsChkPL"
    operation="bsOntologyCChkWSDLOperation"
    xmlns:tns="http://j2ee.netbeans.org/wsdl/
    bsOntologyConsChkProcess/
    bsOntologyCChkWSDL"
    portType="tns:bsOntologyCChkWSDLPortType"
    variable="BsOntologyCChkWSDLOperationIn"/>
    <assign name="AssignInput">
        <copy>
            <from>$BsOntologyCChkWSDLOperationIn.InputPart/
            ns0:getConsChkResult</from>
            <to>$CheckConsistencyIn.parameters/
            getConsChkResult</to>
        </copy>
    </assign>
    <invoke name="ProcessCChk" partnerLink=
    "ProvideOntChkResultPL"
    operation="CheckConsistency" xmlns:tns=
```

```
"http://OnToConsistency.nenad.com/" portType=
"tns:OnToConsistency"
inputVariable="CheckConsistencyIn"
outputVariable="CheckConsistencyOut"/>
<assign name="AssignOutput">
   <copy>
      <from>$CheckConsistencyOut.parameters/
      return</from>
      <to>$BsOntologyCChkWSDLOperationOut.OutputPart
      /ns0:ConsChkResult</to>
   </copy>
</assign>
```

```
<reply name="ReturnOntChkOutcome"
partnerLink="OntologyConsChkPL"
operation="bsOntologyCChkWSDLOperation"
xmlns:tns="http://j2ee.netbeans.org/wsdl/
bsOntologyConsChkProcess/
bsOntologyCChkWSDL"
portType="tns:bsOntologyCChkWSDLPortType"
variable="BsOntologyCChkWSDLOperationOut"/>
</sequence>
</process>
```

4

Effective user selection algorithm for quantized precoding in massive MIMO

Nayan fang[1,2], Jie Zeng[2,*], Xin Su[2], Yujun Kuang[1]

[1]School of Communication and Information Engineering, University of Electronic Science and Technology of China, Chengdu, China
[2]Tsinghua National Laboratory for Information Science and Technology, Tsinghua University, Beijing, China

Abstract

The downlink of a multi-user massive MIMO wireless system is considered, where the base station equipped with a large number of antennas simultaneously serves multiple users. In this paper, an effective user selection algorithm is proposed for quantized precoding in massive MIMO systems. The algorithm aims at minimizing the correlation of precoders among users by relaxing the optimal problem to be convex and solving it using the Primal Newton's Barrier Method. The complexity of the proposed algorithm is relatively low and the performance shown by the numerical results is close to the exhaustive search method. The advantage of the proposed algorithm increasingly shows up as the transmit antennas increase significantly.

Keywords: massive MIMO, user selection, quantized precoding

1. Introduction

Very-large multiple-input multiple-output (MIMO), also called massive MIMO, is a new technique that can potentially offer large network capacities in multi-user scenarios, where the base stations are equipped with a large number of antennas simultaneously serving multiple users. As is known to us, MIMO channels, created by deploying antenna arrays at the transmitter and receiver, offers the opportunity to upscale the spectral and energy efficiencies by order of magnitude, realizing the vision of high-performance green mobile radio. Thus, massive MIMO has become an promising technology to overcome the explosive rate data demand and has attracted much attention [1]. While the advantages of massive MIMO are conceptually straightforward, realizing massive MIMO in the network level is far from simple and several fundamental deadlocks remains.

When the transmitter in base station (BS) equipped with multiple antennas sends different and independent messages or data symbols to each user simultaneously, it promises high capacity and high-quality wireless communication links by spatial multiplexing and diversity. Under the condition that transmitter and receiver have perfect knowledge of the channel state information (CSI),the theoretical sum-capacity of K users grows linearly with the minimum number of

transmit antennas (N_t) and users which is impractical. The inter-user interference should be dealt to achieve the maximize sum-capacity in the system. Here, the CSI at transmitter (CSIT) as an essential component for precoding plays a key role in mitigating the interference. The transmitter need to utilize the CSIT to preprocess or precode the messages before transmission to reduce the inter-user interference and provide multiplexing gain.

CSIT can be acquired by estimating uplink signals in time division duplex (TDD) systems and receiving feedback signals from users in frequency division duplex (FDD) systems. In TDD systems, channel reciprocity can be utilized for pilot training in the uplink to acquire the complete CSIT, but the performance of the system is constrained by the length of pilot and pilot contamination. In FDD systems, the feedback signals are generated based on the fact that each user estimates the CSI via demodulating pilot signals. According to the obtained CSI, the receiver selects the optimal precoder by transversing the whole codebook which is shared by the receiver and the transmitter, then reports the Precoding Matrix Indicator (PMI) to the transmitter via the limited feedback channel. Afer that the transmitter selects a precoder from the codebook as a function of PMI to accomplish the precoding process for the active users.

The performance of a multi-user MIMO(MU-MIMO) system depends on user selection and power allocation schemes under a given quantized codebook, since the

*Corresponding author. Email: zengjie@tsinghua.edu.cn

inter-user interferences restrict the sum rate of the system. The users subset with minimal spatial channel correlation among each other will be scheduled to mitigate the interference. That means the maximum correlation among selected users need to be minimized in all possible user combinations. Several user selection algorithms have been proposed for downlink MU-MIMO system and the basic principle is to maximize the total throughput of the paired users. The exhaustive enumeration with capacity criticism is optimal, but requires infeasible complexity. As a result, several suboptimal user selection algorithms are studied in [3–7]. These suboptimal user selection algorithms can be briefly categorized as capacity-based user selection scheme and Frobenius norm-based user selction scheme. The principle of capacity-based user selection scheme is to select the first user with highest throughput and greedily select the candidate user which provides the highest total throughput together with those selected users which requires lots of singular value decompositions(SVDs)[3, 4]. Meanwhile, the Frobenius norm-based selection algorithms have been proposed which either considers the norm of the channel gain vector[5] or the orthogonality between the selected channel vectors[6] and even jointly considered[7]. But all these mentioned suboptimal selection methods are based on the whole channel matrix which is unreachable at the transmitter in a limited feedback precoding system. Although the Frobenius norm-based user selection can be extended to the limited feedback system, such as the use on IEEE 802.16m and LTE systems, it requires frequent Gram-Schmidt Orthogonalization procdures(GSOs) to calculate null space and aggregate correlation at each iteration. Thus, the complexity of these algorithms are rather high for massive MIMO with hundreds of transmit antennas(proportional to the cube of the number of transmit antennas).

In this paper, we propose an effective user selection algorithm for the quantizied precoding system with the convex optimization theory. In brief, in order to get the approximate orthogonal user subset, we minimize the correlation of precoders obtained by the feedback PMIs. The complexity of our proposed user selection algorithm increases only linearly with the growth of the transmit antennas, and with the extensive attention of massive MIMO technology, this user selection algorithm will have potential application. Simulation results show that the performance of our proposed convex optimization user selection algorithm outperforms the greedy algorithm, and approaches to the performance of the exhaustive search method.

The rest of the paper is organized as follows. In Section II, we briefly describe the MU-MIMO system and summarize the feedback procedures for the user scheduling and precoding in the cellular systems. We

in Section III present the convex optimization user selection algorithm in detail. Section IV shows the simulation results in comparison with the performance of others. And finally, Section V concludes the paper.

2. System Model

2.1. Massive MIMO System Model

In this paper, we focus on the downlink transmission of a massive MIMO system, i.e., the base station (BS) with massive antennas as the transmitter and several user equipment (UE) as the receiver, and we adopt the multi-user MIMO transmission architecture with codebook based precoding mechanism, as shown in Fig1. The BS is equipped with N_t transmit antennas, and N_u UEs with N_r receive antennas each is served by the BS in the cell; and the maximum number of spatial multiplexing users is K which are chosen from N_u distributed users by an user scheduling algorithm. In this paper, we assume $N_r = 1$, $N_t >> K$, and $K \geq 1$. The overall channel is given by $\mathbf{H} = [\mathbf{H}_1^T, \cdots, \mathbf{H}_K^T]^T$, where $\mathbf{H}_k \in \mathbb{C}^{1 \times N_t}$ is the channel matrix between the base station and the kth user. Let $\mathbf{s} = [s_1, s_2, ..., s_K] \in \mathbb{C}^{K \times 1}$, $\mathbf{P} = [\mathbf{W}_{s(1)}, \mathbf{W}_{s(2)}, ..., \mathbf{W}_{s(K)}] \in \mathbb{C}^{N_t \times K}$, and $\mathbf{x} = \mathbf{P}\mathbf{s} \in \mathbb{C}^{N_t \times 1}$ denote the modulation symbol vector,the precoding matrix and the transmit symbol vector, respectively, where s_k and $\mathbf{W}_{s(k)}$ are for the kth user. After $\mathbf{x}$ passing the channel and being added the noise $\mathbf{z}$ where $\mathbf{z}$ is the additive white complex Gaussian noise(AWGN)vector with covariance matrix $\sigma^2 \mathbf{I}$, we will get the received signal $\mathbf{y}$:

$$\mathbf{y} = \mathbf{HPs} + \mathbf{z} \tag{1}$$

For UE k, the received signal y_k is

$$
\begin{aligned}
y_k &= \mathbf{H}_k \mathbf{x}_k + \sum_{i \neq k} \mathbf{H}_k \mathbf{x}_i + z_k \\
&= \mathbf{H}_k W_k s_k + \sum_{i \neq k} \mathbf{H}_k W_i s_i + z_k
\end{aligned}
\tag{2}
$$

For the equation (2), the first item is a desired signal for the kth user and the second term is inter-user interference which caused the significant degradation of the performance. After the channel matrix $\hat{\mathbf{H}}_k$ is obtained through the channel estimation, the received signal y_k is demodulated as:

$$\hat{s}_k = d_k y_k \tag{3}$$

If the interference is unaware to the receiver, which means the interference is treated as part of the noise, the matched filter(MF) is usually adopted:

$$d_k = (\hat{\mathbf{H}}_k \mathbf{W}_k)^* \tag{4}$$

Each user chooses one of codewords from a pre-design codebook and feedbacks the CSI from the

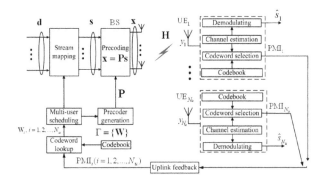

Figure 1. The feedback precoding model of the multi-user MIMO system.

receiver to the transmitter via a limited feedback channel. A codebook containing a list of codewords $\mathbf{W_i}$ which is shared by the the transmitter side and the receiver side is given, with each of codewords reflecting one state of the channel at a certain time, the receiver searches and reports an precoding matrix index within the codebook, called PMI, to the transmitter according to certain rules; If N_u users are waiting to be scheduled at the transmitter, the scheduler will determine $K(< N_u)$ users according to a scheduling algorithm based on the feedback from the users. At the same time, the scheduler will select a precoding matrix $\mathbf{P}$ from the codebook and multiply the signals to accomplish the precoding process for the scheduled K users before transmission.

2.2. Codebook Construction

In Frequency Division Duplexing (FDD) systems, where the fading channel reciprocity cannot be exploited, the performance is achieved by using instantaneous CSI feedback from the users to the BS, between the downlink training phase and the data transmission phase. Hence, the codebook should be pre-designed. Many codebooks have been proposed, such as Kerdock codebook [8], codebooks based on vector quantization [9], Grassmannian packing [10], discrete Fourier transform (DFT)[11], and quadrature amplitude modulation [12]. Codebooks based on vector quantization have taken the channel distribution into account, but have to be re-designed as the change of channel distribution. For uncorrelated channels, the Grassmannian is nearly the optimal codebook, but it has no systematic construction, requiring numerical iterations and high storage. Kerdock codebook, it can provide good system performance with reduced storage requirements and selection computational requirements due to the characteristics of quaternary alphabet and systematic construction. However, in massive MIMO systems, closely-spaced ($\lambda/2$) antenna arrays are favorable to be adopted, and the channels are

usually correlated, hence, the Grassmannian codebook and the Kerdock codebook are not optimized for such correlated channels. Thus, the DFT codebook that is able to trace the substantial correlation is expected to be adopted for massive MIMO system equipped with the closely-spaced antenna array.

2.3. Codeword Search

We can choose the optimal codeword from codebook $\Gamma = \{\mathbf{W}_1, \mathbf{W}_2, ..., \mathbf{W}_N\}$ based on the estimated channel. The codebook is shared by the transmitter and receiver. Codeword selection criteria: the capacity selection criterion is adopted [13], which is to maximize the channel gain as a function of the channel matrix $\hat{\mathbf{H}}$ by traversing the whole codebook Γ:

$$\mathbf{W}_k{}^{\text{opt}} = \mathbf{W}^{\Gamma}(\hat{\mathbf{H}}_k) = \max_{\mathbf{W} \in \Gamma} \left\| \mathbf{W}\hat{\mathbf{H}}_k \right\|_2^2 \qquad (5)$$

where $\mathbf{W}_k{}^{\text{opt}}$ denotes a $N_t \times 1$ matrix of the optimal precoding matrix for the kth user. All users shall report their own requested codewords $\mathbf{W}_i$ to BS.

3. User Selection Algorithm

When multiple users are using the same resources at the same time, there would be severe interference between their signals. Each user should be capable of decoding his respective stream by reducing the interference due to other stream. This can be achieved by pairing users whose procedures are orthogonal to each other in a dataregion and precoding them appropriately at transmitter and post-processing at the receiver so that each user sees only its own information. For the limited feedback systems, the user selection algorithm neither capacity-based user selection algorithm nor Frobenius norm-based user selection algorithm can be employed for massive MIMO, so we propose an effective the user selection method according to the orthogonalization property of precoding matrix. In this section, we formulate the problem of user selection as a constrained convex optimization problem[15] that can be solved efficiently using numerical methods such as interior-point algorithms.

The received signal for user k can be expressed as (2).With formula (4) the demodulated signal can be written as

$$\hat{s}_k = (\mathbf{H}_k \mathbf{W}_k)^* y_k$$
$$= |\mathbf{H}_k \mathbf{W}_k|^2 s_k + (\mathbf{H}_k \mathbf{W}_k)^* \sum_{i \neq k} \mathbf{H}_k \mathbf{W}_i s_i + (\mathbf{H}_k \mathbf{W}_k)^* z_k \qquad (6)$$

The second term of the equation (6) is the inter-user interference. And the interference power of the demodulated signal to the user k is

$$
\begin{aligned}
P_{\text{int}}^k &= \sum_{i \neq k} |(\mathbf{H}_k \mathbf{W}_k)^* \mathbf{H}_k \mathbf{W}_i|^2 \mathrm{E}\left[s_k^* s_k \right] \\
&= \sum_{i \neq k} |(\mathbf{H}_k \mathbf{W}_k)^* \mathbf{H}_k \mathbf{W}_i|^2 \mathrm{E}\left[s_k^* s_k \right]
\end{aligned} \tag{7}
$$

Thus, to minimize P_{int}^k at each step, a candidate user which is orthogonal to previously selected users to be chosen. However, such a user maybe do not exist in practical systems. So under the given precoding technique, we instead select a user with the precoder orthogonal, by greatest extent, to the matrix formed by the precoders of previously selected users. We release our goal to minimize the preceding vector correlation to get the desired user subset S. i.e.

$$
\min \left\| \mathbf{P}_{(S)}^H \mathbf{P}_{(S)} \right\|_1 \tag{8}
$$

where S denotes the desired user subset and $\|\cdot\|_1$ is a matrix norm which returns the value of maximum column sum.

In order to select a suboptimal user subset, we define $\Delta_i (i = 1, \cdots, Nu)$ as the user selection variable for each user as the selection index such that,

$$
\Delta_i = \begin{cases} 1 & i^{th} \ UE selected \\ 0 & otherwise \end{cases} \tag{9}
$$

Now consider a $Nu \times Nu$ diagonal matrix Δ used for user selection at the BS, which has Δ_i as its diagonal entries. The diagonal matrix Δ is represented as

$$
\Delta = \begin{pmatrix} \Delta_1 & & 0 \\ & \ddots & \\ 0 & & \Delta_{Nu} \end{pmatrix}_{Nu \times Nu} \tag{10}
$$

where Δ_i is defined as in (9). Then , the selected users' precoding matrix is $\mathbf{P}\Delta$. Using (8), the modified correlation equation for the massive MIMO can be rewritten as

$$
\min_{\Delta} \left\| (\mathbf{P}\Delta)^H \mathbf{P}\Delta \right\|_1
$$
$$
subject to :
$$
$$
\Delta_i \in \{0,1\} \rightarrow condition1 \tag{11}
$$
$$
trace(\Delta) = \sum_{i=1}^{N_u} \Delta_i = K \rightarrow condition2
$$

The equation in (8) when the substitute $\mathbf{P}_{(S)} = \mathbf{P}\Delta$ is concave in Δ_i. Since the variables Δ_i are constrained to be binary integer, this renders the users selection problem in massive MIMO NP-hard. In order to solve this problem, the concept of linear programming relaxation is used[16]. Linear programming relaxation of the 0-1 integer program is adopting a weaker constraint such that each variable is a real number within the interval [0,1] rather than the constraint that each variable must be 0 or 1 , i.e., for each constraint of the form, one uses a pair of linear constraints $0 \leq \Delta_i \leq 1$ instead of $\Delta_i = \{0, 1\}$ of original integer program. With the linear programming method, we transforms the NP-hard optimization problem of integer programming into a linear program that can be solved in polynomial time. We can prove that the problem with relaxation is an convex optimization problem that the objective function is concave and the constrains is linear inequalities. The solution to the relaxed linear problem can be used to gain information about the solution of original integer problem. Applying this technique, the user selection problem in the massive MIMO can be expressed as

$$
\min_{\Delta} \left\| (\mathbf{P}\Delta)^H \mathbf{P}\Delta \right\|_1
$$
$$
subject to :
$$
$$
0 \leq \Delta_i \leq 1 \rightarrow condition1 \tag{12}
$$
$$
trace(\Delta) = \sum_{i=1}^{N_u} \Delta_i = K \rightarrow condition2
$$

Thus we relaxing a combinatorial optimization problem to a solvable problem with a convex objective function. This optimization problem yields a fractional solution, from which the K largest Δ_i are selected and their indices represent the optimal users.

4. Numerical Results

Based on the selection schemes, we present some simulation results in this section. The simulation procedure follows the system model in Fig 1. Without loss of generality, we take the case of 64 transmit antennas as an example for the performance evaluation of the user selection scheme based on DFT codebook.

For comparison, we show the performances of the massive MIMO system with various user selection schemes: the optimal selection (exhaustive search), the proposed selection scheme, the Frobenius norm-based selection scheme, the greedy selection scheme and the random selection. The greedy selection method is that we iteratively selected the candidate user whose precoding vector is of minimizing correlation together with those selected users. We also shows the theoretical curve with perfect CSIT as the upper bound. Table 1 gives the parameter configuration for the following simulations.

We evaluate the capacity as obtained with our selection algorithm, as well as that obtained by exhaustive search, for practical system parameters. Fig 2 depicts the achievable sum rate of the exhaustive search, the proposed selection scheme,the Frobenius

Table 1. Parameters Configuration

Parameter	Value
Channel model	Winner II
BS antenna setup	64 co-polarized antennas, $\lambda/2$ spacing
UE antenna setup	Single antenna
System frequency (GHz)	2.1
Number of channel realizations	1000
Number of UEs	10, 20
Codebook size (bit)	10

Table 2. The complexity order of different user selection scheme

Schemes	Proposed	Frobenius norm-based	Greedy
Complexity	$o(N_u^{2.5}N_t)$	$o(N_u N_t^3)$	$o(N_u^2 N_t)$

norm-based selection scheme, the greedy and the random selection for six scheduled UEs ($K = 6$) as a function of SNR under 10 UEs. The results indicate that the scheme applied to the DFT-URA deployment has remarkable performance gain compared to the random selection and also outperforms the Frobenius norm-based and greedy scheme. It is because the proposed scheme brings in a better orthogonality of selected UEs than the random selection, suitable for multi-user MIMO in the case where UEs are separated spatially in angular domain. Besides, we show the performance of eight scheduled UEs under 30 UEs in Fig 3 and the performace of greedy scheme is even inferior for the reason that the larger the number of the total user, the more likely it would fall into local optimum.

Optimal selection involves an exhaustive search over all possible $\binom{N_u}{K}$ subsets of UEs requiring $\binom{N_u}{K}K$ additions/multiplications, which grows exponentially with Nu for $K \approx N_u/2$. This can be seen using Stirling's approximation for the factorial. At the same time, the greedy scheme with the complexity of $o(N_u^2 N_t)$. when we use the Primal Newton's Barrier Method(PNB) to solve convex relaxation(12), For the barrier method , the number of Newton steps is upper bounded by $\sqrt{Nu}$ [16]. Each Newton step has a complexity $o(N_u^2 N_t)$. Thus, the total complexity is $o(N_u^{2.5}N_t)$ which is a significant improvement to exhaustive search. We summarize the complexity order of the scheme in table 2. We remark that the complexity of the proposed algorithm is linearly of the number of transmit antennas, which has potential advantages in massive MIMO.

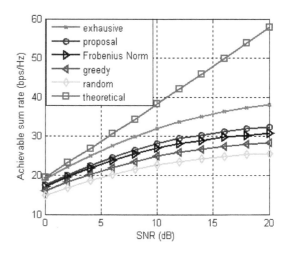

Figure 2. Achievable sum rate with K = 6,Nu=10

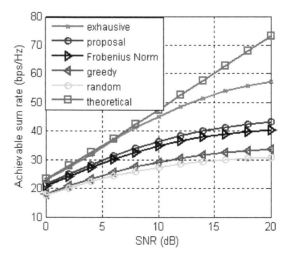

Figure 3. Achievable sum rate with K = 8,Nu=30

5. Conclusion

In this paper, the UE selection scheme for the massive MIMO systems with limited feedback precoding techniques is discussed. Under the codebook design for the URA deployment of the numerous closely-spaced antennas, which would be probably adopted by Massive MIMO, an effective UE selection scheme is proposed. The simulation results show that the proposed scheme outperforms the random selection in terms of the achievable sum rate under the scenario via simulations. The proposed UE selection scheme can contribute to multi-user scheduling for massive MIMO, which would be probably applied in future Beyond 4G systems. Considering the update of 3D channel model, advanced jointing scheduling and antenna selection algorithms should be further studied.

Acknowledgements. This work was supported by Tsinghua-Samsung Joint Research Program, National S&T Major Project (No.2013ZX03001-001), China 863 Project (No. 2012AA011402), Sino-Swedish Strategic Cooperative Program (No. 2012DFG12010) and National Basic Research Program of China(No. 2013CB329002).

References

[1] RUSEK F., PERSSON D., KIONG LAU B. and LARSSON E.G. (October 21, 2011) "Scaling up MIMO:Opportunities and Challenges with Very Large Arrays", *IEEE Signal Processing Magazine.*.

[2] GOLDSMITH, JAFAR S.A. and VISWANATH S. (2003) "Capacity limits of MIMO channels", *IEEE J. Sel. Areas Commun., vol. 21, no. 5, June.*

[3] ZHOU B., BAI B. and LI Y. (2011) "Chordal Distance-Based User Selection Algorithm for the Multiuser MIMO Downlink with Perfect or Partial CSIT", *IEEE International Conference (AINA).*

[4] WANG F., and BIALKOWSKI M. (2011) "A joint design of transmit antenna selection and multiuser scheduling for multiuser MIMO systems employing block diagonalization precoding scheme", *International Conference on Wireless Communications and Signal Processing (WCSP).*

[5] SHEN Z., CHEN R. and ANDREWS J.S. (2006) "Low complexity user selection algorthims for multiuser MIMO systems with block diagonalization", *IEEE Trans. Signal Processing.*

[6] KAVIANI S., and KRZYMIE'N W.A. (Dec. 2008) "Capacity limits of MIMO channels", *IEEE Global Conf. Telecommun.*, pp. 1-5.

[7] ZHANG, GOLKAR Y., SOUSA B. and ELVINO S. (2011) "Efficient User Selection for Downlink Zero-Forcing Based Multiuser MIMO Systems", *IEEE Vehicular Technology Conference (VTC Fall).*

[8] INOUE T., HEATH R.W. (2009) "Capacity limits of MIMO channels", *IEEE Trans. Signal Process*, vol. 57, no. 9, pp. 3711-3716.

[9] ROH J., and RAO B. (2006) "Design and analysis of MIMO spatial multiplexing systems with quantized feedback", *IEEE Trans. Signal Process*, vol.54, no. 8, pp. 2874-2886.

[10] LOVE D.J. and HEATH R.W.(2005) "Limited feedback unitary precoding for spatial multiplexing systems", *IEEE Trans. Inf. Theory*, vol. 51, no.8, pp. 2967-2976.

[11] SCHOBER K., WICHMAN R. and KOIVISTO T. (2011) "MIMO adaptive codebook for closely spaced antenna arrays", *in Proc. European Signal Processing Conference (EUSIPCO)*, pp. 1-5.

[12] RYAN D.J., VAUGHAN I., CLARKSON L., COLLINGS I.B., GUO D. and HONIG M.L. (2007) "QAM codebooks for low-complexity limited feedback MIMO beamforming", *in Proc. IEEE Int. Conf. Commun.*, pp.4162-4167.

[13] LOVE D.J., HEATH R.W. and JR. (2005) "Limited feedback unitary precoding for spatial multiplexing systems", *IEEE Trans. Inf. Theory* vol. 51, no.8, pp. 2967-2976.

[14] S. BOYD and L. VANDENBERGHE (2004) "Convex Optimization", *Cambridge University Press.*

[15] "Linear programming relaxation." http:/en.wikipedia.org/wiki/ *Linear_programming_relaxation,[online:access 2012]*

[16] DUA A., MEDEPALLI K. and PAULRAJ A. (2006) "Recieve antenna selection in MIMO systems using convex optimation", *Wireless Communications.IEEE Transactions.*

5

Hierarchical Codebook Design for Massive MIMO

Xin Su[1], Shichao Yu[1,2], Jie Zeng[1,*], Yujun Kuang[2]

[1]Tsinghua National Laboratory for Information Science and Technology, Tsinghua University, Beijing, China
[2]University of Electronic Science and Technology of China (UESTC), Chengdu, China

Abstract

The Research of Massive MIMO is an emerging area, since the more antennas the transmitters or receivers equipped with, the higher spectral efficiency and link reliability the system can provide. Due to the limited feedback channel, precoding and codebook design are important to exploit the performance of massive MIMO. To improve the precoding performance, we propose a novel hierarchical codebook with the Fourier-based perturbation matrices as the subcodebook and the Kerdock codebook as the main codebook, which could reduce storage and search complexity due to the finite alphabet. Moreover, to further reduce the search complexity and feedback overhead without noticeable performance degradation, we use an adaptive selection algorithm to decide whether to use the subcodebook. Simulation results show that the proposed codebook has remarkable performance gain compared to the conventional Kerdock codebook, without significant increase in feedback overhead and search complexity.

Keywords: Massive MIMO, Kerdock codebook, Fourier-based perturbation, adaptive selection.

1. Introduction

Generally, for the MIMO system, the more antennas the transmitters/receivers equipped with, the higher degree of freedom that the propagation channels can provide, and the higher spectral efficiency, link reliability the system can provide[1]. Therefore, to meet the rapidly increasing demand of high-rate driven by the proliferation of mobile PCs, Smartphones, tablets, etc., and to exploit the precious wireless frequency resources, the massive MIMO system with numerous antennas in the base station (BS) has become an emerging research field. Moreover, in the research of fifth-generation mobile communication system(5G), massive MIMO has become an important research direction.

However, for practical application, the number of antennas should be restricted. Given a fixed number of antennas, we need to explore the massive MIMO system potentials through precoding techniques, which can mitigate the inter-user interference and improve the received SNR. In a closed-loop MIMO frequency division duplexing (FDD) system, the same codebook with a finite set of precoding matrices known to both the transmitter and the receiver should be pre-designed. A larger size codebook can increase the performance of precoding (i.e. Beamforming) at the cost of increasing the uplink feedback overhead, the storage requirement and search complexity. Therefore, it is important to explore better approaches for codebook design to balance the accuracy of precoding and the feedback overhead, and thus optimizing the overall performance.

There is a rich body of literature on the codebook design for traditional MIMO systems. But for the massive MIMO system, they have some disadvantages. Paper [2] has proposed a codebook based on vector quantization, which takes the channel distribution into account, resulting in a need of re-design when the channel distribution changes. The codebook based on Grassmannian packing in [3], [4] is nearly the optimal codebook for uncorrelated channels, but its construction requires numerical iterations, high storage requirement and search computational requirement, and thus restricts its application in massive MIMO. The Discrete Fourier Transform (DFT) codebook with simple systematic construction and appropriate storage requirement has been proposed in [5], but its performance is too low to meet the precoding requirement of massive MIMO, especially under the uncorrelated channels. Compared to these codebooks, the Kerdock codebook [6] can be extended and applied to massive MIMO system easily since its considerable precoding performance with reduced storage and selection computational requirements, besides, it has the characteristics of finite quaternary alphabet and systematic construction.

*Corresponding author. Email: zengjie@tsinghua.edu.cn

In a massive MIMO system, the feedback overhead, storage requirement and codeword searching computational complexity would be significantly increased when the antenna array grows large. To compromise the system performance and feedback overhead, we propose a new hierarchical codebook combining the characteristics of the Kerdock main codebook with the Fourier based perturbation subcodebook. Moreover, to subtract the unnecessary feedback overhead for subcodebook, which has no influence on the precoding performance in some cases, we propose an adaptive selection algorithm to decide when to adopt the hierarchical codebook or the Kerdock codebook only according to the SNR differences among data streams.

The rest of the paper is organized as follows. Section 2 introduces the system model. Section 3 presents the detail design of a hierarchical codebook as well as an adaptive selection algorithm. Simulation results are shown in Section 4. Finally, Section 5 gives the conclusions.

2. System Model

In this paper, we focus on the downlink of a massive MIMO system with N_t transmit antennas at the BS and N_r receive antennas at the user equipment(UE). We adopt the closed-loop MIMO transmission architecture with precoding and feedback mechanism.

Generally, the data processing at the transmitter has a number of steps. Firstly, the data stream which is inputted into the system would be demultiplexed into L substreams, then we get the substream vector $\mathbf{s} = \{s_1, s_2, \cdots, s_{N_r}\}$. L is limited by $1 \le L \le \min(N_t, N_r)$. The condition when $L = 1$ is the case of transmission beamforming, while $L > 1$, we call it multiplexing. Next the data is preprocessed by a precoding matrix $\mathbf{W}$, which is chosen from the codebook stored in the transmitter according to the reported Precoding Matrix Indicator (PMI), the transmitted signal $\mathbf{x}$ shows:

$$\mathbf{x} = \mathbf{W}\mathbf{s} \tag{1}$$

After preprocessed date passing the channel and being added the noise, the signal received by the UE can be expressed as:

$$\mathbf{y} = \mathbf{H}\mathbf{W}\mathbf{s} + \mathbf{z} \tag{2}$$

where $\mathbf{H} \in (N_r \times N_t)$ denotes the fading channel matrix with its entry H_{ij} denoting the channel response from the $j - th$ transmit antenna to the $i - th$ receive antenna, and $\mathbf{z}$ denotes the white complex Gaussian noise (AWGN) vector with covariance matrix $N_0\mathbf{I}_{N_r}$.

Then turn to the hierarchical codebook in limited feedback precoding system. Fig. 1 illustrates the precoding model of hierarchical codebook. It is constructed by two parts, including the main codebook

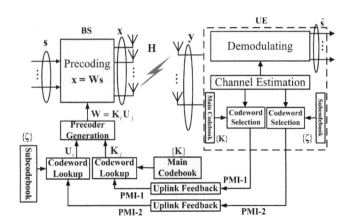

Figure 1. The Limited Feedback Precoding Model

$\mathbf{K} = \{\mathbf{K}_1, \mathbf{K}_2, \cdots, \mathbf{K}_N\}$ as the first part and the subcodebook $\zeta = \{\mathbf{U}_1, \mathbf{U}_2, \cdots, \mathbf{U}_G\}$ as the second part. Assuming that the feedback overhead is limited to B bits, the size of these two parts should satisfy $N \times G = 2^B$. Through receiving the reference signals from the BS and estimating the channel to acquire the Channel State Information (CSI), the UE can select an optimal codeword from $\mathbf{K}$ and ζ respectively according to the CSI and report the corresponding indexes of precoding matrix named as PMI to the BS via the limited feedback channel [7]. Combining the two matrix $\mathbf{K}_i$ and $\mathbf{U}_j$ from the main codebook and the subcodebook respectively, the precoding matrix can be constructed as $\mathbf{W} = \mathbf{K}_i\mathbf{U}_j$. Then the (2) can be rewritten as:

$$\mathbf{y} = \mathbf{H}\mathbf{K}_i\mathbf{U}_j\mathbf{s} + \mathbf{z} \tag{3}$$

At the receiver, the UE obtains the estimated channel matrix through channel estimation and demodulates the received signal by employing various methods, such as Zero-Forcing (ZF) or Minimum Mean Square Error (MMSE), etc.

3. Design of the Novel Hierarchical Codebook

For limited feedback precoding, codebook design is crucial. With the characteristics of quaternary alphabet and systematic construction, Kerdock codebook has relatively low storage requirement and selection computational requirement, which are critical factors in massive MIMO system. However, it is not sufficient since the number of precoding matrices based on the Kerdock is limited. Paper [11] has proved that adding the unitary perturbation to codebook could improve the performance of the codebook with the linear ZF or MMSE receiver. In order to enhance the performance of the massive system, we propose a hierarchical codebook which adopts Kerdock as the first part of the codebook,

Table 1. Codebooks for $L = 1 \sim 4$

L	1	2	3	4
Codeword $\mathbf{K}_i$		$\mathbf{Y}_{k_i}^{\{L_i\}}$		
Index i		$32k_2 + k_1$		
k_1		0-31		
k_2	0-31	0-15	0-7	0-7
Size(bits)	10	9	8	8

Note:$L_i = Lk_2, Lk_2 + 1, ..., Lk_2 + L - 1$

and Fourier-based perturbation matrices as the second part. Moreover, based on the proposed hierarchical codebook, we proposed an adaptive selection algorithm to decide when to use the subcodebook.

3.1. Main codebook design

The basic idea of the Kerdock codebook design is using the feature of Mutually Unbiased Bases (MUB) to construct the precoding matrix, and with the main feature that all the elements of the matrix in the codebook are composed of ± 1 and $\pm j$.

An MUB is the set of bases satisfying the mutually unbiased property. Supposing $\mathbf{S} = \{\mathbf{s}_1, \mathbf{s}_2, \cdots, \mathbf{s}_M\}$ and $\mathbf{Q} = \{\mathbf{q}_1, \mathbf{q}_2, \cdots, \mathbf{q}_M\}$ are two $M \times M$ orthonormal bases. If the column vectors drawn from $\mathbf{S}$ and $\mathbf{Q}$ satisfy $|\langle \mathbf{s}_i, \mathbf{q}_j \rangle| = 1/\sqrt{M}$, we can say that they have the mutually unbiased property [8].

The Kerdock codebook has several construction methods such as Sylvester-Hadamard construction and power construction [6]. In this paper, we adopt the Sylvester-Hadamard construction. The construction method is presented as follows:

Firstly, we generate a number of diagonal matrices $\mathbf{D}_n$ with the size of $N_t \times N_t$, for $n = 0, 1, 2 \cdots, N_t - 1$, $N_t = 2^B$, i.e., N_t should be the power of 2. The details about their generation can be referred to [9].

Then we construct the corresponding orthogonal matrix:

$$\mathbf{Y}_n = \frac{1}{\sqrt{N_t}} \mathbf{D}_n \hat{\mathbf{H}}_{N_t} \quad (4)$$

where $\hat{\mathbf{H}}_{N_t}$ is the $N_t \times N_t$ Sylvester-Hadamard matrix:

$$\hat{\mathbf{H}}_{N_t} = \hat{\mathbf{H}}_2 \otimes \hat{\mathbf{H}}_2 .... \quad (5)$$

where $\hat{\mathbf{H}}_2 = \begin{pmatrix} 1 & 1 \\ 1 & -1 \end{pmatrix}$

Finally, we need to construct the codebook by selecting unique column combinations from each $\mathbf{Y}_n$ according to the number of substreams, i.e. L. Taking $N_t = 32$ as an example, the codebooks for $L = 1 \sim 4$ are shown in Table 1.

Besides, having the estimated channel matrix, the codeword selection is an important issue. In this paper, for the main codebook, we choose the Minimum

Singular Value Selection Criterion (MSV-SC) to select the optimal codeword from $\mathbf{K}$.

For the beamforming, the beamformer that minimizes the probability of symbol error for maximum ratio combining receiver is expressed as:

$$\hat{\mathbf{f}}[i] = \arg \max_{\mathbf{f} \in \mathbf{K}} \|\mathbf{H}[i]\mathbf{f}\|_2^2 \quad (6)$$

where $\mathbf{f}$ denotes a $N_t \times 1$ matrix.

For spatial multiplexing with a ZF receiver, the minimum singular value selection criterion is expressed as:

$$\hat{\mathbf{F}}[i] = \arg \max_{\mathbf{F} \in \mathbf{K}} \lambda_{\min}\{\mathbf{H}[i]\mathbf{F}\} \quad (7)$$

where λ_{min} denotes the minimum singular value of the argument. This selection criterion approximately maximizes the minimum substream Signal-to-Noise Ratio (SNR).

3.2. Fourier-based perturbation subcodebook design

Several methods for constructing hierarchical codebooks have been proposed in [10]. Rotation-based perturbation is simple but it demands a large number of matrices to gain high performance. In this paper, we employ a novel construction method with Fourier based perturbation matrices since it can track the short-term change of the channel and deal with the SNR imbalance among the substreams.

Because of the high throughput of massive MIMO system, our main consideration of subcodebook design is the bit error ration (BER) improvement. The unitary perturbation can enhance the performance of the codebook with linear receivers like ZF and MMSE. Using the linear ZF receiver, the SNR of the $k - th$ data stream can be expressed as:

$$SNR_k(\mathbf{H}, \mathbf{K}_p) = \frac{1}{N_0[(\mathbf{K}_p^H \mathbf{H}^H \mathbf{H} \mathbf{K}_p)^{-1}]_{kk}} \quad (8)$$

Then we add the perturbation matrix $\mathbf{U}_p$ to the precoding matrix: $\mathbf{W}_p = \mathbf{K}_p \mathbf{U}_p$ ($\mathbf{K}_p$ is the main codebook). We know that the BER performance of a precoding system is dominated by the performance of the substream with the lowest SNR. So the best condition is that every substream has an equivalent SNR. Using ZF receiver, the lowest SNR can be expressed as:

$$SNR_{\min}^{ZF}(\mathbf{H}, \mathbf{W}_p) = \min_{1 \leq k \leq L} SNR_k^{ZF}(\mathbf{H}, \mathbf{W}_p)$$
$$= \frac{1}{\max_{1 \leq k \leq L} (N_0[\mathbf{U}_p^H (\mathbf{K}_p^H \mathbf{H}^H \mathbf{H} \mathbf{K}_p)^{-1} \mathbf{U}_p]_{kk})} \quad (9)$$

We can see that the SNRs of the substreams are determined by the diagonal elements of the matrix $(\mathbf{U}_p^H \mathbf{F}_p^H \mathbf{H}^H \mathbf{H} \mathbf{F}_p \mathbf{U}_p)^{-1}$. For pursuing better BER performance, we should balance the SNR of every substream. The optimal matrix $\mathbf{U}_{opt}$, which can make the SNRs of different substreams approximately equal, would provide a room for balancing the SNRs of the substreams to maximize the overall BER performance. And we should note that the unitary perturbation can affect the performance of the codebook with linear receivers (ZF, MMSE), but it has no impact on the ML receiver.

Next is the method to construct the subcodebook matrix $\mathbf{U}_g$. The Fourier based perturbation matrices, used in the second part of the codebook, can be designed as follows:

$$\mathbf{U}_g = \Lambda_g \mathbf{D}_L, g = 0, 1, ..., G - 1 \tag{10}$$

where $\mathbf{D}_L$ is the normalized DFT matrix to be given in (12), and Λ_g is the rotation diagonal matrix to be given in (13), L is the number of layers and G is the total number of $\mathbf{U}_g$.

$$\mathbf{D}_L = \{d_{kl}, k, l = 0, ..., L - 1\}, d_{kl} = \frac{1}{\sqrt{M}} \exp(\frac{j2\pi kl}{M}) \tag{11}$$

$$\Lambda_g = diag(1, \exp(\frac{j2\pi g}{LG}), \cdots, \exp(\frac{j2\pi(L-1)g}{LG})) \tag{12}$$

Finally, like main codebook, we should discuss the codeword search scheme for subcodebook. From the design of the subcodebook, we can see that the search of the subcodebook should rely on the search of main codebook in addition to the channel matrix. Since we construct the subcodebook based on BER improvement, the selection criterion for the second codeword can be expressed as:

$$\mathbf{U}_{opt} = \arg \min_{\mathbf{U} \in \zeta} \max_{1 \leq k \leq L} ([\mathbf{U}^H (\mathbf{I}_L + \frac{1}{N_0} \mathbf{K}_p^H \mathbf{H}^H \mathbf{H} \mathbf{K}_p)^{-1} \mathbf{U}]_{kk}) \tag{13}$$

3.3. Adaptive selection

In this paper, we have proposed a novel hierarchical codebook in addition. The design of the hierarchical codebook with Fourier based perturbation is to balance the SNRs of different substreams and make the SNRs of different data substreams in the precoding system approximately equal. So when the SNR of each substream have big difference, massive MIMO system would have remarkable performance gain through using subcodebook. However, under the condition where the SNRs of different substreams have little difference, the second part of the codebook(subcodebook) makes little sense on the precoding performance. Thus, it is unnecessary to feedback the PMI of the second part. For this reason, we proposed an adaptive hierarchical selection algorithm according to the substream SNRs. In order to mark the codebook we use, we define a bit

Table 2. Simulated Codebook Type Selection Algorithm

Function GetCTI $(SNR[], L)$
$//L$: the number of substreams
$//\varepsilon$: the fixed threshold set previously

$//CTI$: codebook type indicator
$CTI = 0$
for $i = 0 : L - 1$
    for $j = 0 : L - 1$
        if $SNR[i] - SNR[j] > \varepsilon$ **then**
            $CTI = 1$
            break
        end if
    end for
end for
return CTI

variable named Codebook Type Indicator (CTI). Firstly, we compute the difference between any two SNRs of spatial multiplexing streams. Then if all the differences is lower the fixed threshold that we set previously, the CTI will be set as 0, which means we will only use the Kerdock codebook; otherwise, we will use the hierarchical codebook with Fourier based perturbation. Through this method, we can save some feedback overhead. The pseudo code of this adaptive algorithm is shown in Table 2.

4. Numerical Results

In this section, we present some simulation results in the configuration of $N_t = 32, N_r = 3$. The independent and identically distributed (i.i.d.) Rayleigh flat fading channels are assumed. Besides, receiver is able to acquire the perfect channel information, and the ZF method is adopted for equalization. The size of the proposed hierarchical codebook is denoted as $[N, G]$, where N and G are the size of the main codebook and the perturbation subcodebook respectively. For this antenna deployment, we have $N = 1024, G = 8$.

Fig. 2 and Fig. 3 show the curves of BER performance for different codebooks under the conditions with little and great SNR differences of different data substreams, respectively. The SNR in the x-axis denotes the average SNR of all the streams. For the purpose of comparison, we also present the BER performance under the ideal condition with the perfect CSI, as well as under the worst condition whithout precoding.

From Fig. 2, we can see that the proposed codebook has little BER performance gain compared with the original Kerdock codebook, since in most cases only the main codebook is used due to little SNR differences among the substreams. While in Fig. 3, we add

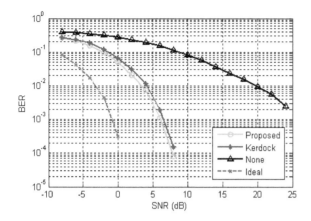

Figure 2. The BER Performance of Different Codebooks (little SNR differences)

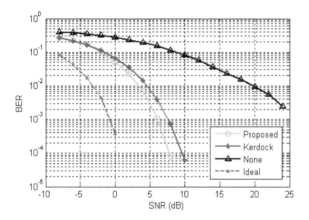

Figure 3. The BER Performance of Different Codebooks (great SNR differences)

different powers of noise for different substreams. It is shown that the proposed codebook has significant BER performance gain since the Fourier-based perturbation subcodebook make a difference.

5. Conclusion

In this paper, we proposed a hierarchical codebook for the massive MIMO system. We adopt the Kerdock codebook as the main codebook of the hierarchical codebook due to its low requirement for storage and selection computational complexity. To balance the SNR differences among different data substreams, thus getting higher precoding performance, we adopt the Fourier-based perturbation matrices as the subcodebook. In addition, for the purpose of compromise

between the system performance and feedback overhead, we proposed an adaptive selection algorithm to reduce the feedback overhead and search complexity without noticeable precoding performance degradation. Simulation results show that compared to the Kerdock codebook, the proposed codebook has significant performance gain, especially under the condition with large SNR differences among substreams.

Acknowledgements. This work was supported by National S&T Major Project (No.2013ZX03001-001), China 863 Project (No. 2012AA011402), Sino-Swedish Strategic Cooperative Programme (No. 2012DFG12010) and National Basic Research Program of China(No. 2013CB329002).

References

[1] Rusek F.,Persson D., Lau B. K., and Tufvesson F., "Scaling up MIMO: opportunities and challenges with very large arrays," *IEEE Signal Processing Magazine*, pp. 1-44, Oct. 21, 2011.

[2] Inoue T.and Heath R. W., "Kerdock codes for limited feedback precoded MIMO systems," *IEEE Trans. Signal Process.* vol. 57, no. 9, pp. 3711-3716, 2009.

[3] Roh J.and Rao B., "Design and analysis of MIMO spatial multiplexing systems with quantized feedback," *IEEE Trans. Signal Process* vol. 54, no. 8, pp. 2874-2886, 2006.

[4] Love D. J.and Heath R. W., "Limited feedback unitary precoding for spatial multiplexing systems," *IEEE Trans. Inf. Theory*, vol. 51, no. 8, pp. 2967-2976, 2005.

[5] Love D. J.and Strohmer T., "Grassmannian beamforming for multiple-input multiple-output wireless systems," *IEEE Trans. Inf. Theory*, vol. 49, no. 10, pp. 2735-2747, 2003.

[6] Raghavan V., Heath R. W., and Sayeed A. M., "Systematic Codebook Designs for Quantized Beamforming in Correlated MIMO Channels," *IEEE Journal on Selected Areas in Commun.*, vol. 25, no. 7, pp. 1298-1310, 2007.

[7] Ryan D. J., Vaughan I., Clarkson L.and Honig M.L., "QAM codebooks for low-complexity limited feedback MIMO beamforming," *in Proc. IEEE Int. Conf. Commun.*, pp. 4162-4167, 2007.

[8] 3GPP TS 36.211: Evolved Universal Terrestrial Radio Access (E-UTRA); Physical channels and modulation.

[9] Klappenecker A. and Roetteler M., "Constructions of mutually unbi-ased bases," *Finite Fields Appl.*, pp. 137-144, 2004.

[10] Heath R. W., Strohmer T. and Paulraj A. J., "On quasi-orthogonal signatures for CDMA systems," *IEEE Trans. Inf. Theory*, vol. 52, no. 3, pp. 1217-1226, 2006.

[11] Huang Y., Xu D. and Yang L., "Limited Feedback Precoding Based on Hierarchical Codebook and Linear Receiver," *IEEE Trans, on Wireless Commun.*, vol. 7, no. 12, pp. 4843-4848, 2008.

6

Salus: Kernel Support for Secure Process Compartments

Raoul Strackx*, Pieter Agten*, Niels Avonds,†, Frank Piessens*

iMinds-DistriNet - KU Leuven, Celestijnenlaan 200A, 3001 Heverlee, Belgium

Abstract

Consumer devices are increasingly being used to perform security and privacy critical tasks. The software used to perform these tasks is often vulnerable to attacks, due to bugs in the application itself or in included software libraries. Recent work proposes the isolation of security-sensitive parts of applications into protected modules, each of which can be accessed only through a predefined public interface. But most parts of an application can be considered security-sensitive at some level, and an attacker who is able to gain in-application level access may be able to abuse services from protected modules.

We propose Salus, a Linux kernel modification that provides a novel approach for partitioning processes into isolated compartments sharing the *same* address space. Salus significantly reduces the impact of insecure interfaces and vulnerable compartments by enabling compartments (1) to restrict the system calls they are allowed to perform, (2) to authenticate their callers and callees and (3) to enforce that they can only be accessed via unforgeable references. We describe the design of Salus, report on a prototype implementation and evaluate it in terms of security and performance. We show that Salus provides a significant security improvement with a low performance overhead, without relying on any non-standard hardware support.

Keywords: Privilege separation, principle of least privilege, modularization

1. Introduction

Both desktop and mobile devices are increasingly being used to perform security and privacy critical tasks, such as online banking, online tax declarations and purchasing goods from online stores. The software to perform these tasks either runs inside a web browser, or is written as a standalone application. In both cases, the software is often vulnerable to attack, either due to bugs in the application itself or due to bugs in included software libraries or in the runtime environment used to execute the application (e.g. the browser).

Because of their widespread use and potentially high-impact nature, such applications form an interesting target for cybercriminals. Past research has focused on defending against specific attack vectors such as buffer overflows [1–4], format string vulnerabilities [5] and non-control-data attacks [6]. Even though many of these defense mechanisms are applied in practice, successful attacks against high-value applications are still common.

To provide stronger security guarantees, recent research efforts have shifted from trying to defend entire applications against every possible attack to providing strong isolation of sensitive parts of an application with a minimal trusted computing base (TCB). Cryptographic keys of an application, for example, can be isolated in a protected module that has complete control over its own secrets; the module can only be accessed via its public interface. Accessing the cryptographic keys directly at assembly level is prevented by the security architecture. Thus, an attacker that has successfully exploited a vulnerability in the non-security sensitive part of the application still cannot access the cryptographic keys.

A large number of security architectures providing such protection mechanism have been proposed in this field, including software implementations using hardware virtual machine support [7, 8], trusted computing primitives [9], implementations based on system management mode [10] and even completely hardware-based solutions [11–13]. Recent research papers by Intel indicate that hardware support for these security architectures will also become available on mainstream x86 platforms in the near future [14–16].

In practice, isolating security-sensitive parts of an application is difficult since most program logic can be considered security-sensitive at some level [17]. A too

*firstname.lastname@cs.kuleuven.be
†niels.avonds@gmail.com

coarse-grained approach will result in bloated modules that may contain vulnerabilities and that are too big to be formally verified [18]. Minimum-sized modules on the other hand, can provide strong and easily verifiable guarantees, but may need to expose insecure interfaces to interact with other modules. This is a common problem of module-isolating security platforms, both in software as in hardware. Application developers are trapped in a catch-22 with possibly severe security consequences. In the recent DigiNotar attack [19], for example, the root CA's private cryptographic key was securely stored in a hardware security module (HSM), but its insecure interface enabled attackers to sign arbitrary certificates.

In order to improve upon these shortcomings, we acknowledge that almost every part of an application performs security-sensitive operations. To reduce chances of a successful attack, we propose to partition the *entire* application into compartments and implement a non-hierarchical access control mechanism between compartments. Compartments not only provide provable secure isolation of stored private data (as modules in related work do), but are also able to confine software vulnerabilities to the compartments they occur in by (1) restricting the types of system calls that they are allowed to issue, (2) enabling authentication of calling and called compartments and (3) enabling compartments to only service requests made through unforgeable references, reducing the impact of insecure interfaces. By separating likely attack vectors from attack targets and placing them into different compartments, an attacker would need to exploit vulnerabilities in multiple compartments to reach her goal.

Each compartment resides in its own chunk of memory, consisting of a *public section* containing the code of the compartment and a *private section* storing sensitive data (e.g. cryptographic keys or passwords). Only when executing the public section of a compartment can the private section of that compartment be accessed. To force other compartments to use a compartment's public interface, execution can only enter the public section via well-defined code entry points and, if required by the compartment, unforgeable references. As an additional protection measure and to support the principle of least privilege [20], compartments have the ability to restrict the types of system calls they are allowed to perform. Once a compartement drops a system call privilege, it cannot be re-acquired. This further reduces the impact of compromised compartments. The compartments of a single process all run in the same address space, providing a lightweight programming model that enables legacy applications to be ported easily and incrementally.

Consider, as an example, an X.509 certificate signing application consisting of a parser, a validator, a signer

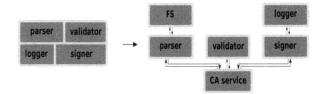

Figure 1. Salus' compartmentalization enables strong isolation of security-sensitive data *and* contains possibly vulnerable code. Multiple vulnerable compartments need to be exploited to attack the system successfully.

and a logging component (Figure 1). When run as a single monolithic application, a vulnerability in any one of these components can lead to the compromise of the entire application. When placing each of these components in a separate compartment under Salus, components can only call each other through their well-defined interfaces using unforgeable references and each component can authenticate both its callers and its callees. This restricts the flow of data and control between compartments to predefined patterns and raises the bar for a successful attack significantly. Consider as an example an attacker who exploited a vulnerability in the parser. In order for her to sign arbitrary certificates, she would either need to provide specially crafted credentials for the submitted certificate that would not cause the "Validator" to raise flags, or she would need to gain direct access to the "Signer" compartment by exploiting another vulnerability in the "CA Service" compartment to leak the unforgeable reference.

Furthermore, by combining unforgeable references and restricting the system calls that can be issued by a compartment, we can provide fine-grained access control to the kernel. Consider as an example the parser and assume that it reads its signing requests directly from the file system. At development time, there are two options. Option 1 is to grant the compartment access to the open/close and read/write system calls. In that case an attacker who exploited a vulnerability in the parser can inspect the entire file system with the application's privileges. The second option provides stronger security guarantees by revoking the parser compartment all system call privileges and only providing it with an unforgeable reference to a file system compartment (FS in Figure 1). This newly added compartment tightly restrict access to a single folder or file type and only provides the parser access to the files it approves. Having almost unrestricted access to the file system itself, a vulnerability in the FS compartment would enable an attacker to launch similar attacks as in option 1. However, given that this compartment is likely to be several orders of magnitude smaller than the parser compartment, the probability that such an exploitable vulnerability can be found is limited. Such constructs are a well-known advantage of capability systems [21–23].

Concretely, we make the following contributions in this paper:

- We present a novel approach for partitioning processes into compartments with support for strong isolation of sensitive data *and* containment of vulnerabilities. To the best of our knowledge, Salus is the first solution that simultaneously (1) reduces the impact of insecure compartment interfaces, (2) enables compartments to restrict the types of system calls they are allowed to perform and (3) executes compartments in the same address space allowing legacy applications to be ported easily without having to marshall in- and output messages.

- We report on a prototype implementation of Salus in the Linux kernel.

- We evaluate the security of our approach and the performance of our prototype.

This paper is an extended version of a conference paper published at SecureComm 2013 [24]. This journal version gives a substantially extended description of the Salus system, and in addition adds support for unforgeable references to the compartment model. The remainder of this paper is structured as follows: in Section 2 we define our attacker model and describe our desired security properties. In Section 3 we provide a high-level overview of Salus, before presenting our prototype implementation in Section 4. Finally we evaluate our approach in Section 5, discuss related work in Section 6 and conclude in Section 7.

2. Attacker Model & Security Properties

We consider an attacker able to inject and execute malicious shellcode in vulnerable compartments, for example, by exploiting a buffer overflow vulnerability. Our system must defend against such attacks in the following way:

- The exploitation of a compartment must not affect the security of compartments other than those that explicitly trust the compromised compartment.

- Once a compartment is exploited, an attacker is only able to call other compartments via their proper interfaces *iff* it received a reference to those compartments. Simply guessing the compartment's virtual address is not sufficient.

- An exploited compartment may still interact with other compartments and pass compartment

references. Called compartments however, will check the types of received arguments and will refuse to call other compartments with an incorrect type.

- Attackers are explicitly allowed to create new compartments. There is thus no guarantee that compartments requesting protection can be trusted. Hence, Salus must isolate compartments from one stakeholder from those of another, possibly malicious, stakeholder.

- An attacker should not be able to execute system calls that have been revoked.

Kernel-level and physical attacks are considered out of scope. Regarding the cryptographic primitives used, we assume the standard Dolev-Yao model [25]: An attacker can observe, intercept and adapt any message. Moreover, an attacker can create messages, for example by duplicating observed data. However, the cryptographic primitives used cannot be broken.

3. Overview of the Approach

This section presents a high-level overview of Salus. Section 3.1 describes the memory access control mechanisms on which Salus is based. Section 3.2 presents the services Salus provides to protected applications and section 3.3 shows how these services are used in a typical life cycle of a compartmentalized application. Authenticated communication between compartments and unforgeable references to compartments are discussed in sections 3.4 and 3.5 respectively. Finally we discuss how new and legacy applications can be compartimentalized in section 3.6.

3.1. Compartments of Least Privilege

Structure of a Compartment. The basic layout of a compartment, shown in Figure 2, is a virtual memory region divided into two sections: a public section and a private section. The *public* section contains the compartment's code and any data that should be read accessible by other compartments of the same application. This section can never be modified after initialization, which enables other compartments to authenticate the compartment based on a cryptographic hash of the public section (see Section 3.4). The start of the functions that make up the compartment's public interface are marked as entry points. Execution of the compartment can only be entered through these memory locations (see Table 1).

The *private* section contains the compartment's private data, which consists of application-specific security-sensitive data (e.g. cryptographic keys) as well as data relevant to the correct execution of the compartment, such as the runtime call stack. The data

from\to	Entry pnt.	Public section	Private section	Unprot. mem.
Entry pnt.	- - -	- - x	- - -	- - -
Public section	r - x	r - x	r w -	r w x
Private section	- - -	- - -	- - -	- - -
Unprot. mem/ other compartment	r - x	r - -	- - -	r w x

Table 1. The memory access control model enforces, for example, that a compartment's private section (4th column) can only be read-write accessed from the public section of the same compartment (3rd row)

Figure 2. Salus' memory access control model enables the creation of compartments that provide strong isolation guarantees to sensitive data. Secure communication primitives reduce the impact of an insecure interface.

in the private section is read and write accessible[1] from within the compartment, but completely inaccessible for code executing outside of the compartment. Note that since each compartment has its own private call stack, intercompartmental function call arguments and return addresses must be passed via CPU registers (as opposed to passing them using the runtime stack).

Applications can still have a memory region that is not part of any compartment. This region is termed *unprotected memory* and is read/write accessible from any compartment. All compartments of the same application run in the same address space, which facilitates the compartmentalization of legacy applications. Nonetheless, fine-grained compartmentalization of a large code base can still require significant developer effort. Therefore, Salus enables applications to be compartmentalized incrementally by storing code and/or data in unprotected memory. While unprotected memory does not provide any of the security guarantees of

compartments, it does provide an incremental upgrade path for legacy applications.

As an example of a compartment, consider a single compartment providing a certificate signing service. The compartment provides two functions as part of its public interface (see Figure 2). The first function, set_key, allows setting the cryptographic key used to sign certificates. This key is stored as the m_key variable in the private section. The second function, sign_cert, handles the actual signing requests. Salus' memory access control model ensures that only these two functions are executable; any attempt to jump to another memory location in the compartment will fail. Similarly, any attempt to directly read or write the cryptographic key in the private section from unprotected code or from another compartment will be prevented. Only after calling a valid entry point will read and write access to the private section be enabled, making the cryptographic key only accessible while the compartment is being executed. When the function is terminated, execution returns to the caller and read/write access to the compartment's private section will again be disabled.

Special care is required when execution returns to a compartment after a call to another compartment. Execution must resume at the return location, which is the instruction right after the call instruction in the caller compartment. This location however does not typically correspond to an entry point and hence would cause a memory access violation according to Salus' memory access control model (see Table 1). Compartments can implement a *return entry point* to avoid this access violation. Right before calling another compartment, the return location is placed on the top of the calling compartment's private stack while the location of the return entry point is passed to the callee in a register. When the intercompartmental call has finished, execution flow jumps to the return entry point where the return location is retrieved from the compartment's stack and jumped to. Note that a return entry point is a software implementation and follows the same access rights as any other entry point.

---

[1] By preventing code execution in the private section, the chances that an attacker is able to successfully exploit a vulnerability in a compartment, is reduced significantly. We acknowledge that this restriction may hinder applications that rely on generated code (e.g., JITed applications). Support for such applications could be easily added; at creation-time the creator should specify whether the new compartment's private section should be executable. As we believe this is a special case, we will not consider it for the remainder of the paper.

Restriction of Privileges. Salus provides two important primitives to limit the impact of a compromised compartment. The first primitive is caller and callee authentication. By authenticating callers and callees, a compartment can limit its interaction to trusted compartments only. Although this does not protect against trusted compartments that have been compromised, it does significantly limit the capabilities of an attacker after a successful exploit. For instance, the "signer" compartment of the CA signing service displayed in Figure 1, may only accept calls from the "CA service" compartment. As such, an attacker who successfully exploited a vulnerability in the parser may attempt to call the signing compartment, but the latter will refuse to service the attacker's service request.

The second primitive allows compartments to disable specific system calls for any code executed from within their public section. Once a system call is disabled, it cannot be re-enabled. By carefully partitioning an application into compartments, each of which should disable any system call it doesn't need, the impact of the exploitation of a vulnerable compartment is minimized. Note that much more fine-grained solutions exist than restricting complete system calls [26]. However, we focus on providing strong compartmentalization primitives that can be used as a building blocks for finer-grained privilege restriction mechanisms.

3.2. Provided Services

To enable compartmentalization of applications, Salus provides runtime support of the following services:

Create After code is loaded into memory, this service can be used to create a new compartment. Given a memory location and size for the compartment to create, Salus will enable memory protection for this region and will return a system-wide unique ID for the new compartment. Note that our attacker model explicitly allows the creation of new compartments by an attacker.

Destroy A compartment can only be destroyed by the compartment itself. After destruction, the memory access protection is disabled. Hence, a compartment should overwrite any private data before destruction.

Request compartment ID and layout To support secure communication, Salus provides a service to request the ID and layout (i.e. the size and locations of the public and private sections and the available entry points) of a compartment covering a given memory location. If there is no compartment at the specified location, the service returns an error code. This service is used as a primitive in compartment authentication.

Request caller ID To support caller authentication, Salus provides a service to request the ID and layout of the compartment that called an entry point of the current compartment.

Disable system call To limit the impact of the exploitation of a compartment, unused system calls can be disabled. To prevent an attacker from gaining system call privileges by creating a new compartment, compartments inherit system call privileges from their parent.

3.3. Life Cycle of a Compartmentalized Application

Compartmentalized applications can be started as any other application. After the (trusted) operating system or loader loads the application into memory and starts its execution, the application can create the required compartments. Finally, execution can jump to the compartment containing the application's main function. Compartments can be created at any point during the application's execution, for example, at the time a new (compartmentalized) plugin is loaded.

Creation of Compartments. Figure 3a shows the process of setting up a compartment. As the first step of setting up a new compartment, the application allocates (unprotected) memory and loads the compartment's code. Next, the application enables protection of this memory region, by calling Salus' creation service. Note that there is no guarantee that the new compartment's code has been loaded correctly into memory, since the creator might have been compromised already. However, any tampering with the code will be detected when the compartment tries to communicate with another compartment, as will be explained in Section 3.4.

When a new compartment is created, Salus clears the first byte of the private section. This serves as a flag to indicate to the compartment that it should initialize itself when its service is first requested. As part of its initialization, a compartment should clear the private memory locations it will use. This prevents an attacker from crafting a private section by setting it up in unprotected memory locations where a new compartment will later be created. Initialization code should typically also disable the system calls that will not be used during further execution of the compartment.

Destruction of Compartments. The destruction of a compartment, shown in Figure 3b, can only be initiated by the compartment itself. This ensures that compartments can clear their private section (which may contain sensitive data), before the memory protection is lifted. In addition, trusted communication endpoints

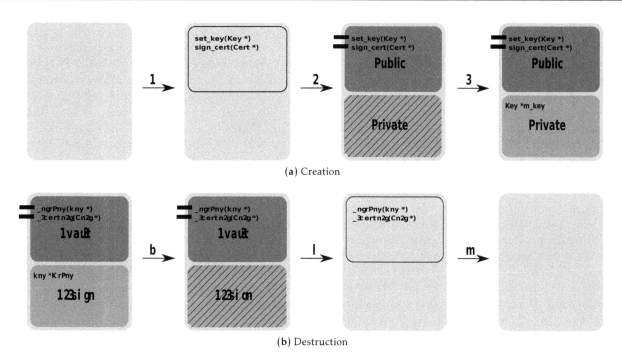

(a) Creation

(b) Destruction

Figure 3. The life cycle of a protected compartment from creation (3a) to destruction (3b)

could be notified of the compartment's imminent destruction. After destruction, the unprotected memory area of the destructed compartment can be freed.

3.4. Secure Communication

Salus' memory isolation mechanism provides strong guarantees that sensitive data in the private section can only be accessed by code in the public section [27–30]. Reconsidering our certificate signing service as an example (see Figure 1), we can prove that the signing key will never leave its compartment. But an attacker with access to the compartment's interface is still able to sign arbitrary certificates. Salus can limit the feasibility of such attacks in two ways: (1) by enforcing both caller and callee authentication, and (2) by requiring that callers have an unforgeable reference to the compartment at hand, which means that guessing the location of a compartment is insufficient to access it. In this section we will focus on authentication of compartments. While we will only discuss authentication of calling and called compartments, a similar approach can be applied when locations of other compartments are passed as arguments. In Section 3.5 we will show how compartments can enforce that they can only be called through unforgeable references.

Security Report. Authenticating a compartment consists of verifying whether that compartment adheres to a trusted *security report* of that compartment. A security report of a compartment consists of:

The cryptographic hash of its public section This allows any code to verify that the public section of the compartment has not been tampered with: the cryptographic hash should be recalculated at runtime and be compared to the known-good value stored in the security report. This protects against an attacker who is able to modify the public section of a compartment during its creation, before memory protection is enabled (see Section 3.3).

The layout of the compartment When a creation request originates from unprotected memory, the request itself may have been tampered with. An attacker could, for instance, specify an incorrect private section size for the compartment to create. This may result in the use of unprotected memory that should be under Salus' protection. By storing the known-good layout of the compartment in the security report, any code can verify that the layout was not tampered with during creation of the compartment.

A cryptographic signature In order to have integrity protection and authentication of the security report, it is digitally signed by its issuer. Each compartment can decide independently whether or not to trust a certain issuer, which opens up the opportunity to integrate compartments from different parties into a single application. Since the cryptographic signature provides integrity protection, security reports can simply be stored in unprotected memory.

Authentication of Called Compartments. When exchanging sensitive information between compartments, caller and callee must authenticate each other *before* sensitive data is exchanged.

To authenticate a compartment to be called, its ID must first be obtained using Salus' 'request compartment ID' service. Next, the callee's security report must be acquired. For this a central service where each compartment registers to on initialization, can be used. Given the callee's ID, the service should return the (location of the) corresponding security report. Note that this service need not be trusted, as any tampering with the information returned will be detected during the next steps. Once the security report has been obtained, it should be validated by checking the cryptographic signature and by checking that the issuer is trusted. Each compartment should contain a list of trusted security report issuers. Next, the callee compartment's layout should be requested from Salus and a hash of the public section should be calculated. The layout and the hash must be compared to the values listed in the security report. This completes the authentication and allows the caller to securely call one of the callee's public functions.

When calling a compartment that has already been authenticated in the past, a re-validation must occur because the callee may have been destroyed since the last interaction. A full authentication using the security report on every call would be very time consuming, so to reduce the performance impact, Salus allows compartments to be re-authenticated quickly based on their ID. Salus ensures each compartment has an ID that is unique on the system until the next reboot. Hence, a re-authentication can simply consist of requesting the ID of the compartment to be called (using the 'request compartment ID' service) and checking that it is the same as during the initial authentication. Using unique identifiers has the added benefit that code can distinguish between different instances of the same compartment.

Authentication of Calling Compartments. To enable compartments to limit use of their (possibly insecure) interface to trusted caller compartments, Salus provides primitives for caller authentication. For a compartment to authenticate its caller, it can first request the caller's ID and memory location (using the 'request caller ID' service) and proceed to authenticate the caller using the same steps as described above.

3.5. Unforgeable references

Salus' access control mechanism and supporting services enable authentication of both callers and callees. Unfortunately, in some situations this does not suffice. Let's reconsider the CA service from Section 1 as an example but now assume that it

receives signing requests over a network. Figure 4 displays how the application can be partitioned into different compartments. A compartment Listener listens for incoming network connections and spawns a new CAConnection compartment for every connecting client. This compartment is in charge for all future communication with the client. This is similar to a Socket object in an object-oriented language. When a connection is established, clients must provide login credentials and a certificate request. In order to isolate vulnerabilities, CAConnection hands off incoming messages to a compartmentalized parser. If messages parsed correctly, the parser returns Credentials and Request compartments to the CAConnection compartment, or an error code if parsing failed. Once all data is collected, the CAService is called. Based on the provided Credentials and Request compartments, it will authenticate the client credentials, verify that the client is allowed to request a certificate for the specified domain and finally instruct the Signer (not displayed) to sign the certificate request.

By compartmentalizing the Parser, we wish to isolate possible vulnerabilities. Unfortunately, in this setup an attacker able to exploit a vulnerability in the parser may still be able to request certificates for domains that she does not own. The problem arises when the parser returns Request and Credentials compartments to CAConnection. Even though CAConnection is able to authenticate the Parser, it cannot verify that the received Request and Credentials compartments are based on the actual data passed to the parser. An attacker who successfully exploited a vulnerability in the parser may be able to scan[2] the entire memory and steal a Credentials compartment belonging to a different network connection.

To remedy the problem, we propose using unforgeable references to compartments. *Only* compartments with an unforgeable reference to a compartment have the *capability* to access it. Thus, even if a compartment was compromised, it cannot access or pass references to other compartments that it finds in memory. In our example, a compromised parser may still find a Credentials compartment in memory, but it is infeasible that it can guess the correct nonce (i.e., it cannot create a correct *unforgeable reference* to it). Even a compromised parser can thus not return "stolen" credentials. This results in a strict separation between different connections.

While unforgeable references in higher programming languages are easily enforceable by a type system, we

---

[2]An in-application level attacker may scan the entire memory in a number of ways. For example, by using Salus' service to request the layout of a compartment for likely compartment locations until a non-error result is returned, or by reading the entire program memory for telltale signs of entry points.

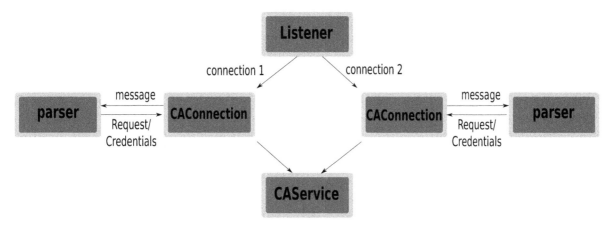

Figure 4. By enforcing that compartments can only be accessed via unforgeable references, stronger security guarantees can be guaranteed. Even if an attacker is able to exploit a vulnerability in a parser, she will be unable to access Request/Credentials compartments belonging to another connection.

cannot apply the same approach. An attacker able to exploit a vulnerability in a compartment has assembly-level access and can simply scan the entire memory area to access other compartments. Instead we propose establishing unforgeable references as (location, nonce) tuples. Newly created compartments must be assigned a cryptographic nonce, which can serve as a key to access the compartment's public interface. If and only if a caller provides the correct nonce, will a call to the compartment be serviced. This approach has four advantages: (1) with a sufficiently large nonce, it is computationally infeasible to forge references, (2) references can be stored in the secret section of compartments, just like any other reference, (3) compartments can implement unforgeable references using the default Salus services, and (4) both standard and unforgeable references can exist in the same application. Section 4.4 describes in detail how compartments can implement support for unforgeable references.

3.6. Writing Compartmentalized Applications

Writing safe compartments is a non-trivial task; each compartment should keep track of it's own stack, callbacks to unprotected memory should return through a specific return entry point, etc. To ease the creation of such compartments, we developed a C compiler and linker that takes care of such considerations. Application developers can simply annotate functions indicating that they are entry points, reside in unprotected memory or are located in another compartment.

Unfortunately, our tool does not solve all problems at hand. The developer is still in charge to ensure that sensitive data stored in a compartment is never accessed from unprotected memory or by another compartment. The difficulty in ensuring this depends heavily on the

programming language used and the quality of the source code. Applications written in C may not be very structured. Each function may allocate memory regions and pass pointers implicit (e.g., stored in allocated memory, or type casted as an integer) or explicit (e.g., as arguments) to other functions. Compartmentalizing such legacy applications may be difficult, but given that all compartments execute in the same address space, an incremental path exists. Developers may place functions that operate on the same sensitive data in the same compartment, while initially still storing the data in unprotected memory. When all functions are placed in the compartment and sensitive data is thus only accessed by a single compartment, it can safely be allocated inside the compartment. Tools such as logging access right violations during development instead of stopping the application (as proposed by [31]) may be helpful in this process but manual inspection of code is still required.

Object-oriented languages on the other hand, may already enforce strict data encapsulation; data may only be accessed through the object's public interface. In such cases each class may be compiled as a separate compartment but to minimize overhead caused by crossing protection boundaries, multiple classes may be placed together in a single compartment.

4. Implementation

Access rights to compartment sections depend on the value of the program counter. For instance, only if execution is in the public section of a compartment, will the private section of that compartment be read/write accessible. This program counter-based memory access scheme is at the core of Salus' protection mechanism. Enforcing this scheme purely in software would have a huge performance impact as every memory access has to be checked. A pure hardware implementation of the scheme is possible [12, 13], but prohibits its use on commodity, off-the-shelf PC platforms. The approach taken for Salus combines the best of both alternatives, by using the key insight that memory access rights for compartments only need to change when execution crosses a compartment border. This allows Salus to use the standard memory management unit (MMU) to enforce the memory protection scheme.

A prototype for Salus has been implemented as a Linux kernel modification. Section 4.1 describes how the program counter-based access control mechanism is implemented in this prototype. Section 4.2 describes the API Salus provides to processes and Section 4.3 lists the Linux system calls that had to be modified in order to provide a secure implementation of the protection mechanism.

4.1. Program Counter–Based Access Control

By aligning compartment sections to pages, the standard MMU found on any recent commodity computer can be applied to enforce the required memory protection scheme. After a compartment is created (e.g. from unprotected memory), the MMU access rights for the pages of the new compartment are set up according to Table 1: the public section is world-readable while the private section is isolated completely.

When execution tries to enter a compartment (e.g., because of a call instruction), a page fault is generated by the MMU. Based on the memory location addressed and the access type (read, write or execute), Salus determines whether a valid entry point was called and, if necessary, modifies the access rights of the calling and called compartments' public and private sections, according to Table 1. Access rights of pages unrelated to the two involved compartments are not modified, which minimizes the number of page faults and access right modifications, thereby reducing the overall performance impact.

Because unprotected memory is always readable, writable and executable, no page fault is generated when execution returns from a compartment to unprotected memory. To restore the access rights of the exited compartment, the compartment itself must issue a system call to Salus.

Since all threads of the same process normally share the same page tables, our approach cannot guarantee the required security properties in case of multiple threads. However, this is not a fundamental limitation of our model. Support for multithreaded applications can be added by modifying the kernel in order to provide each thread with a separate set of page tables. All threads have identical virtual-to-physical mappings, but with different access rights depending on the currently executing compartment in each thread.

Compartments also must be multithreading-aware and provide a separate stack per thread. Our prototype currently does not support multithreading.

The Linux page fault handler was modified to implement these access right modifications. To keep track of a process' compartments, the Linux process descriptor data structure was extended with a list of comp_struct structures. Each comp_struct describes a single compartment and contains:

- The (virtual) start address and length of the public and private sections

- The compartment's unique ID

- The compartment's saved stack pointer

- A list of the compartment's remaining system call privileges

4.2. System Call API

The following new system calls were implemented in the Linux kernel. These system calls represent the API Salus provides to processes.

void salus_create(void* start, uint len_pub, uint len_priv) Before a new compartment is created, the list of existing compartments is checked to ensure that the new compartment will not overlap with any existing ones. New compartments must also not overlap with the kernel or have their memory pages mapped to files. When these checks succeed, a new compartment is created and added to the current process' compartment list. It receives the same system call privileges as its parent.

void salus_destroy(void) Since compartments can only be destroyed from within their own public section, this system call does not require any arguments. This system call restores the original memory access rights on the memory region occupied by the executing compartment and then removes the compartment from the current process' compartment list.

struct comp_layout* salus_layout(void* addr) This system call returns the ID and memory layout of the compartment covering a given memory location. It can be implemented by simply iterating over the current process' compartment list until a matching compartment is found. A null pointer is returned when there is no compartment covering the given address.

struct comp_layout* salus_caller(void) This system call returns the ID and memory layout of the compartment that last called an entry point of the current compartment. A `null` pointer is returned when the current compartment was last called from unprotected memory.

void salus_syscall_disable(uint syscall_id)
This system call disables further use of the specified system call, by removing it from the list of system call privileges in the `comp_struct` of the current compartment. Once a system call is revoked, it cannot be re-acquired.

void salus_return(void* addr) Before execution returns from a called compartment back to its caller (i.e. unprotected memory or another compartment), the access rights of the called compartment's pages need to be restored. This system call performs this access rights modification and then continues execution at the specified address.

4.3. Conflicting System Calls

Some existing system calls in the Linux kernel conflict with Salus' compartmentalization. Additional security checks had to be inserted for these conflicting system calls.

mprotect The `mprotect` system call can be used to change the access rights of pages in memory. Additional checks were added to prevent this system call from modifying the access rights of compartments.

mmap Existing system calls such as `mmap` or `mremap` modify the virtual address space of a process. An attacker could abuse these system calls to map a compartment's private section to a file, for instance. When the compartment then writes sensitive information to the newly mapped pages, this information may leak to an attacker. We prevent this attack by verifying that a compartment is mapped correctly before it is called. These checks were also added to the `salus_layout` API call.

personality In Linux, each process has a *personality*, which defines the process' execution domain. The personality includes, among other settings, a `READ_IMPLIES_EXEC` bit, which indicates whether read rights to a memory region should automatically imply executable rights as well. For compartments this would result in world-executable public sections, nullifying the use of designated compartment entry points. Therefore, Salus enforces that this bit is disabled for compartmentalized processes.

fork The `fork`, `vfork` and `clone` system calls can be used to create a new process or thread.

As these processes or threads share parts of their page tables, the elevated access rights of the private section of a called compartment, affects all processes/threads and enable its access from unprotected memory. While these system calls could be modified to create copies of the page tables leading to the same virtual-physical address translation but with different access rights, our research prototype currently does not support this. Linux' existing `CLONE_VM` and `VM_DONTCOPY` flags are used to prevent compartments being mapped in the new process or thread. Checks were also added to the `madvice` system call, since it can be used to modify the `VM_DONTCOPY` flag.

4.4. Unforgeable references

Implementing support for unforgeable references consists of two steps: (1) newly created compartments must generate a cryptographic nonce, and (2) whenever a compartment is called, it must check whether the caller did indeed have the capability to access it.

The first step can be achieved in two ways. One option is to modify Salus' `salus_create` service call (see Section 4.2). After creating the compartment, the kernel generates a new cryptographic nonce and stores it at a specific location in the compartment's private section. Finally the `salus_create` service call returns the (location, nonce) tuple as the unforgeable reference.

Alternatively, newly created compartments can be taken ownership of on a first-call basis, by providing a `take_ownership` entry point that generates and returns an unforgeable reference on its first call. Only the first compartment that requests ownership will be provided with the unforgeable reference, subsequent calls to this entry point will be rejected. While malicous compartments may "steal" newly created compartments by taking ownership as soon as possible, they do not gain any additional power, since compartments are created from unprotected memory and hence do not possess any sensitive information that may leak to an attacker. Listing 1 shows a sample implementation of the `take_ownership` entry point in pseudo code.

In the second step, a called compartment must check whether the caller did indeed have the capability to access the compartment. To perform this check, the caller must pass the cryptographic nonce of the unforgeable reference to the called entry point. If and only if the provided nonce is identical to the nonce stored in the compartment's private section, will the call be serviced. Otherwise an error value will be returned. Note that the compartment is able to specify for every entry point whether or not it requires the nonce to

```
1  take_ownership:
      if ( nonce != 0 )
3        return -1;
      else
5      {
          nonce = gen_nonce ( ) ;
7          return nonce ;
      }
```

Listing 1: An implementation of the take_ownership entry point

access it. The take_ownership entry point, for example, will never require a capability.

5. Evaluation

The effectiveness of Salus' protection mechanisms is evaluated in Section 5.1 and its performance impact is discussed in Section 5.2.

5.1. Security Evaluation

To evaluate Salus' security, we make a distinction between memory-safe and memory-unsafe compartments. A memory-unsafe compartment can be exploited by an attacker using low-level attack vectors such as buffer overflows [1–4], format string vulnerabilities [5] or non-control data attacks [6]. A memory-safe compartment does not contain such vulnerabilities, for instance because it was written in a memory-safe language or simply because the compartment doesn't contain any memory-safety bugs.

Since memory-safe compartments cannot be exploited directly, the only attack vector against them is through exploitation of another compartment in the same address space. However, recent research [27–30] has shown that memory protection mechanisms such as those offered by Salus, are able to provide full source code abstraction. This means that, even when other compartments have been successfully exploited, an attackers' capabilities are limited to interacting with the memory-safe compartment through its public interface. A carefully constructed interface can thus effectively limit the attack surface of a compartment. But in many cases, creating a secure interface is still a challenging problem [32]. Recall the example of a certificate signing compartment introduced in Section 3.1: even if the private cryptographic key is never exposed, an attacker could potentially still use the compartment's interface to sign arbitrary certificates [19]. By taking advantage of Salus' support for caller/callee authentication however, the risk of such an attack can be minimized by only servicing requests from compartments that would issue them as part of the normal operation of the application (e.g. in Figure 1, the signer compartment should only accept requests from the validator compartment).

Memory-unsafe compartments may still contain vulnerabilities that can be exploited by attackers. Even though Salus does not prevent such attacks, compartmentalization can still provide significant security benefits. Firstly, high-risk components can be identified and be placed in separate compartments. Effective but high-overhead countermeasures [33, 34] can be used to harden such compartments. By only applying these countermeasures to likely vulnerable compartments, their performance impact remains limited.

Secondly, Salus' ability to provide unforgeable references and it's ability to restrict access to system calls, can be used to enforce fine-grained access policies. Enabling a compartment to issue open/close and read/write system calls, essentially provides it access to the entire file system[3]. Alternatively, small, safe compartments can be created that provide similar support but may limit access to a specific folder. Since the compartment cannot issue open system calls herself, it can only access the file system through the received "capability" compartment (see Section 1 for an example).

Thirdly, compartmentalization can automatically thwart certain types of attacks. For instance, limiting entrance of compartments to valid entry points significantly reduces the chance of an attacker finding enough gadgets to successfully execute a return-oriented-programming (ROP) attack [35, 36].

Fourthly, compartmentalization can be used as a building block for new countermeasures. For instance, a custom loader could be implemented that loads compartments at different locations in memory for every program execution. This is similar to address space layout randomization (ASLR) [37], but can be applied at a much finer-grained level.

Finally, even when a compartment has been successfully exploited, Salus can still limit the impact of the attack. Because Salus provides entry point enforcement, caller/callee authentication and system call privilege containment, an attacker will likely have to compromise multiple vulnerable compartments before reaching her intended target. This significantly increases the effort an attacker must take to successfully exploit the application. The ability to confine attackers to the exploited compartment even allows implementing a tightly controlled sandbox where user-provided machine code can be executed safely.

5.2. Performance Evaluation

To evaluate the performance of Salus, we performed micro- and macrobenchmarks. All tests were run on

---

[3]Of course this is restricted by the access rights the application is executing in

Type	CPU cycles	Relative
Function Call	5,944	1
System Call	193,970	32.63
Compartment Call	4,024,227	677.02

Table 2. Compartment access overhead

Concurrency	Vanilla	Salus	Relative perf.
1	109.11	96.54	-11.52 %
2	165.56	153.62	-7.21 %
4	184.31	164.78	-10.60 %
8	199.98	175.35	-12.32 %
16	206.82	181.00	-12.48 %
32	207.78	181.50	-12.65 %
64	206.64	180.35	-12.72 %
128	206.49	180.97	-12.36 %

Table 3. Requests per second of an SSL-enabled webserver where every SSL session is protected in its own compartment, for an increasing number of clients.

a Dell Latitude E6510. This laptop is equipped with an Intel Core i5 560M processor running at 2.67 GHz and contains 4 GiB of RAM. A Ubuntu Server 12.04 distribution with (modified) Linux 3.6.0-rc5 x86_64 kernel was used as the operating system.

System-wide impact. To show that legacy applications not using the modularization technique are not impacted by our changes to the Linux kernel, we ran the SPECint 2006 benchmark. All tests finished within ±0.4% compared to the vanilla kernel.

Microbenchmarks. To measure the overhead caused by switching the access rights, we created a microbenchmark that measures the cost of a call to a secure compartment and compare it to the cost of calling a regular function and calling a system call. The compartment used in the benchmark immediately returns to the caller. The system call and function behave similarly.

Table 2 displays the results of this microbenchmark. Calling a compartment is about 677 times slower compared to calling a regular function. This overhead is attributed to the need to modify the access rights of pages. Compared to calling a system call, the compartment is only 20 times slower. Due to these high costs, there is a trade-off to be made between a low number of compartment transitions and small compartments with additional security guarantees.

Secure Web Server. As a macrobenchmark, we compartmentalized an SSL-enabled web server based on an example provided by PolarSSL library[4]. For every new connection a new compartment is created, securing session keys even in the event that an attacker is able to inject shellcode in the compartment providing its own SSL session.

The secure compartment was built using the PolarSSL cryptographic library and a subset of the diet libc library. A simple static 74-byte page is returned to the clients over an SSL-connection protected by a 1024-bit RSA encryption key.

We used the Apache Benchmark to benchmark this web server for an increasing number of clients that are concurrently requesting pages. The results are shown in Table 3. The performance overhead tops at 12.72%

and is mainly attributed to the many compartment boundaries crosses during the SSL negotiation phase.

Compartmentalized parser. As input files are often under the control of an attacker and sanitation of their content can be difficult, parsers are a likely attack vector for many applications. As a second benchmark, we isolated the decompressing function of gzip (GNU zip). While disabling unused system calls for the entire process would result in similar security guarantees, we are interested in the impact of repeated compartment crossings in a parser setting. Applications that place their parser and the rest of the application in different compartments, would incur a similar overhead as only one additional compartment boundary needs to be crossed.

To benchmark the application, we created input files with randomized content, ranging from 16 KiB to 64 MiB in size, compressed them and measured the time taken to decompress the files with the hardened application. The application was run 100 times on each file. File I/O used a buffer of 32 KiB and the output was redirected to the null device. Figure 5 displays the results.

Given the relatively high overhead of a call to a compartment and the low computation cost of the decompressing function, it is unsurprising that for small input files the overhead can be as high as 21.9%. When the input size is increased however, the overhead drops steadily to -0.5% for 64 MiB input files, even though also the number of compartment-border crossings increases from 8 to 8200. We attribute this significant drop in overhead to the increased amount of slow disk I/O that needs to be performed as the input file size gets bigger, an effect that we predict to see in most parser-like compartments. The small performance gain of 0.5% can be attributed to cache effects.

The way an application is partitioned will have a significant impact on performance. Applications should be compartmentalized in logical blocks where each compartment has direct access to most of its required

---

[4]https://polarssl.org/

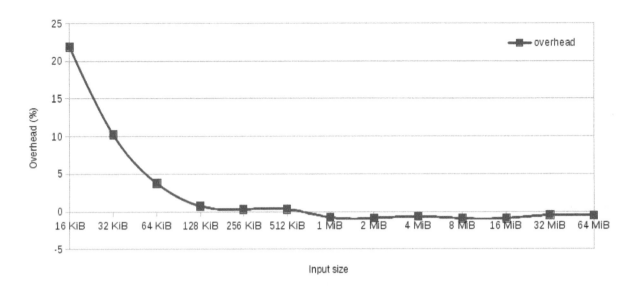

Figure 5. Salus' performance overhead on the gzip macro benchmark drops significantly as the input file size increases.

data. Once a logical block has finished, control and all data should be passed to the next compartment, reducing the number of inter-compartment calls. Smaller, heavily protected compartments such as an SSL compartment, provide strong security but may impact performance more significantly when called repeatedly. This makes the performance impact of compartmentalization difficult to predict. Therefore we advocate for automatic partitioning tools that analyze the application's call graph and information flow to reduce the number of compartment crosses and help the programmer decide which compartments should be hardened.

6. Related Work

Various security measures have been proposed to harden applications. Many of them aim to protect against very specific vulnerabilities such as buffer overflows [1–4], format string vulnerabilities [5] or non-control data attacks [6]. While these countermeasures make it significantly more difficult for an attacker to compromise software applications, they cannot offer complete protection. Static verification of source code [38], in contrast, is able to provide such hard security guarantees, but typically comes at a significant economic cost in terms of programming and verification effort.

Singaravelu et al. [17] proposed to isolate security-sensitive parts of applications in complete isolation from the rest of the system. Many research proposals have since been filed based on this principle. Each of them provides some way of executing modules in isolation, relying on a trusted code base ranging from only a few thousands of lines of code [7, 8] to only the protected modules themselves and a

small runtime library [9, 10]. More recently, specially tailored hardware support has been proposed in academia [11–13] and industry [14–16]. While these research prototypes offer provable security to the sensitive data that they protect [27–30], they do not attempt to reduce the impact of a vulnerability elsewhere in the code by executing modules with the least amount of privileges possible [20]. An attacker who successfully gains control over the platform is still able to interact with other protected modules unrestrictedly.

Other work focuses on confining possible software vulnerabilities. Early work focused on reducing the size of the kernel itself [39], where process privileges are managed by capabilities. Recently Watson et al. [26] proposed applying a similar idea to partition applications themselves, where capabilities can be granted to each created partition. As partitions live in their own process, interaction takes place through remote procedure calls and passed data must be marshalled. Salus avoids these drawbacks by executing compartments in the same address space and unprotected memory can be used to gradually partition legacy applications (see section 3.6).

Provos et al. [40] and Brumley et al. [41] propose separating sensitive applications into a privileged monitor and one or multiple slave components. Monitor and slaves communicate through system sockets and thus also require arguments to be marshalled. Subsequent work by Provos [42] argues for finer grained access policies for system calls. Bittau et al. [31] also propose splitting applications into compartments (called sthreads) executing with least privilege. Developers can tag memory locations and a security policy enforces that a compartment can only access memory locations with a matching tag. When an sthread requires more privilege operations, it can request so by calling a callgate. A security policy enforces which callgates an sthread can call. Salus' unforgeable references enable a much more flexible security policy. Compartments can be provided temporary access to system resources by encapsulating them in a compartment. As all interaction to the resource passes by this compartment, the caller's access rights can easily be revoked at a later point in time [43].

Native Client (NaCl) [44, 45], which builds upon the concepts of software fault isolation [46], takes another approach and attempts to completely sandbox x86 code. Accesses to the environment from within a sandbox are tightly controlled by runtime facilities. While NaCl focuses on downloaded, untrusted binary code, it could be used to partition entire applications. Interaction between two NaCl partitions is provided through a service similar to Unix domain sockets, making porting existing legacy applications a challenging undertaking. Salus on the other hand can provide a similar tightly controlled sandbox by placing such partitions in one compartment while the remaining legacy application is placed in another. A specially created wrapper can ensure that all system call privileges are revoked before execution control is given to the sandboxed code. There are however two major differences compared to NaCl. First, Salus only impacts performance when compartment boundaries are crossed. NaCl on the other hand places constraints on the binary code itself, resulting in a varying performance impact. Second, Salus employs a non-hierarchical separation of privilege, allowing compartments to be completely isolated from other compartments (possibly provided by other vendors) while compartments of the same vendor can co-operate easily.

Finally, our earlier work [8, 11] is the most related to Salus. It also employs a program-counter based access control mechanism, but assumes a safe interface. Therefore it has the same limitation as other research prototypes [7, 9, 10] that provide strong isolation of sensitive data: it does not reduce the possible impact of exploited vulnerabilities.

7. Conclusion

Protected-module architectures isolate sensitive parts of applications. They guarantee that sensitive data can only be accessed via a well-defined interface. In practice, however, it is hard to isolate security-sensitive parts, as most code in an application is sensitive up to some level. As a result, modules of such platforms may need to provide unsafe interfaces; an attacker may not access the sensitive data directly, but access to the provided interface may still lead to unwanted behavior.

We presented Salus, a new security architecture providing strong isolation guarantees of both sensitive data and software vulnerabilities. Salus significantly reduces the impact of unsafe interfaces by (1) supporting the authentication of compartments and (2) enabling compartments to enforce that they can only be accessed through unforgeable references. This allows likely attack vectors and targets to be placed in different compartments, such that an attacker must successfully attack multiple compartments before an attack target can be reached.

Acknowledgement. This work has been supported in part by the Intel Lab's University Research Office. This research is also partially funded by the Research Fund KU Leuven, and by the EU FP7 project NESSoS. With the financial support from the Prevention of and Fight against Crime Programme of the European Union (B-CCENTRE). Raoul Strackx holds a PhD grant from the Agency for Innovation by Science and Technology in Flanders (IWT). Pieter Agten holds a PhD fellowship of the Research Foundation - Flanders (FWO).

References

[1] Aleph One: Smashing the stack for fun and profit. Phrack magazine 7(49) (1996)

[2] Erlingsson, Ú.: Low-level software security: Attacks and defenses. In Aldini, A., Gorrieri, R., eds.: Foundations of Security Analysis and Design IV. Volume 4677 of Lecture Notes in Computer Science. Springer-Verlag (2007) 92–134

[3] Strackx, R., Younan, Y., Philippaerts, P., Piessens, F., Lachmund, S., Walter, T.: Breaking the memory secrecy assumption. In: Proceedings of the Second European Workshop on System Security. EuroSec'09, New York, NY, USA, ACM (2009) 1–8

[4] Younan, Y., Joosen, W., Piessens, F.: Code injection in C and C++ : A survey of vulnerabilities and countermeasures. Technical Report CW386, Department of Computer Science, KULeuven (2004)

[5] Cowan, C., Barringer, M., Beattie, S., Kroah-Hartman, G., Frantzen, M., Lokier, J.: Formatguard: automatic protection from printf format string vulnerabilities. In: Proceedings of the 10th conference on USENIX Security Symposium. SSYS'01, Berkeley, CA, USA, USENIX Association (2001) 1–9

[6] Chen, S., Xu, J., Sezer, E.C., Gauriar, P., Iyer, R.K.: Non-control-data attacks are realistic threats. In: Proceedings of the 14th conference on USENIX Security Symposium. Volume 14 of SSYM'05., Berkeley, CA, USA, USENIX Association (2005) 177–192

[7] McCune, J.M., Li, Y., Qu, N., Zhou, Z., Datta, A., Gligor, V., Perrig, A.: TrustVisor: Efficient TCB reduction and attestation. In: Proceedings of the IEEE Symposium on Security and Privacy. S&P'10, Washington, DC, USA, IEEE Computer Society (May 2010) 143–158

[8] Strackx, R., Piessens, F.: Fides: Selectively hardening software application components against kernel-level or process-level malware. In: Proceedings of the 19th ACM

conference on Computer and Communications Security. CCS'12, New York, NY, USA, ACM (October 2012) 2–13

[9] McCune, J.M., Parno, B., Perrig, A., Reiter, M.K., Isozaki, H.: Flicker: An execution infrastructure for TCB minimization. In: Proceedings of the ACM European Conference in Computer Systems. EuroSys'08, New York, NY, USA, ACM (April 2008) 315–328

[10] Azab, A., Ning, P., Zhang, X.: SICE: a hardware-level strongly isolated computing environment for x86 multi-core platforms. In: Proceedings of the 18th ACM conference on Computer and communications security. CCS'11, New York, NY, USA, ACM (2011) 375–388

[11] Strackx, R., Piessens, F., Preneel, B.: Efficient Isolation of Trusted Subsystems in Embedded Systems. In Jajodia, S., Zhou, J., eds.: Security and Privacy in Communication Networks (SecureComm'10). Volume 50 of Lecture Notes of the Institute for Computer Sciences, Social Informatics and Telecommunications Engineering., Springer Berlin Heidelberg (2010) 344–361

[12] Noorman, J., Agten, P., Daniels, W., Strackx, R., Herrewege, A.V., Huygens, C., Preneel, B., Verbauwhede, I., Piessens, F.: Sancus: Low-cost trustworthy extensible networked devices with a zero-software trusted computing base. In: 22nd USENIX Security Symposium. SSYM'13, USENIX Association (August 2013)

[13] Owusu, E., Guajardo, J., McCune, J., Newsome, J., Perrig, A., Vasudevan, A.: OASIS: on achieving a sanctuary for integrity and secrecy on untrusted platforms. In: Proceedings of the 2013 ACM SIGSAC conference on Computer & communications security. CCS'13, New York, NY, USA, ACM (2013) 13–24

[14] Anati, I., Gueron, S., Johnson, S., Scarlata, V.: Innovative technology for CPU based attestation and sealing. In: Proceedings of the 2nd International Workshop on Hardware and Architectural Support for Security and Privacy. Volume 13 of HASP'13., New York, NY, USA, ACM (2013)

[15] Hoekstra, M., Lal, R., Pappachan, P., Phegade, V., Del Cuvillo, J.: Using innovative instructions to create trustworthy software solutions. In: Proceedings of the 2nd International Workshop on Hardware and Architectural Support for Security and Privacy. HASP'13, New York, NY, USA, ACM (2013) 11

[16] McKeen, F., Alexandrovich, I., Berenzon, A., Rozas, C.V., Shafi, H., Shanbhogue, V., Savagaonkar, U.R.: Innovative instructions and software model for isolated execution. In: Proceedings of the 2nd International Workshop on Hardware and Architectural Support for Security and Privacy. HASP'13, New York, NY, USA, ACM (2013) 8

[17] Singaravelu, L., Pu, C., Härtig, H., Helmuth, C.: Reducing TCB complexity for security-sensitive applications: three case studies. In: Proceedings of the 1st ACM SIGOPS/EuroSys European Conference on Computer Systems. EuroSys'06, New York, NY, USA, ACM (2006) 161–174

[18] Garfinkel, T., Pfaff, B., Chow, J., Rosenblum, M., Boneh, D.: Terra: A virtual machine-based platform for trusted computing. In: Operating Systems Review. Volume 37 of OSR'03., New York, NY, USA, ACM (2003) 193–206

[19] Hoogstraten, H., Prins, R., Niggebrugge, D., Heppener, D., Groenewegen, F., Wettinck, J., Strooy, K., Arends, P., Pols, P., Kouprie, R., Moorrees, S., van Pelt, X., Hu, Y.Z.: Black Tulip - report of the investigation into the DigiNotar certificate authority breach. Technical report, FoxIT (2012)

[20] Saltzer, J., Schroeder, M.: The protection of information in computer systems. In: Proceedings of the IEEE. Volume 63., IEEE (1975) 1278–1308

[21] Carter, N.P., Keckler, S.W., Dally, W.J.: Hardware support for fast capability-based addressing. In: Proceedings of the Sixth International Conference on Architectural Support for Programming Languages and Operating Systems. ASPLOS'94, New York, NY, USA, ACM (1994) 319–327

[22] Woodruff, J., Watson, R.N.M., Chisnall, D., Moore, S.W., Anderson, J., Davis, B., Laurie, B., Neumann, P.G., Norton, R., Roe, M.: The CHERI capability model: Revisiting RISC in an age of risk. In: Proceedings of the 41st International Symposium on Computer Architecture. ISCA'14 (2014)

[23] Dennis, J.B., Van Horn, E.C.: Programming semantics for multiprogrammed computations. Communications of the ACM **9** (March 1966) 143–155

[24] Avonds, N., Strackx, R., Agten, P., Piessens, F.: Salus: Non-hierarchical memory access rights to enforce the principle of least privilege. In Zia, T., Zomaya, A., Varadharajan, V., Mao, M., eds.: Security and Privacy in Communication Networks (SecureComm'13). Volume 127 of Lecture Notes of the Institute for Computer Sciences, Social Informatics and Telecommunications Engineering., Springer International Publishing (September 2013) 252–269

[25] Dolev, D., Yao, A.C.: On the security of public key protocols. In: IEEE Transactions on Information Theory. Volume 29., Piscataway, NJ, USA, IEEE Press (September 1983) 198–208

[26] Watson, R.N., Anderson, J., Laurie, B., Kennaway, K.: Capsicum: practical capabilities for UNIX. In: Proceedings of the 19th USENIX Security symposium. SSYM'10, Berkeley, CA, USA, USENIX Association (2010)

[27] Agten, P., Strackx, R., Jacobs, B., Piessens, F.: Secure compilation to modern processors. In: Proceedings of the 25th Computer Security Foundations Symposium. CSF'12, Los Alamitos, CA, USA, IEEE Computer Society (2012) 171–185

[28] Patrignani, M., Clarke, D.: Fully abstract trace semantics of low-level isolation mechanisms. In: Proceedings of the 29th Annual ACM Symposium on Applied Computing. SAC'14, ACM (March 2014) 1562–1569

[29] Patrignani, M., Agten, P., Strackx, R., Jacobs, B., Clarke, D., Piessens, F.: Secure compilation to protected module architectures. In: Accepted for publication in Transactions on Programming Languages and Systems (TOPLAS), New York, NY, USA, ACM

[30] Patrignani, M., Clarke, D., Piessens, F.: Secure Compilation of Object-Oriented Components to Protected Module Architectures. In Shan, C.c., ed.: Proceedings of the 11th Asian Symposium on Programming Languages and Systems (APLAS'13). Volume 8301 of Lecture Notes

in Computer Science., Springer International Publishing (2013) 176–191

[31] Bittau, A., Marchenko, P., Handley, M., Karp, B.: Wedge: Splitting applications into reduced-privilege compartments. In: Proceedings of the 5th USENIX Symposium on Networked Systems Design and Implementation. NSDI'08, Berkeley, CA, USA, USENIX Association (2008) 309–322

[32] Longley, D., Rigby, S.: An automatic search for security flaws in key management schemes. Computers & Security 11(1) (1992) 75–89

[33] Younan, Y., Philippaerts, P., Cavallaro, L., Sekar, R., Piessens, F., Joosen, W.: Paricheck: an efficient pointer arithmetic checker for c programs. In: Proceedings of the 5th ACM Symposium on Information, Computer and Communications Security. ASIACCS '10, New York, NY, USA, ACM (2010) 145–156

[34] Akritidis, P., Costa, M., Castro, M., Hand, S.: Baggy bounds checking: An efficient and backwards-compatible defense against out-of-bounds errors. In: Proceedings of the 18th conference on USENIX security symposium. SSYM'09, USENIX Association (2009) 51–66

[35] Shacham, H.: The geometry of innocent flesh on the bone: return-into-libc without function calls (on the x86). In: Proceedings of the 14th ACM conference on Computer and communications security. CCS '07, New York, NY, USA, ACM (2007) 552–561

[36] Checkoway, S., Davi, L., Dmitrienko, A., Sadeghi, A.R., Shacham, H., Winandy, M.: Return-oriented programming without returns. In: Proceedings of the 17th ACM Conference on Computer and Communications Security. CCS'10, New York, NY, USA, ACM (2010) 559–572

[37] Bhatkar, S., DuVarney, D.C., Sekar, R.: Address obfuscation: An efficient approach to combat a broad range of memory error exploits. In: Proceedings of the 12th USENIX security symposium. Volume 12 of SSYM'03., Berkeley, CA, USA, USENIX Association

(2003) 105–120

[38] Jacobs, B., Piessens, F.: The verifast program verifier. CW Reports CW520, Department of Computer Science, K.U.Leuven (August 2008)

[39] Liedtke, J.: Toward Real Microkernels. Communications of the ACM 39(9) (1996) 77

[40] Provos, N., Friedl, M., Honeyman, P.: Preventing privilege escalation. In: Proceedings of the 12th Conference on USENIX Security Symposium. SSYM'03, Berkeley, CA, USA, USENIX Association (2003)

[41] Brumley, D., Song, D.: Privtrans: Automatically partitioning programs for privilege separation. In: Proceedings of the 13th Conference on USENIX Security Symposium. Volume 13 of SSYM'04., Berkeley, CA, USA, USENIX Association (2004)

[42] Provos, N.: Improving host security with system call policies. In: Proceedings of the 12th Conference on USENIX Security Symposium. SSYM'03, Berkeley, CA, USA, USENIX Association (2003)

[43] Miller, M., Yee, K.P., Shapiro, J.S.: Capability myths demolished. Technical Report SRL2003-02, Johns Hopkins University (2003)

[44] Yee, B., Sehr, D., Dardyk, G., Chen, J.B., Muth, R., Ormandy, T., Okasaka, S., Narula, N., Fullagar, N.: Native client: A sandbox for portable, untrusted x86 native code. In: Proceedings of the 30 IEEE Symposium on Security and Privacy. S&P'09, IEEE (2009) 79–93

[45] Sehr, D., Muth, R., Biffle, C., Khimenko, V., Pasko, E., Schimpf, K., Yee, B., Chen, B.: Adapting software fault isolation to contemporary CPU architectures. In: Proceedings of the 19th USENIX Security Symposium. SEC'10 (2010)

[46] Wahbe, R., Lucco, S., Anderson, T.E., Graham, S.L.: Efficient software-based fault isolation. In: Proceedings of the fourteenth ACM symposium on Operating systems principles. SOSP '93, New York, NY, USA, ACM (1993) 203–216

7

Trust in social computing
The case of peer-to-peer file sharing networks

Heng Xu[1], Tamara Dinev[2], Han Li[3],*

[1]College of Information Sciences and Technology, Pennsylvania State University, University Park, PA 16802, USA; [2]Barry Kaye College of Business, Florida Atlantic University, Boca Raton, FL 33431, USA; [3]School of Business, Minnesota State University, Moorhead, MN 56563, USA

Abstract

Social computing and online communities are changing the fundamental way people share information and communicate with each other. Social computing focuses on how users may have more autonomy to express their ideas and participate in social exchanges in various ways, one of which may be peer-to-peer (P2P) file sharing. Given the greater risk of opportunistic behavior by malicious or criminal communities in P2P networks, it is crucial to understand the factors that affect individual's use of P2P file sharing software. In this paper, we develop and empirically test a research model that includes trust beliefs and perceived risks as two major antecedent beliefs to the usage intention. Six trust antecedents are assessed including knowledge-based trust, cognitive trust, and both organizational and peer-network factors of institutional trust. Our preliminary results show general support for the model and offer some important implications for software vendors in P2P sharing industry and regulatory bodies.

Keywords: network-based community, peer-to-peer (P2P) file sharing, risks, social computing, trust

1. Introduction

New applications and services that facilitate user collective action and social interaction with rich data exchange have been driving a dramatic evolution of the Web [1, 2]. Examples include blogs, wikis, social bookmarking, user-driven ratings, peer-to-peer (P2P) networks, photo and video sharing communities, and online social networks. In the literature, these applications or services have been variously referred to as *social computing*, which reflects the increased role of computing in social structures, in empowering individual users and communities and not just institutions [1, 3]. Social computing platforms share several features that differentiate them from traditional organizational computing and content sharing. Specifically, these platforms tend to be decentralized, dynamic, and flexibly structured in terms of how information is gathered and distributed [1, 3].

Social computing is seen as having a profound impact on the global economy both through impacting the social structure [4] and the technology development as a whole

[2]. In terms of social structure, individuals increasingly take cues from one another rather than from public or private organizations such as corporations, media, religion, and political institutions. Charron *et al.* [4] point to several important tenets driven by the social computing: innovation shift from top-down to bottom-up; value shift from ownership to experience; power shift from institutions to communities. In terms of IT developments, social computing seeks to improve social software that can facilitate interaction either between groups or individuals and computing tools.

The above-mentioned innovation, value, and power shift from organizations (which represent an identifiable and law-bound entity) to communities informs also a shift in risk and trust perceptions and their importance to the community members. As Parameswaran and Whinston [1] pointed out, social computing platforms empower 'individual users with relatively low technological sophistication in using the Web to manifest their creativity, engage in social interaction, contribute their expertise, share content, collectively build new tools, and disseminate information and propaganda' (p. 763). In these platforms users

*Corresponding author. Email: li@mnstate.edu

routinely engage with a large number of user communities with whom they have little or no prior interaction. This exposes users to an even greater risk of opportunistic behavior by malicious or criminal communities that can make use of the anonymity, fault tolerance, robustness, and low cost of online communities to build platforms for illegitimate interaction, communication, and data exploration [1]. Given that users face realistic concerns pertaining to social computing, we seek in this paper to understand what steps can be taken to increase users' trust perceptions and reduce their risk perceptions so as to encourage legitimate interactions in social computing platforms.

Trust is a crucial enabling factor in relations where there are uncertainty, interdependence, risk, and fear of opportunism [5, 6]. Little is known, however, regarding how trust beliefs are formed and developed in terms of using social computing applications, and what individual, organizational, and community factors influence the trust formation. Following the conceptual and integrative development of trust in the field of Information Systems, we develop and empirically test a research model that incorporates multiple, interrelated factors contributing to the formation of trust beliefs in the context of P2P networks. With specific reference to P2P networks as a social computing platform, such environment facilitates the development of communities through the creation of 'architectures that allow peer-wise communication and social action'. In developing the research model, we identify new trust-building mechanisms, namely peer-network situational normality and peer-network structural assurances. These peer-network structures would be especially suited for network-based virtual community such as P2P networks, constituted by individuals with unstructured and non-static relationships, interacting together in a community [7].

Moreover, it has been noted that the openness of P2P networks that renders them advantageous can also be a liability in terms of attack vulnerability [8]. Given these potential liabilities, 'in the social platforms, reputation and trust will be key determinants' in their usage [1, p. 774]. Such concerns certainly support the study of trust and perceived risk in P2P networks. In current research, we developed a trust-risk model and empirically tested our model with a survey of 136 experienced and voluntary P2P users in a large university in Singapore. By rigorously specifying the antecedents of trust beliefs, our objective is to conceptually clarify and verify the multiple factors that inform trust formation in social computing context. In what follows, we first describe the theoretical foundation that guides the development of the research model. Then we develop the research hypotheses that identify factors included in the process wherein individuals form trusting beliefs. This is followed by the research methodology and findings. The paper concludes with a discussion of the results, the practical and theoretical implications of the findings, and directions for future research.

2. Theoretical foundation and research hypotheses

2.1. P2P networks

Peer-to-peer (P2P) networks consist of nodes communicating directly for the purpose of exchanging content files, with no centralized governing node [9]. Each participant in the P2P network can behave either as a client, receiving files, or as a server, sending files, or both [see 8 for a review]. Users need to run specific P2P sharing software on their local computers in order to participate in the P2P network. P2P sharing software, in this study, is defined as an application running P2P sharing technology dedicated for searching, downloading, and sharing digital resources among peer users (peers), such as file sharing software like Gnutella and KaZaA, and CPU sharing software like SETI@home. While the term P2P is widely associated with sharing of music and movies that often involve copyright violations, its scope is far wider which can be used to exchange any digitized content [8]. Parameswaran and Whinston view P2P software as 'social software taken to the extreme, bypassing limitations of the browser interface and the DNS (Domain Name System) addressing, radically decentralized, and relying almost exclusively on collective action by users at the edge' (p. 765).

Given the current existence of various uncertainties in P2P sharing such as resource piracy [10], computer attack by malicious peers [11], and privacy invasion [12–14], it is crucial to understand the factors that will affect individual's usage of P2P sharing. In this study, we only focus on examining the voluntary use of P2P sharing software, free of charge. Non-voluntary use of P2P sharing often happens within organizations and such a case is not the focus of current research. Furthermore, we focus on those P2P sharing networks where all peers are equal and anonymous[1] [12], without central administrators or power peers who have the capability to control other peers. Throughout the rest of the paper, we use the following terms: *resources* referring to files or computer hardware being shared out; *resources download* referring to retrieving resources from other users' computer over the Internet; *sharing resources* referring to allowing others to access and download the shared resources; *vendor* referring to the producer of P2P sharing software; *peer* is used in exchange with the term *user; peer network*, or *P2P sharing*

---

[1]Reiter and Rubin (1999) conceptualized three degrees of anonymity: (i) type, which states sender or receiver anonymity; (ii) adversary, or who is trying to break the anonymity, and (iii) degree, which may range from absolute privacy (imperceptible presence) to possible innocence, to exposed (to the adversary), to provably exposed (to others). In P2P sharing, peer anonymity is referred to as a peer's identity hidden from other peers (type), but with the possibility of being exposed to a malicious peer (adversary and degree).

network, is the network of peers running the same P2P sharing software.

2.2. Risks in P2P file sharing

Risk has been generally defined as the uncertainty resulting from the potential for a negative outcome [15] and the possibility of another party's opportunistic behavior that can result in losses for one self [16, 17]. Perceived risks affect an individual's intention and actual usage of a technology especially in a high uncertain environment, such as online shopping [18]. An individual's calculation of risk involves an assessment of the likelihood of negative consequences as well as the perceived severity of these consequences [19]. The negative perceptions related to risk may affect an individual emotionally, materially, and physically [20].

In the P2P sharing context, a user could be exposed to uncertainties related to three sources: *peers* (including the user herself), the *vendor of the P2P sharing software,* and *the Internet.* The user, therefore, may perceive that there is some probability of suffering a loss when downloading or sharing resources in the P2P network. For example, a peer may find her computer overloaded or attacked by malicious peers (peer-related performance risk); she may face legal suit or even jail (legal risk) when she shares pirated resources with other peers [10]; she may by mistake share her entire hard disk or other principal data repository as material available to others (peer-related privacy risk). Moreover, the user may not be informed of her online activities being disclosed to third parties by the software vendor (vendor-related privacy risk) [13]. A peer may find the software's performance is not as good as expected, or hard to find and download her intended resources (vendor-related performance risk). Furthermore, the data transmission over Internet incurs potential channel risk as the attacker might be an eavesdropper that can observe some or all messages sent and received over the Internet [21]. Without proper control of the risk in P2P file sharing, a user may choose not to use the software due to high risks [22]. For example, Pavlov and Saeed [23] reported that the deteriorated performance of Gnutella software often causes failed download and leads users to give up the usage. In this study, we define a user's perceived risks in using P2P sharing software as the user's perceived probability of suffering a loss when downloading or sharing resources in the P2P network.

2.3. Trust

Trust has received a great deal of attention from scholars in the disciplines of social psychology [24], sociology [25], management [26], and marketing [27]. In examining the published literature on trust, various definitions of trust have been proposed in many different ways. Nevertheless, across disciplines there is consensus that trust is a crucial enabling factor in relations where there are uncertainty, interdependence, risk, and fear of opportunism [5, 6]. 'The need for trust only arises in a risky situation', and trust could be an effective mechanism to reduce the complexity of human conduct in situations where people have to cope with uncertainty [28]. Trust involves at least two entities in relation to each other—a trustor and a trustee. In e-commerce, the consumer is usually seen as the trustor, the party who places him or herself in a vulnerable situation; and the e-vendor is the trustee, the party in whom trust is placed and who has the opportunity to take advantage of the trustor's vulnerability [29].

Before engaging in a discussion of trust, it is helpful to delineate the differences between trust belief and trust intention. Trust belief, on the one hand, is the trustworthiness perception of certain attributes specific to a *trustee*, while trust intention, on the other hand, is the psychological state of a *trustor*, i.e. trustor's intention to engage in trust-related behaviors with a specific trustee. Even though efforts have been devoted to differentiating trust belief from trust intention [see 30, 31], most researchers adopted the conceptualization of trust as a set of specific *trust beliefs* in e-commerce studies [32, 33]. Consequently, this study has adopted the conceptualization of trust as three specific beliefs that are utilized most often [31, 33, 34]: competence (ability of the trustee to do what the trustor needs), benevolence (trustee caring and motivation to act in the trustor's interests), and integrity (trustee honesty and promise keeping).

In the context of P2P file sharing, because of the absence of proven guarantees that the vendors and other users or third parties will not engage in harmful opportunistic behaviors, trust is crucial in helping users overcome their perceptions of uncertainty and risk [18, 35]. Research has shown that vendors' trustworthiness attributes are important to users. Lee [36] found that the four out of the top 10 most important features in P2P sharing software rated by users are related to trust, including 'stability', 'reliability' (competence), 'can exit nicely' (integrity), and 'gives error message' (benevolence). Tsivos *et al.* [13] also proposed that P2P sharing systems should have built-in self-regulatory characteristics to reduce the complexity of uncertainty, including characteristics such as stopping queries that are bound to match too many files and eliminating duplicate packets from overzealous users. Following trust definition in e-commerce and other contexts [16, 32, 37], we define *trust of P2P vendors* as a set of *specific beliefs* dealing primarily with the integrity, benevolence, and competence of vendors. Lack of trust and high risks (e.g. security risks and legal risks in terms of copyright infringement) have seriously undermined the development of consumer-friendly P2P business initiatives [8].

Three types of trust antecedents will be examined in this research: institution-based trust (specifically, structural assurance beliefs and situational normality beliefs), knowledge-based trust (specifically, direct knowledge of or experiential interaction with a trustee), and cognition-based trust (specifically, reputation categorization process). Figure 1 presents our research model. The following sections develop and elaborate the key constructs and the theoretical rationale for the causal relationships among the constructs in the research model.

2.4. Knowledge-based trust

Familiarity with vendors comes from prior first-hand experience. It is suggested that familiarity builds trust in *a priori* trustworthy party [38] and validated in e-commerce context [32]. Familiarity with vendors is different from situational normality because the latter does not involve the knowledge about the actual vendor [32]. In the context of P2P file sharing, familiarity with a vendor, e.g. refers to how knowledgeable a user is about the procedures and techniques for performing P2P sharing activities.

It is suggested that trust in an *a priori* trustworthy party grows as the trust-relevant knowledge is accumulated from experience with the other party [24]. In e-commerce, familiarity with e-vendors is found to lead to higher trust beliefs in vendors [32]. In the context of P2P file sharing, trust-relevant knowledge that is derived from prior experiences, such as the procedures and techniques for performing P2P sharing activities, should help the development of trust in the software vendor. Therefore, we hypothesize:

H1: Familiarity with the vendor of P2P sharing software positively affects trusting beliefs.

2.5. Cognitive trust

Cognition-based trust is formed *via* categorization processes in which individuals place more trust in people similar to themselves and assess trustworthiness based on second-hand information and on stereotypes [30, 32]. Prior research considers reputation as an important subcomponent of the cognitive trust and suggests that a trustor may categorize a trustee as trustworthy or untrustworthy based on the reputation of the trustee

[30]. The reputation categorization process infers that a trustee with a good reputation is believed to be trustworthy [30]. Therefore, if the trustee has a good reputation, trustor will quickly develop trusting beliefs about the trustee, even without first-hand knowledge or direct experiential information [30, 39]. Thus, in the context of P2P sharing, we predict that vendors with a good reputation are seen as trustworthy and those with a bad reputation as untrustworthy.

H2: Reputation of the vendor of P2P sharing software positively affects trusting beliefs.

2.6. Institutional trust

Institution-based trust means that 'one believes that the necessary impersonal structures are in place to enable one to act in anticipation of a successful future endeavor' [30, p. 478]. Among the above-mentioned trust antecedents, institution-based trust is consistently found to have positive impacts on the development of trust in e-vendors for both experienced [32] and inexperienced users [18, 31, 40]. Institutional trust is one's perception of the existence of guarantees, safety nets, or other impersonal structural conditions to facilitate achieving the expected outcomes [41, 42].

Prior research examining the institutional trust in information systems has mostly focused on a single institutional context in the electronic market environment—the organizational context [33]. However, when studying trust in social computing, we believe more complex institutional contexts should be considered. For example, in a P2P environment, two sets of structures are involved in forming users' trust beliefs—the organizational structures and the peer-network structures. We refer the organizational structures to a user's perceptions of the institution environment of a P2P sharing network. Influential factors in forming users' trust beliefs include the organizational resources and procedures, vendor guarantee such as the code of conduct of P2P United [43], the association of P2P software vendors like BearShare, Grokster, and eDonkey, which regulates member vendors in terms of user privacy, security, and respect for copyright laws. In the light of frequent calls for self-regulations among P2P sharing vendors [23], it is important for us to examine the impacts of these organizational structures.

We further believe that the peer-network structures would be the other important factor in forming users' trust beliefs. In a P2P network, trust of a peer is hardly developed because trust is often applicable to a relationship with another *identifiable* party [44] and a peer can easily hide her identity from others. Thus, peers' behaviors such as free-riding shared resources can deteriorate the performance of a P2P sharing network and thus negatively impact others' sharing activities [23, 45]. This is not surprising as in P2P networks, both cooperative and non-cooperative

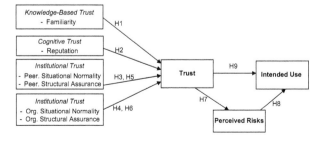

Figure 1. Research model.

behaviors are facilitated given the decentralized nature of such networks [46]. Moreover, there are reports that P2P networks are inserted with low-quality or damaged versions of music files for various purposes [47, 48]. Also P2P networks are criticized being utilized for exchanging pirated resources among some peers. These uncertainties in peer-network may expose users to various risks and drive them to withdraw from the use of P2P sharing software [22]. Consequently, we propose a new dimension in institutional trust, namely *peer-network structure*, relating to one's perceptions that other users on the same peer-network appear to be normal or favorable and the P2P sharing actions are likely to incur low risk.

Existing research on institutional trust has focused on the organizational structure in the context of electronic markets [5, 31, 32, 49, 50]. However, as a specific attribute of network-based virtual communities in social computing, peer-network structure should be explicitly conceptualized as one type of institutional trust that is distinct from the organizational structure. When the network of relationships in network-based virtual community is unstructured and non-static, it is especially important for participants in such network to recognize that the peer network they are interacting with is of low-risk [cf. 7]. At the same time, peer-network structure is distinct from knowledge-based familiarity with a P2P vendor, or cognitive-based trust, in two ways: (i) the peer-network structure is the perception about collective peers who may not be identifiable and (ii) the perceptions of peer-network structure can be derived from first-hand experiences of using the particular P2P sharing software, second-hand information such as news from media, or a combination of both. In current research, we believe a thorough investigation of both peer-network and organizational institutional contexts will improve our understanding of trust-risk model in social computing.

There are two components of institutional trust discussed in the literature: (i) situational normality, defined as the belief that the situation appears to be normal or favorable and success is likely [51] and (ii) structural assurances, defined as the belief that success is likely because such contextual conditions as promises, contracts, regulations, and guarantees are in place [30].

Situational normality. Situational normality stems from the belief that the environment is in proper order [25] and success is likely because the situation is normal or favorable [51, 52]. Situational normality could be related to a greater trust belief because it assures people that everything in the setting is as it ought to be [30, 42] and thus their interactions with others in this setting are in accordance with what they consider to be anticipated [32]. When people face unanticipated or abnormal situations, they are uncomfortable and tend to not trust others in this kind of setting [39]. Empirical studies in e-commerce context have generally supported the positive impacts of situational normality on trust [31, 32] and operationalized situational normality by referring to the trustee being studied, e.g. a specific online vendor or a particular web site. In current research, we operationalize two situational normality constructs: organizational and peer-network situational normality. Situational normality in the peer-network context is operationalized with collective peers who share or download resources on a P2P network. Users are more likely to have positive trusting beliefs if they believe that majority of peers are interacted in a predictable and reliable manner. Therefore, we hypothesize:

H3: Peer-network situational normality positively affects trusting beliefs.

Situational normality in the organizational context is operationalized with vendors of P2P sharing software. Users are more likely to have positive trusting beliefs in a P2P vendor if they observe that the P2P sharing software has a typical user interface, a set of expected procedures, and a typical set of functionalities for P2P sharing activities based on their knowledge and experiences of other similar P2P sharing software. For example, a number of P2P software encourages users to share their resources by offering some rewards, e.g. the more being shared, the faster download a peer can enjoy. As a result, a user would expect such a rewarding mechanism to be built in the P2P sharing software and tend to build trust into the vendor if the vendor provided such functionality. Therefore, we hypothesize:

H4: Organizational situational normality positively affects trusting beliefs.

Structural assurances. Structural assurances refer to the beliefs that structures like regulations, guarantees, and legal resources could guide, empower, and constrain the conduct of individuals and organizations [30, 31, 53]. Examples of structural assurances built into the Web environment could include regulatory or watchdog agencies, legal resources, seals of approval, explicit privacy policy statements, guarantees, affiliation with respected companies, and special interest groups such as consumer or trade associations [5, 32, 54]. Similar to situational normality, two structural assurance constructs were operationalized in current research: organizational and peer-network structural assurances. In a peer-network context, techniques such as reputation building, prevention of pirated resources from being injected into P2P network, and risk reduction mechanisms like anti-flooding and anti-attack have been proposed and implemented into P2P sharing software [55, 56]. These peer-network structures can prevent opportunistic behaviors of peers [57] and thus can build user confidence and trust in the P2P systems and their vendors. Thus, users who perceive high peer-network structural assurances would attribute this to the competence and integrity of the system and thus increase trust in the vendor. Therefore, we hypothesize:

H5: Peer-network structural assurance positively affects trusting beliefs.

In the context of e-commerce, it has been found that organizational structural assurance could limit the firm's ability to behave in negative ways, allowing consumers to form and hold beliefs about expectations of positive outcomes [58]. When violation occurs, these structures could provide mechanisms of voice and recourse [30, 58], which could create strong incentives for firms to refrain from opportunistic behavior and behave appropriately. For example, industry self-regulation body such as P2P united, created code of conduct which regulates member vendors in areas such as users' privacy, security, and respect for copyright laws. Users should be more inclined to trust vendors who are members of P2P United due to the statements of guarantees. Besides these, safety guards such as vendor's privacy statement could also lead to higher trust in vendors. Hence, we hypothesize:

H6: Organizational structural assurance positively affects trusting beliefs.

2.7. Trust, perceived risk, and intended use

The effect of trust on risk reduction has been empirically supported in e-commerce context [18, 21, 35, 59]. Trust could reduce information complexity and lower the perceived risk of a transaction. It has been established in e-commerce that trust in an e-vendor reduces the level of perceived risk [18]. Based on these findings, in the context of P2P sharing, we propose that trusting beliefs in vendor's attributes such as competence, benevolence, and integrity should lower users' risk perceptions in P2P sharing. With the trusting belief in the vendor's capabilities, a user may perceive a lower level of risk such as privacy invasion, free-riding, virus attack, and injection of pirated resources, etc. Hence, we hypothesize:

H7: Trusting beliefs reduce perceived risks in P2P file sharing.

Along the line of Theory of Reasoned Action [60, 61], risk perception viewed as the negative antecedent belief, and trust viewed as the positive antecedent belief, could both affect a person's attitude that in turn influence a person's behavioral intention [18]. Empirical evidence supports the above expectations of the negative relationship between perceived risk and behavioral intention, and the positive relationship between trust and behavioral

intention in e-commerce context [62, 63]. We suggest that the same logic can be extended to P2P sharing context and thus we hypothesize:

H8: Perceived risks in P2P file sharing decrease intended use of P2P sharing software.
H9: Trusting beliefs increase intended use of P2P sharing software.

3. Research method

3.1. Instrument development

Measurement items were developed based on procedures advocated by Churchill [64] and Moore and Benbasat [65]. As far as possible, constructs were adapted from existing measurement scales used in prior studies to fit the context of P2P file sharing where necessary. All the constructs are operationalized as reflective constructs, and adapted from prior trust literature with modifications to reflect the specific context of the P2P sharing in the survey questions. *Intended use* was measured with three items asking the extent to which users would reuse the P2P sharing software [33]. Measures of *perceived risks* were based on the measures used in Pavlou and Gefen [33], adapted to refer to the expectation that a high potential for loss would be associated with the use of P2P sharing software. *Trusting beliefs* were measured with three items that were directly taken from Gefen *et al.* [32]. The measures for *reputation* were developed based on a review of reputation-based trust [31, 50]. These items generally referenced a vendor having an overall good reputation [31]. *Familiarity* was measured by three items based on Gefen *et al.* [32]. In terms of *institutional trust*, measurement for organizational situational normality and structural assurance was adapted based on the measurement of trustworthy attributes of a vendor in Gefen *et al.* [32]; measurements for peer-network situational normality and structural assurance were adapted from the measurements of institution-based trust in McKnight *et al.* [31]. All items in the questionnaire were anchored on 7-point Likert scale. Appendix A presents the final questions measuring each construct in this study.

3.2. The survey

To examine the effects of perceived risk in P2P sharing and trust in vendors on the intention to use P2P sharing software, a survey technique was employed. Email addresses of 600 undergraduate students were randomly collected from an online learning system at a large university in Singapore. Invitation emails explained the purpose of the study and stated that only those who have prior experience in P2P sharing were eligible to participate in the online survey. Also included in the invitation emails was the URL link to the Web-based survey questionnaire. The respondents were told that their anonymity would be assured and the results would be reported only in aggregate. As an incentive for participation, three monetary awards of Singapore dollar $40 per person[2] were raffled among the participants.

To ensure that the data are collected among experienced users of P2P sharing software, respondents were requested to complete the online questionnaire by answering the questions regarding the recent P2P sharing software which they used for resource searching, download, or sharing. Respondents were also required to indicate the name of that P2P sharing software and the usage frequency during past three months. Questionnaires from respondents who had not indicated the previous usage of P2P sharing software were discarded. A total of 136 responses were resulted. The mostly used software applications from the respondents were KaZaA (72%), BitTorrent (12%), Emule (9%), and Shareaza (6%).

4. Data analysis

A second-generation causal modeling statistical technique—partial least squares (PLS), was used for data analysis in this research for three reasons. First, PLS is widely accepted as a method for testing theory in early stages, while LISREL is usually used for theory confirmation [66]. Thus PLS is more suitable for our exploratory study. Second, PLS is well suited for highly complex predictive models [67]. Prior research that applied PLS [68] has claimed that it is best suited for testing complex relationships by avoiding inadmissible solutions and factor indeterminacy. This makes PLS suitable for accommodating the relatively complex relationships among various constructs in this research. Third, PLS has the ability to assess the measurement model within the context of the structural model, which allows a more complete analysis of interrelationships in the model.

4.1. Testing the measurement model

The measurement model was evaluated by examining the relationships between the constructs and the indicators. Such examinations may include the test of the convergent and discriminant validity of constructs. Three tests are used to determine the convergent validity [69]: reliability of questions, the composite reliability of constructs, and the average variance extracted by constructs. Reliability of these questions was assessed by examining the loading of each question on the construct. In order for the shared variance between each question and the construct to exceed the error variance, the reliability score for the question should be at least 0.707 [70]. Given that all questions had reliability scores above 0.707 (see Table 1),

the questions measuring each reflective construct had adequate reliability. Composite reliabilities of constructs with multiple indicators exceeded Nunnally's [71] criterion of 0.7 while the average variances extracted for these constructs were all above 50% and Cronbach's alphas were also all higher than 0.7. Overall, the above test results indicate that the convergent validity of all constructs is adequate.

Discriminant validity is the degree to which measures of different constructs are distinct [72]. To test discriminant validity, the squared correlations between constructs (their shared variance) should be less than the average variance extracted for a construct. Table 2 reports the descriptive statistics and the results of discriminant validity, which is checked by comparing the diagonal to the non-diagonal elements. All items fulfilled the requirement of discriminant validity.

4.2. Testing the structural model

After establishing the validity of the measures, we tested the structural paths in the research model using PLS. We conducted hypothesis tests by examining the sign and significance of the path coefficients. A jack-knife resampling technique was applied to estimate the significance of the path coefficients. Given that each hypothesis corresponded to a path in the structural model, support for each hypothesis could be determined based on the sign (positive or negative) and statistical significance for its corresponding path. Figure 2 shows a graphical display of the results of hypothesis testing. The explanatory power of the structural model is assessed based on the amount of variance explained in the endogenous construct (i.e. intended use). The structural model could explain 33.5% of the variance for intended use. This greatly exceeded 10%, which was suggested by Falk and Miller [73] as an indication of substantive explanatory power.

As shown in Figure 2, all hypotheses were supported except H6 (Organizational Structural Assurance → Trust). In support of H1 and H2, the results indicate that familiarity with the vendor of P2P sharing software and reputation of the vendor of P2P sharing software were positively related to trusting beliefs. H3 and H4 postulate the influences of peer-network situational normality and organizational situational normality on trusting beliefs. In support of H3 and H4, the positive relationships between peer-network and organizational situational normality and trusting beliefs were found significant. Regarding the influences of structural assurances on trusting beliefs, peer-network structural assurance was positively related to trusting beliefs (H5 was supported); but organizational structural assurance did not have significant impact on trusting beliefs (H6 was not supported). Trusting beliefs were negatively related to perceived risks (H7 was supported); perceived risks were found to be negatively related to intended use

---

[2]The reward was framed in Singapore dollars. One Singapore dollar was around 59 US cents at the time of experiment.

Table 1. Psychometric properties of the measurement model.

Construct indicators	Factor loadings	Composite reliability	Cronbach's alpha	Variance extracted
Intention to use (INT)				
INT1	0.983	0.986	0.873	0.959
INT2	0.984			
INT3	0.970			
Perceived risk (RISK)				
RISK1	0.976	0.949	0.889	0.862
RISK2	0.889			
RISK3	0.918			
Trust in P2P Vendor (TRU)				
TRU1	0.939	0.892	0.837	0.735
TRU2	0.805			
TRU3	0.822			
Reputation (VR)				
VR1	0.898	0.916	0.886	0.783
VR2	0.859			
VR3	0.898			
Familiarity (FV)				
FV1	0.921	0.912	0.899	0.776
FV2	0.896			
FV3	0.822			
Organizational structural assurances (OSA)				
OSA1	0.912	0.880	0.811	0.712
OSA2	0.890			
OSA3	0.715			
Organizational situational normality (OSN)				
OSN1	0.717	0.858	0.878	0.670
OSN2	0.820			
OSN3	0.908			
Peer-network structural assurances (PSA)				
PSA1	0.905	0.929	0.865	0.814
PSA2	0.924			
PSA3	0.877			
Peer-network situational normality (PSN)				
PSN1	0.889	0.913	0.850	0.778
PSN2	0.862			
PSN3	0.894			

Table 2. Discriminant validity.

	VR	FV	OSA	OSN	PSA	PSN	TRU	RISK	INT
VR	0.885								
FV	0.604	0.881							
OSA	0.547	0.585	0.844						
OSN	0.379	0.185	0.300	0.812					
PSA	0.467	0.364	0.422	0.261	0.902				
PSN	0.433	0.378	0.460	0.309	0.378	0.882			
TRU	0.590	0.578	0.470	0.361	0.458	0.467	0.857		
RISK	0.100	0.119	0.335	0.287	0.022	0.106	−0.351	0.928	
INT	0.253	0.086	0.008	0.513	0.128	0.260	0.389	−0.217	0.979

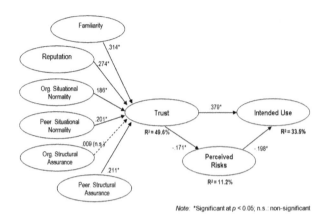

Figure 2. Structural model.

(H8 was supported); trusting beliefs were positively related to intended use (H9 was supported).

5. Discussion

Social computing focuses on how users may have more autonomy to express their ideas and participate in social exchanges in various ways, one of which may be P2P file sharing. While technical issues are still relevant from a social computing perspective, this research examines P2P file sharing from a more social perspective to understand how factors influencing individuals in groups or networks affect their online behaviors. The results show that experienced users' continued use of P2P sharing software depends on both trust beliefs in the software vendor and risk perceptions associated with P2P sharing. Most prior research on trust has reported knowledge-based factors, cognitive factors, and institutional factors as significant determinants of trusting beliefs. Our results confirmed the effects of these trust antecedents in a social computing context. In addition, this research has differentiated two components of institutional trust: the organizational structures and the peer-network structures. Our results showed that the organizational structures were overshadowed when peer-network structures, knowledge-based trust, and cognitive trust were also assessed. Specifically, the proposed organizational structural assurances did not have impact on trust in P2P vendor. A possible explanation for this could be that the survey participants are not aware of the existence of any legal protection or not familiar with the industry's self-regulation body. In fact, none of the survey respondents lastly used a P2P sharing software that is provided by a member vendor in P2P United [43].

Our preliminary findings have several practical implications in the P2P landscape. First, this study highlights the important roles of P2P vendors and peer network in building effective online communities. As discussed earlier, building trustworthy dependency on other peers is difficult due to high uncertainties of peers. Peer-to-peer

vendors, therefore, should contribute to provide functions into the sharing software to build a safe, effective, and stable P2P network that can lead to users' perception of peer-network situational normality and structure assurance [74]. P2P vendors should also actively take efforts in addressing such issues like free-riding, content piracy, malicious computer attack, rather than passively leaving these issues as they were and merely playing a role as software provider. This is also supported by the call for the self-regulation in P2P industry by researchers [8] and by lawmakers [56]. In practice, more and more P2P software vendors are implementing such mechanisms, like providing peers with incentives for opening more shares in exchange for faster download speed, and offering the accounting functions in the software to protect against malicious users. Second, reputation of the vendor and user familiarity with the vendor should also be actively promoted, e.g. via promoting the new features and procedures of the software through mass media. Third, both organizational and peer-network situational normality, such as a typical user interface, and effective mechanisms like free-riding prevention and anti-flooding, are important strategies for trust building.

Organizational structural assurances were shown not important in trust building by our data analysis. This insignificant effect is probably due to the participants' lack of knowledge about available organizational safeguards. One interpretation of this result is that, users of social computing applications increasingly take trust cues from one another or from communities rather than from organizational sources. As such, user communities are increasingly driving innovations and communications from the bottom-up, and the information flow, economic value, and power are starting to shift from organizations to user communities. This interpretation is in line with the trends discussed in the Introduction per researchers' observations [2, 4].

6. Conclusion

P2P networks do not merely implement Web-based interfaces but also they design architectures that allow peer wise communication and social action; that is, P2P networks imply communities as well as sophisticated enabling technologies [3]. This exploratory study seeks to understand what steps can be taken to increase users' trust beliefs and reduce their risk perceptions so as to encourage legitimate interactions in social computing platforms. Our results can also be applied to users who do not have initial experiences, as the information about the vendor's trustworthiness and the level of perceived risk can be passed on to and propagated among initial users and affect their adoption of P2P sharing software [75]. Although the data generally support the proposed model, caution must be exercised when generalizing these findings. This study was conducted in Singapore, care must be taken when generalizing these findings to consumers in other social, economic, and cultural environments, and future research should attempt to replicate this study in other countries, especially those in North America and in Europe, to further validate the research model. Most P2P sharing networks are running globally over the Internet, and the legal risks presented in one country may be absent in other countries. How to effectively prevent global P2P users from sharing and downloading copyright violated materials? Who will play the most important role in regulating the usage behavior in using P2P sharing software? These would be fruitful questions for future research.

Through the causal modeling of the antecedents affecting use intentions, our findings provide preliminary empirical support to understand trust and risk issues in the context P2P file sharing networks. Nevertheless, since some characteristics of this study may limit the generalizability of our findings, several avenues for future work remain. We hope this study makes a modest contribution to stimulating further research in the field of the P2P file sharing networks.

Appendix A.

Measurement items (measured on 7-point Likert-type scale).

Intended use (INT): (Pavlou and Gefen [33])
INT1   I intend to continue using the P2P file sharing software to search for, download or share resources.
INT2   I predict I would continue using the P2P file sharing software.
INT3   I plan to continue using the P2P file sharing software.

Perceived risk (RISK): (Pavlou and Gefen [33])
RISK1   There is a high potential for loss involved in using the P2P file sharing software.
RISK2   There is a considerable risk involved in using the P2P file sharing software to search for, download and/or share resources.
RISK3   My decision to use the P2P file sharing software is risky.

Trust in P2P vendor (TRU): (Gefen *et al.* [59])
TRU1   I believe the vendor is honest.
TRU2   I believe the vendor cares about its users.
TRU4   I believe the vendor is reliable.

Reputation (VR): (McKnight *et al.* [5])
VR1   The vendor has a reputation for being honest.
VR2   The vendor has a reputation for being concerned about the users.
VR3   Most users think that this vendor has a reputation for being fair.

Familiarity (FV): (Gefen *et al.* [59])
FV1   I am familiar with the vendor through resource searching and download by the P2P file sharing software.
FV2   I am familiar with the vendor through sharing resources to other peers by the P2P file sharing software.
FV3   I am familiar with the vendor through reading magazine/newspaper articles or ads.

Peer-network structural assurances (PSA): (McKnight *et al.* [5])
PSA1   The peer-network has enough safeguards to make me feel comfortable transferring and sharing resources with other peers.
PSA2   I feel assured that the technological structures adequately protect me from peers' opportunistic behaviors.
PSA3   I feel assured that other peers cannot free ride on my shared resources.

Peer-network situational normality (PSN): (McKnight *et al.* [5])
    Based on your experiences with peers whom you have downloaded resources from, or shared resources to, among those peers:
PSN1   Most peers are in general predictable and consistent regarding their behaviors.
PSN2   Most peers are trustworthy in transferring and sharing resources with other peers.
PSN3   Most peers are reliable to download resources from, and/or share resources to.

Organizational structural assurances (OSA): (Gefen *et al.* [59])
OSA1   I feel assured that downloaded resources are legal because the vendor provides statements of guarantees that all shared resources are legal.
OSA2   I feel safe using the P2P sharing software because the vendor is on the list of P2P United.
OSA3   I am comfortable searching, downloading or sharing resources because of the regulatory and technological structures built by the vendor.

Organizational situational normality (OSN): (Gefen *et al.* [59])
Based on your experiences with other similar P2P sharing software...

(continued on next page)

Appendix A. Continued.

Intended use (INT): (Pavlou and Gefen, 2004)	
OSN1	The mechanisms built into the software to encourage peers to download/share resources are typical of other similar P2P file sharing software.
OSN2	The steps required to search for, download and share resources are typical of other similar P2P file sharing software.
OSN3	The approach used by the software to encourage peers to download/share resources is the type of approach most similar P2P sharing software employs.

Acknowledgements. The authors would like to thank Hao Wang and Audrey Lim at the Pennsylvania State University for their assistance on an earlier version of this paper.

References

[1] PARAMESWARAN, M. and WHINSTON, A.B. (2007) Social computing: an overview. *Commun. Assoc. Inf. Syst.* **19**: 762–780.

[2] IP, R. and WAGNER, C. (2008) Weblogging: a study of social computing and its impact on organizations. *Decis. Support Syst.* **45**(2): 242–250.

[3] WANG, F.Y., CARLEY, K.M., ZENG, D. and MAO, W. (2007) Social computing: from social informatics to social intelligence. *IEEE Intell. Syst.* March/April: 79–83.

[4] CHARRON, C., FAVIER, J. and LI, C. (2006) *Social Computing: How Networks Erode Institutional Power, and What to Do About It* (Forester Research). Retrieved 26 May 2011, from http://www.forrester.com/Research/Document/Excerpt/0,7211,38772,00.html.

[5] MCKNIGHT, D.H. and CHERVANY, N.L. (2002) What trust means in e-commerce customer relationships: an interdisciplinary conceptual typology. *Int. J. Electron. Com.* **6**(2): 35–59.

[6] HOFFMAN, D.L., NOVAK, T. and PERALTA, M.A. (1999) Information privacy in the marketspace: implications for the commercial uses of anonymity on the Web. *Inf. Soc.* **15**(2): 129–139.

[7] DHOLAKIA, U.M., BAGOZZI, R.P. and PEARO, L.K. (2004) A social influence model of consumer participation in network- and small-group-based virtual communities. *Int. J. Res. Marketing* **21**(3): 241–263.

[8] HUGHES, J., LANG, K.R. and VRAGOV, R. (2008) An analytical framework for evaluating peer-to-peer business models. *Electron. Com. Res. Appl.* **7**(1): 105–118.

[9] SCHODER, D. and FISCHBACH, K. (2003) Peer-to-peer prospects. *Commun. ACM* **46**(2): 27–29.

[10] MCGUIRE, D. (2004) *Lawmakers Push Prison for Online Pirates.* Retrieved 26 May 2011, from http://www.washingtonpost.com/wp-dyn/articles/A40145-2004Mar31.html.

[11] DINGLEDINE, R., FREEDMAN, J.M. and MOLNAR, D. (2001) Accountability. In ORAM, A. [ed.] *Peer-to-Peer: Harnessing the Power of Disruptive Technologies* (Cambridge, MA: O'Reilly & Associates), 271–339.

[12] REITER, M.K. and RUBIN, A.D. (1999) Anonymous Web transactions with crowds. *Commun. ACM* **42**(2): 32–38.

[13] TSIVOS, P., WHITLEY, A.E. and HOSEIN, I. (1999) An exploration of the emergence, development and evolution of regulatory characteristics of information systems. In *Proceedings of the Twentieth International Conference on Information Systems (ICIS)* (Charlotte, NC), 813–816.

[14] BORLAND, J. (2003) *Fingerprinting P2P pirates.* Retrieved 26 May 2011, from http://news.com.com/2100-1023_3-985027.html.

[15] HAVLENA, W.J. and DESARBO, W.S. (1991) On the measurement of perceived consumer risk. *Decis. Sci.* **22**(4): 927–939.

[16] GANESAN, S. (1994) Determinants of long-term orientation in buyer–seller relationships. *J. Marketing* **58**: 1–19.

[17] YATES, J.F. and STONE, E.R. (1992) Risk appraisal. In YATES, J.F. [ed.] *Risk-Taking Behavior* (Chichester, UK: John Wiley & Sons), 49–85.

[18] JARVENPAA, S.L. and TRACTINSKY, N. (1999) Consumer trust in an Internet store: a cross-cultural validation. *J. Comput. Mediated Commun.* **5**(2): 1–35.

[19] PETER, J.P. and TARPEY, S.L.X. (1975) A comparative analysis of three consumer decision strategies. *J. Consum. Res.* **2**(1): 29.

[20] MOON, Y. (2000) Intimate exchanges: using computers to elicit self-disclosure from consumers. *J. Consum. Res.* **26**: 323–339.

[21] STEWART, K.J. (2003) Trust transference on the world wide Web. *Organ. Sci.* **14**(1): 5–17.

[22] PEW-INTERNET, AMERICAN-LIFE-PROJECT (2004) *The State of Music Downloading and File-Sharing Online.* Retrieved 26 May 2011, from http://www.pewinternet.org/~/media/Files/Reports/2004/PIP_Filesharing_April_04.pdf.pdf.

[23] PAVLOV, V.O. and SAEED, K. (2003) A resource-based assessment of the gnutella file-sharing network. In *Proceedings of 24th Annual International Conference on Information Systems (ICIS)* (Seattle, WA), 85–95.

[24] LEWICKI, R. and BUNKER, B.B. (1995) Trust in relationships: a model of trust development and decline. In BUNKER, B.B. and RUBIN, J.Z. [eds.] *Conflict, Cooperation, and Justice* (San Francisco, CA: Jossey-Bass), 133–173.

[25] LEWIS, J.D. and WEIGERT, A.J. (1985) Trust as a social reality. *Soc. Forces* **63**(4): 967–985.

[26] LANE, C. and BACHMANN, R. (1996) The social constitution of trust: supplier relations in Britain and Germany. *Organiz. Stud.* **17**(3): 365–395.

[27] MOORMAN, C., DESPHANDE, R. and ZALTMAN, G. (1993) Factors affecting trust in market research relationships. *J. Marketing* **57**(1): 81–101.

[28] LUHMANN, N. (1988) Familiarity, confidence, trust: problems and alternatives. In GAMBETTA, D.G. [ed.] *Trust* (New York: Basil Blackwell), 94–107.

[29] GRABNER-KRÄUTER, S. and KALUSCHA, E.A. (2003) Empirical research in online trust: a review and critical assessment. *Int. J. Hum. Comput. Stud. Special Issue on 'Trust and Technology'* **58**(6): 783–812.

[30] McKNIGHT, D.H., CUMMINGS, L.L. and CHERVANY, N.L. (1998) Initial trust formation in new organizational relationships. *Acad. Manage. Rev.* **23**(3): 472–490.

[31] McKNIGHT, D.H., CHOUDHURY, V. and KACMAR, C. (2002) Developing and validating trust measures for e-commerce: an integrative typology. *Inf. Syst. Res.* **13**(3): 334–359.

[32] GEFEN, D., KARAHANNA, E. and STRAUB, D.W. (2003) Trust and TAM in online shopping: an integrated model. *MIS Q.* **27**(1): 51–90.

[33] PAVLOU, P.A. and GEFEN, D. (2004) Building effective online marketplaces with institution-based trust. *Inf. Syst. Res.* **15**(1): 37–59.

[34] BHATTACHERJEE, A. (2002) Individual trust in online firms: scale development and initial test. *J. Manage. Inf. Syst.* **19**(3): 211–241.

[35] KOLLOCK, P. (1999) The production of trust in online markets. *Adv. Group Processes* **16**: 99–123.

[36] LEE, J. (2003) An end-user perspective on file-sharing systems. *Commun. ACM* **46**(2): 49–53.

[37] GIFFIN, K. (1967) The contribution of studies of source credibility to a theory of interpersonal trust in the communication process. *Psychol. Bull.* **68**(2): 104–120.

[38] LUHMANN, N. (1979) *Trust and Power* (Chichester, UK: John Wiley & Sons).

[39] LI, X., HESS, T.J. and VALACICH, J.S. (2008) Why do we trust new technology? A study of initial trust formation with organizational information systems. *J. Strategic Inf. Syst.* **17**(1): 39–71.

[40] GRAZIOLI, S. and WANG, A. (2001) Looking without seeing: Understanding unsophisticated consumers' success and failure to detect Internet deception. In *Proceedings of the International Conference on Information Systems (ICIS)* (New Orlean, LA), 193–204.

[41] SHAPIRO, D.L., SHEPPARD, B.H. and CHERASKIN, L. (1992) Business on a handshake. *Negotiation J.* **3**: 365–377.

[42] ZUCKER, L.G. (1986) Production of trust: institutional sources of economic structure, 1840–1920. In STAW, B.M.A and CUMMINGS, L.L. [eds.] *Research in Organizational Behavior* (Greenwich, CT: JAI Press), 53–111.

[43] P2P UNITED. (2004) *P2P United: Fighting for the Future of Peer-to-Peer Technology*. Retrieved 26 May 2011, from http://www.ftc.gov/bcp/workshops/filesharing/presentations/eisgrau.pdf.

[44] MAYER, R.C., DAVIS, J.H. and SCHOORMAN, F.D. (1995) An integration model of organizational trust. *Acad. Manage. Rev.* **20**(3): 709–734.

[45] SAMANT, K. (2003) Free riding, altruism, and cooperation on peer-to-peer file-sharing networks. In *Proceedings of 24th Annual International Conference on Information Systems (ICIS)* (Seattle, WA), 914–920.

[46] KWOK, S.H. and YANG, C.C. (2004) Searching the peer-to-peer networks: the community and their queries. *J. Am. Soc. Inf. Sci. Technol.* **55**(9): 783–793.

[47] BORLAND, J. (2002) *Start-Ups Try to Dupe File-Swappers* (CNET News.Com). Retrieved 26 May 2011, from http://news.com.com/2100-1023-943883.html.

[48] LEVINE, D. (2002) *Not the Real Slim Shady*. Retrieved 26 May 2011, from http://dir.salon.com/story/tech/feature/2002/06/10/eminem_mp3/index.html.

[49] LEE, K.C., KANG, I.W. and McKNIGHT, D.H. (2007) Transfer from offline trust to key online perceptions: an empirical study. *IEEE Trans. Eng. Manage.* **54**(4): 729–741.

[50] PENNINGTON, R., WILCOX, H.D. and GROVER, V. (2003) The role of system trust in business-to-consumer transactions. *J. Manage. Inf. Syst.* **20**(3): 197–226.

[51] BAIER, A. (1986) Trust and antitrust. *Ethics* **96**: 231–260.

[52] GARFINKEL, H. (1963) A conception of, and experiments with, 'trust' as a condition of stable concerted actions. In HARVEY, O.J. [ed.] *Motivation and Social Interaction* (New York: Ronald Press), 187–238.

[53] SHAPIRO, S.P. (1987) The social control of impersonal trust. *AJS* **93**: 623–658.

[54] PALMER, J.W., BAILEY, J.P. and FARAJ, S. (2000) The role of intermediaries in the development of trust on the WWW: The use and prominence of trusted third parties and privacy statements. *J. Comput. Mediated Commun.* **5**.

[55] ORAM, A. (2001) *Peer-to-Peer: Harnessing the Power of Disruptive Technologies* (Cambridge, MA: O'Reilly & Associates).

[56] BORLAND, J. (2003) *Senators Ask P2P Companies to Police Themselves* (CNET News.com). Retrieved 26 May 2011, from http://news.com.com/2100-1028_3-5110785.html.

[57] WALDMAN, M., CRANOR, F.L. and RUBIN, A. (2001) Trust. In ORAM, A. [ed.] *Peer-to-Peer: Harnessing the Power of Disruptive Technologies* (Cambridge, MA: O'Reilly & Associates), 242–270.

[58] JOHNSON, L.J. and CULLEN, B.J. (2002) Trust in cross-cultural relationships. In GANNON, M.J. and NEWMAN, K.L. [eds.] *The Blackwell Handbook of Cross-Cultural Management* (Oxford, UK and Malden, MA: Blackwell), 335–360.

[59] GEFEN, D., RAO, V.S. and TRACTINSKY, N. (2003) The conceptualization of trust, risk and their relationship in electronic commerce: the need for clarifications. In *Proceedings of the 36th Hawaii International Conference on System Sciences (HICSS)* (Big Island, HI), 192–201.

[60] AJZEN, I. (1985) From intentions to actions: A theory of planned behavior. In KUHL, J. and BECKMANN, J. [eds.] *Action Control: From Cognition to Behavior* (New York, NY: Springer Verlag), 11–39.

[61] AJZEN, I. (1991) The theory of planned behavior. *Organiz. Behav. Hum. Decis. Processes* **50**: 179–211.

[62] GEFEN, D. (2002) Customer loyalty in e-commerce. *J. Assoc. Inf. Syst.* **3**: 27–51.

[63] PAVLOU, P.A. (2003) Consumer acceptance of electronic commerce: integrating trust and risk with the technology acceptance model. *Int. J. Electron. Com.* 7(3): 69–103.

[64] CHURCHILL, G.A. (1979) A paradigm for developing better marketing constructs. *J. Marketing Res.* **16** (February): 64–73.

[65] MOORE, G.C. and BENBASAT, I. (1991) Development of an instrument to measure the perceptions of adopting an information technology innovation. *Inf. Syst. Res.* **2**(3): 173–191.

[66] FORNELL, C. and BOOKSTEIN, F.L. (1982) Two structrual equation models: lisrel and pls applied to customer exit-voice theory. *J. Marketing Res.* **19**(11): 440–452.

[67] CHIN, W.W. (1998) The partial least squares approach to structural equation modeling. In MARCOULIDES, G.A. [ed.] *Modern Methods for Business Research* (London: Psychology Press), 295–336.

[68] KIM, D. and BENBASAT, I. (2006) The effects of trust-assuring arguments on consumer trust in Internet stores: application of Toulmin's model of argumentation. *Inf. Syst. Res.* **17**(3): 286–300.

[69] COOK, M. and CAMPBELL, D.T. (1979) *Quasi-Experimentation: Design and Analysis Issues for Field Settings* (Boston, MA: Houghton Mifflin).

[70] CHIN, W.W. (1998) The partial least squares approach to structural equation modeling. In MARCOULIDES, G.A. [ed.] *Modern Methods for Business Research* (Mahwah, NJ: Lawrence Erlbaum Associates), 295–336.

[71] NUNNALLY, J.C. (1978) *Psychometric Theory* (New York: McGraw-Hill), 2nd ed.

[72] CAMPBELL, D.T. and FISKE, D.W. (1959) Convergent and discriminant validation by the multitrait–multimethod matrix. *Psychol. Bull.* **56**(1): 81–105.

[73] FALK, R.F. and MILLER, N.B. (1992) *A Primer for Soft Modeling* (Akron, OH: University of Akron Press).

[74] ARINGHIERI, R., DAMIANI, E., VIMERCATI, S.D.C.D., PARABOSCHI, S. and SAMARATI, P. (2006) Fuzzy techniques for trust and reputation management in anonymous peer-to-peer systems. *J. Am. Soc. Inf. Sci. Tech.* **57**(4): 528–537.

[75] SONG, J. and WALDEN, A.E. (2003) Consumer behavior in the adoption of peer-to-peer technologies: an empirical examination of information cascades network externalities. In *Ninth Americas Conference on Information Systems (AMCIS)* (Tampa, FL), 1801–1810.

8

Single and Multiple UAV Cyber-Attack Simulation and Performance Evaluation

Ahmad Y. Javaid[1,*], Weiqing Sun[2], Mansoor Alam[1]

[1]2801 W. Bancroft St., EECS Department, College of Engineering, The University of Toledo, Toledo, Ohio, USA
[2]2801 W. Bancroft St., ET Department, College of Engineering, The University of Toledo, Toledo, Ohio, USA

Abstract

Usage of ground, air and underwater unmanned vehicles (UGV, UAV and UUV) has increased exponentially in the recent past with industries producing thousands of these unmanned vehicles every year. With the ongoing discussion of integration of UAVs in the US National Airspace, the need of a cost-effective way to verify the security and resilience of a group of communicating UAVs under attack has become very important. The answer to this need is a simulation testbed which can be used to simulate the UAV Network (UAVNet). One of these attempts is - UAVSim (Unmanned Aerial Vehicle Simulation testbed) developed at the University of Toledo. It has the capability of simulating large UAV networks as well as small UAV networks with large number of attack nodes. In this paper, we analyse the performance of the simulation testbed for two attacks, targeting single and multiple UAVs. Traditional and generic computing resource available in a regular computer laboratory was used. Various evaluation results have been presented and analysed which suggest the suitability of UAVSim for UAVNet attack and swarm simulation applications.

Keywords: UAV Cyber-security, performance evaluation, simulation, testbed

1. Introduction

With applications in almost every field, UAVs have become really popular for applications which were limited by human element. Until a few years ago, primary focus of development was military in nature but their use in other real world civil applications are on a rapid increase. With applications like pizza delivery (Pizza Hut), local package delivery (Amazon), agricultural chemical deployment, ecological surveys [1–4], industries and academia are using UAVs for their research, businesses, etc, and there are much more applications to be thought of. Without doubt, their importance in the military domain has increased several folds in the recent past due to their impact on human effectiveness and safety. Another important point to be noted is the delay in inclusion of civil and other kinds of UAVs in the National Airspace System (NAS) due to several issues including communication security [5].

Increased attack attempts in recent past on such mobile cyber-physical systems (CPS) are alarming and have raised concerns over their use, especially with increasing autonomy level [6, 7]. Keeping this in the mind, the authors noticed the need of cost-effective and safe virtual simulation testbed environment for testing the accurate implementation of various security related technologies in an Unmanned Aerial System (UAS). Addressing various environment variations, such as weather, loss of connectivity and contested communication are some of the most important aspects of such a simulation testbed due to dependency of UAV control on communication and its security. Therefore, we focus on two basic types of attack - one targeting a single UAV and second, targeting multiple UAVs in the mission area.

The rest of the paper is organized to provide background on related and our previous works in sections 2 and 3 respectively. Section 4 provides more details about UAVSim covering its design and various features. Section 5 describes all the performance analysis done and related results and inferences. Section 6 concludes the paper and discusses possible future enhancements to the work.

2. Related Work

In this section, we discuss some of the recent advances and works related to simulation or actual hardware based evaluations. As these UAV related issues are addressed by policy makers and bureaucrats, the need of a secure and safe UAV system stays unquestionable for military as well as civil applications due to safety and privacy threats imposed by their compromise.

Therefore, several researchers have been working on development of different kinds of simulation testbeds in order to validate safe states of these systems and check possibility of moving into an unsafe state. These simulation testbeds can be classified in four major categories based on the resources they employ.

2.1. Software based Single UAV Simulation

Software simulation testbeds are purely based on well-known software platforms and do not employ any kind of hardware. Testbeds developed using Matlab/Simulink [8], FlightGear [9], JSBSim/FlightGear [7, 10] and Matlab/FlightGear [11] are some of the recent outcomes of research in this area. All these simulation testbeds have focused on testing a single-UAV model instead of modeling its behavior in presence of other UAVs in the real world.

2.2. Software–Hardware based Single UAV Simulation

Some other simulation testbeds using hardware along with software, have also been developed where the hardware might be actual UAVs [12], robots [13, 14], or just laptops [15, 16]. A very recent work of this type [10] focuses on analytical and component based simulation and analysis. In this work, the area of focus for cyber attacks is sensor compromise of various degrees.

2.3. Software based Multiple UAV Simulation

This class of simulation testbeds are also solely based on software platforms and these are developed in-house as well. One of the most important works in this class, SPEEDES (Synchronous Parallel Environment for Emulation and Discrete Event Simulation) [17], simulates a swarm of UAVs on a high performance parallel computer so that it can match the speed and communication rate of a real UAVNet. Another recent work of this class, DCAS (distributed cyber attack simulator) [18], presents a distributed simulation framework for modeling cyber attacks and the evaluation of security measures. DCAS is based on Portico, an open source HLA (high-level architecture) simulation engine. Limitation of this work is that it is for a generic wired or wireless network and does not include mobile components. On the other hand, UAVSim addresses these limitations and incorporates various mobility models, mobile radio propagation models, mobile ad-hoc routing protocols, etc.

2.4. Software–Hardware based Multiple UAV Simulation

This class of simulation testbeds are primarily based on software platforms but real or emulated UAVs can also be used within them. Rather than being commercial, these are mostly developed in-house for research purposes, specifically for UAVs. The only testbed that could be found in this category, $C3UV$ [19], Center for Collaborative Control of Unmanned Vehicles at UC Berkeley, has been constantly updated by their researchers since 2004. Over the years, the $C3UV$ team has incorporated multiple-UAV simulation on parallel computing environment along with the capability of using real UAVs. This kind of testbed, despite all its achievements, involves huge expenses in terms of high performance parallel computing hardware and optional use of real UAVs. While in UAVSim, cost involved is quite less and once positive results are achieved, the tested mechanism can be directly implemented in real UAVs only if required.

3. Our Previous Work

After studying all the important works done until now and their limitations, UAVSim was designed and developed keeping in mind the primary objective of UAVNet security simulation. Initially, UAV system model was defined to represent the system approximately so that a software model could be created. An analytical threat and vulnerability analysis was also performed and attack impacts were demonstrated using FlightGear simulation software [6]. Further, an independent simulation module (called UAVSim) was developed and a few cyber-attacks, such as, Jamming and DDoS (Distributed Denial of Service) were implemented using the base simulation engine of OMNeT++. One of the major features developed in this phase was an interactive GUI for beginner level users. Various simulation results and related insights were presented and the accuracy of UAVSim was demonstrated [20]. This work also describes the technical details of the developed software simulation testbed. In continuation, advanced features like multi-user support, server based centralized simulation, etc., using ubiquitous computing infrastructure, were added to UAVSim and the testbed performance was analysed in different scenarios for different modes of operations for DDoS attack [21].

In this paper, we extend the analysis for DDoS attack with increased number of concurrent users and present detailed analysis for Jamming attack as well. Primary reason behind selection of these two attacks for our performance analysis is the huge computational resource requirement for simulation of both of these attacks. Most cyber attacks which aim to take control over the subject, do not involve large amount of data transmission, instead, these attacks only require minimum data transmission in terms of some unique command and control messages. Therefore, if high computing resource consumption attacks (from the testbed perspective) can be simulated, it would prove the testbed's capability to simulate all other attacks which will consume less resources on the underlying computing infrastructure.

4. UAVSim: Design and Features

As discussed in Section 2, the primary focus of developing a simulation testbed has been simulating the behavior of a single UAV to check its proper functioning. Nowadays, use of large number of UAVs in various applications demands for their performance test in an existing swarm of aircrafts, especially when the US Government is working on integrating UAVs in the US National Airspace.

4.1. Testbed requirements

There are other important requirements to be met to make such a simulation testbed more useful. Testing of security measures in terms of hardware as well as software should be supported. Impact evaluation, on system components and overall performance must be supported as well. The testbed should allow use of various UAV models, developed in UAVSim as well as other popular software. In order to make it available to UAV-experts, who are not technically sound, the testbed should have an interactive and easy to use GUI. For advanced users, an advanced GUI can also be an option. The environment designed should be also verified and validated in order to correctly simulate the UAV model. One of the most important aspects of communication should be addressed and the UAV should be treated as a network of components which replicates the component communication behavior. Other environment variations, such as contested communication, collaborative control, mobility models and mission paths should be modeled and addressed.

4.2. Design

Keeping the above requirements in mind, UAV component level modeling, individual simulation, attack classification and attack modeling were performed [6] and later a software simulation testbed, UAVSim, for simulations of all sizes of UAV networks was developed and in-depth design was presented in [20]. Preliminary performance evaluation was done in [21] which covered simulations for the DDoS attack with maximum number of users limited to 6. UAVSim is developed using the open source network simulator OMNeT++ and one of its independently developed open source modules called INET for mobility and related protocols. For satellite communication, another open source component called OS3 (Open source satellite simulator) has been used. Network design and higher level code is coded in NED, a language specifically designed for OMNeT++ while the lower level functionality is coded

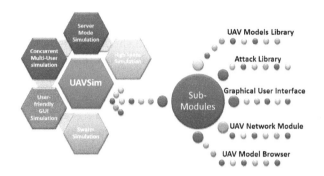

Figure 1. UAVSim: Modes of operations and various components

in C++ [22]. Figure 1 shows the design of UAVSim in the right half. The left half shows various modes of simulations.

4.3. Features

Security simulation being the primary focus and one of the most important features of UAVSim, several attacks have been implemented in the attack library of the testbed. Further, basic and advanced models of popular UAVs have been designed. The interactive GUI allows beginner and intermediate users to vary various parameters while advanced users can directly make changes to the simulation configuration files. Apart from supporting mobile wireless communication and UAV component level modeling capability, UAVSim also supports detailed network analysis at lower levels of the protocol stack. Further, attacks targeting different layers can also be designed, launched and tested in UAVSim. One of the most important features of UAVSim from user perspective is its user-friendly design and its ability to work on generic computing environment. Figure 1 summarizes the important features and modules of UAVSim.

User-friendly GUI Simulation. The simulation testbed supports both command line and graphical user interface. We have developed a custom GUI for UAVSim which lets basic users select possible options for some parameters. Users do not get a lot of independence in the basic GUI. While, the advanced users can edit all other parameters as well using the configuration file in the simulation project. Although the GUI might cost some resource, it definitely can be counted as one of the performance parameters as the testbed has been designed to be used for all levels of users, basic, intermediate or advanced.

Server Mode Simulation. In order to enhance performance, a high performance computer can also be utilized in our simulation testbed. The connection details to a server or high performance computer can be set using the GUI by the administrator or the person setting up the testbed for initial use. It should be noted that the core testbed simulation files should be installed on the server prior to this setup and ssh should be enabled on the high performance computer to enable seamless communication and execution.

High Speed (No-GUI) Simulation. While the testbed has a well-designed GUI, the aim of providing a non-GUI option was to enhance the performance. There is an option of express command mode execution as well, which prints the minimum required simulation statistics in order to let the user know that the simulation is running and the computer is not frozen. Using this option, the simulation can be run at the maximum speed and thus gives the best performance. This mode was primarily designed for Server mode simulation because the communication with the server might slow down execution. Nevertheless, this mode can be used on the desktop mode as well as server mode.

Concurrent Multi-User Simulation. The testbed also provides a multi-user option which allows multiple users to concurrently run their simulations through their individual machines. This option utilizes the Server Mode of the testbed. As mentioned before, if the testbed needs to be used for high speed simulation or, by several users at the same time, a non-GUI server option is available. One of the most important prerequisites to use this option is the connection oriented access availability on the server to all the user accounts. This is necessary in order to enable independent simulation for each user. The core simulation modules need to be installed on the server while users remotely connect to the server using UAVSim. The UAVSim, once configured with the server and connection details, automatically connects to the server and displays results in a console window. It should be noted that the multi-user simulation is only available in non-GUI option.

Swarm Simulation. Although the simulation testbed was primarily developed for UAVNet security simulation, it also supports UAV swarm simulation. This feature enables users to test the network behavior when large numbers of UAVs are used for any specific application. The use can be commercial, civil or military in nature but in case of swarms, usually it should be a sensor based application with a large number of sensors. The performance for swarm simulation using a large number of nodes has also been evaluated.

4.4. Attack Anatomy

Here we have described the design and implementation of the two selected attacks in brief. A detailed explanation is not really necessary because of their well known anatomy. As mentioned earlier, the focus was to select two resource intensive attacks - one which attacks a single UAV and second, which attacks multiple UAVs - in order to measure the performance accurately. DDoS attack was chosen as the attack which will target a single UAV while a Jamming attack will target multiple UAVs in the mission area. Both of these attacks are discussed below.

DDoS Attack - Single Target. The DDoS (Distributed Denial of Service) attack aims at loss of communication through network congestion. This is achieved by making the host appear unavailable to other hosts in the network, mostly, due to the increase in response time and almost 100% packet drop. The reason behind huge packet drop and response time is large number of adversary hosts sending frequent requests to the host being attacked, the requests might be PING, SYN or any other kind of packet demanding an acknowledgment.

This attack has been implemented in UAVSim using a number of attack nodes, which can be defined by the user based on the total number of UAVs in the network. Although it has been proved experimentally that even a single host is capable of launching a successful DDoS attack using a PING packet because of its small payload size [23]. In order to implement the DDoS attack, we have used the traditional way of transmitting spoofed packets to a single host from several attack nodes. All attack nodes behave similar to regular UAV hosts and are assigned the IP addresses of the same range in order to make them indistinguishable from other trustworthy UAVs. During the simulation analysis, we have varied this number to check the success rate of attack in different scenarios. All attack nodes transmit packets to a single UAV host in order to make it unreachable and thus, launch a successful attack on a single target. Approximate time taken to successfully launch this attack is only few seconds for all simulations and packet loss for the attacked node reaches 99.9% in less than 2 seconds.

Jamming Attack - Multiple Target. Any kind of radio signal based communication can be interrupted using Jamming, which involves transmission of noise in the mission area. This attack results in loss or corruption of packets. The noise usually spans over all the frequencies and prevents communication at any frequency. If the attack node has a powerful transmitter, a signal can be generated that will be strong enough to overwhelm the targeted signals and disrupt communications. The most common types of signal used in an jamming attack are random noise and pulse [24]. Jamming equipment is readily available in the market as well as on online shopping websites like *amazon* and *eBay*. In addition, jamming equipment can be mounted from a location remote to the target networks. This attack can not be handled by most of modern wireless devices and is relatively easier to launch.

This attack has been implemented in UAVSim by creating several attack nodes which send noise signals to all the hosts in a round robin fashion over different frequencies. The number of these attack nodes can be varied in the simulation. Transmitting random signals to all UAV hosts will launch a multiple target attack as aimed. Various techniques have been developed in the recent past to take care of jamming attacks, most of them expect the attack to be in a particular signal frequency and therefore, most of these methods use frequency hopping and spread spectrum communication to counter Jamming attack [25]. That is why we have implemented total-frequency band jamming attack. Although post-2010, several researchers have proposed various anti-jamming encoding, encryption, etc., we have not addressed those techniques in our jamming attack implementation. Total time for successful completion of this attack takes a little more time than DDoS attack and is about 5 seconds for most simulations while packet loss for all hosts reach above 90% for all nodes.

5. Performance Analysis and Results

The primary focus of this paper is to demonstrate the usefulness of the simulation testbed even with regular computing infrastructure. Usually, in an academic or research setup, where resources are constrained, purchasing expensive high end computing infrastructure might be quite difficult or even impossible. Therefore, the testbed should allow users to use it for any kind of UAV network in a cost-effective manner. As mentioned in Section 2, various works which allow simulation of UAV swarms for various purposes, use quite high-end computing facilities and are not available to the public. On the contrary, our simulation testbed is designed to work with already existing simulation engine and components which are open source and thus, free to use. At the same time, this testbed does not need expensive machines to get results. Clearly, one might have to compromise on computation time.

Another point to note is the expected increase in simulation run times for Jamming attack. DDoS works on principle of sending huge number of packets to one node causing congestion and stopping it from communicating with others. On the other hand, Jamming works by transmitting noise on all frequencies so that communication is jammed due to noise traffic on the wireless channel. Therefore, in order to implement DDoS attack, lesser number of attack nodes are required as only single node working at a frequency needs to be jammed. On the contrary, Jamming requires all the frequencies to be jammed for a successful attack.

5.1. Simulation Setup

Keeping in mind the primary goal of performance evaluation, it is necessary to have a clear understanding of what kind of Hardware or Software environment has been used in order to make sure that performance claims are accurate. Here we have described the hardware, software as well as simulation testbed setup used for our simulations.

Hardware Setup. The PC being used for simulations has a Intel® Core^{TM} i7-3770 CPU (1 × 3.40 GHz 4-core, L2/L3 Cache: 1 MB/8 MB) and a system memory of 8.0 GB while the server used during the server mode simulation has a Intel® Xeon® Processor E5-2630 (2 × 2.30 GHz 6-core, L2/L3 Cache: 1.5/15 MB) and a system memory of 64.0 GB. Apart from these, for use in multi-user concurrent simulations, a few generic laptops (more or less 2-3 years old) were used and their configurations are not listed due to negligible computation taking place in those systems and hence, no impact on overall testbed performance.

Software Setup. Both of the systems defined in previous paragraph, the PC and the Server machines, run Ubuntu version 12.04 LTS. Needless to say, the Server runs the x64 server version while the PC is running the x64 desktop version. Both has the OMNeT++ version 4.2.2 with the INET version 2.2 and CNI_OS3 version 1.0. As mentioned before, our simulation module UAVSim makes use of OMNeT++ and these two open-source plugins to accurately simulate a UAVNet.

Testbed Setup. All simulations were 300 seconds long while actual time taken to finish this simulation were observed. As established before in [26], actual time to attack a time-sensitive military system such as a missile defense system is only a couple of seconds. Even in our simulations, attacks were launched right after the simulation started and it was noted that the attacks were successful in only a couple of seconds. The most basic UAV Model has been used for our simulation as using more advanced models detailing various sub-modules would clearly increase simulation times. Advanced UAV models can be used while implementing more complex confidentiality or integrity compromising attacks. Frequency for UAV communication is fixed at 5 GHz for Single Target attack scenario while it varies between the range of 5-15 GHz for use in Multiple Target attack scenario.

Several cases were evaluated for different types of simulation. For Case I, running time and swarm behavior analysis, we have used both the number of attack nodes as well as the number of UAVs. In Case I_a, the number of UAVs was varied from 50 to 500 and in Case I_b, number of attack nodes was varied from 2 to 20. In case of Jamming attack, run time increased to days after 350 nodes, therefore, that was the limit

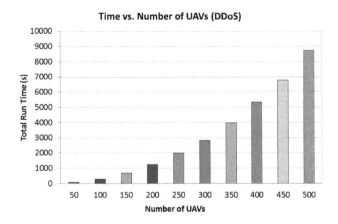

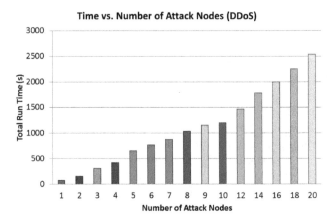

Figure 2. Run time variation with increasing number of UAVs for DDoS attack (same frequency of all hosts)

Figure 3. Run time variation with number of attack nodes for DDoS attack

for Jamming attack. Further, for checking multi-user behavior, two separate analysis were done with separate cases. Case II, where performance of swarm behavior with multiple concurrent user was evaluated. Case II_a, where 50 nodes were used for swarm behavior analysis while concurrent number of users was increased from 1 to 8. Case II_b, number of nodes was increased to 100. Case III, where performance of the testbed was evaluated for attack simulation. Case III_a being 5 attack nodes and Case III_b being 10 attack nodes, keeping the number of UAV nodes as 10 for both the scenarios, III_a and III_b. These three cases will be referred to during the discussion and analysis.

5.2. DDoS Attack – Single Target

This subsection covers the results for all the simulations for Distributed DOS attack. The various simulations were done for the 3 above mentioned cases, namely Case I, II and III. We have analysed the effect of increasing number of UAVs, attack nodes, concurrent users in Server Mode operation and use of GUI.

Number of UAVs. For this analysis, we have used Case I_a (number of UAVs varied). As mentioned earlier, Case I_a involves use of regular UAV nodes in order to ascertain the UAVSim capability for UAV swarm simulation other than the primary capability of UAVNet security simulation. It is clear from Fig. 2 that the testbed run-time varies exponentially with the increasing number of UAV nodes for this attack. Clearly, the run time is directly proportional to the powers of each 50 nodes and thus, is easily predictable for higher number of UAVs. Looking at the simulation times, it can be argued that large number of UAVs can be used for swarm simulations using single frequency scenarios.

Number of Attack Nodes. For the second analysis, we have used Case I_b (varying number of attack nodes). Fig. 3 depicts the performance for simulations with

increasing number of attack nodes while the number of UAVs is kept constant as 10. Looking at the trend obtained, it is understood that the variation is exponential with respect to the number of attack nodes and instead of multiples of 50, here we have multiples of 2, therefore, large number of attack nodes may not be used for security simulations. Keeping in mind the number of attack nodes which can be simulated in reasonable time, using large value (more than 50) for this variable is neither possible nor required.

Graphical User Interface. The third performance metric was the use of GUI (graphical user interface) which displays the network animation. It is understood that having a GUI displaying the network animation and various network statistics during a CPU intensive operation might impact the performance. Therefore, we used Case I_b, where we varied the number of attack nodes and measured the speed of simulation for GUI and non-GUI options. Fig. 4 show the results obtained for GUI and non-GUI options on the server as a blue dashed line and a black dotted line. The red dashed line shows the percentage difference between the two modes with respect to the lower value (non-GUI option).

As shown in Fig. 4, the non-GUI run time follows a non-linear polynomial trend with respect to the number of attack nodes. The percentage change between GUI and non-GUI options for a DDoS attack is not more than 7% for all cases with most cases being between 2 – 5%. Therefore, it can be said that the performance is not much affected by use of GUI for this particular attack.

Number of Concurrent Users – Single Frequency Swarm Scenario. The fourth performance test was done varying number of concurrent users accessing the simulation framework in the server mode option. As mentioned earlier, the server based simulation works only in non-GUI mode to enhance execution performance and reduce the server to PC communication. In our earlier

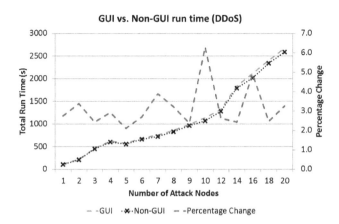

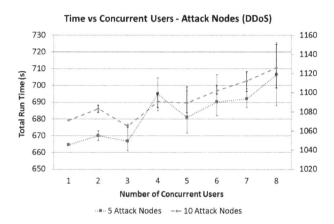

Figure 4. Run time variation with GUI and Non-GUI options, and the percentage change in two options for DDoS Attack

Figure 6. Run time variation of security simulation with increasing number of concurrent users in server mode operation for DDoS attack

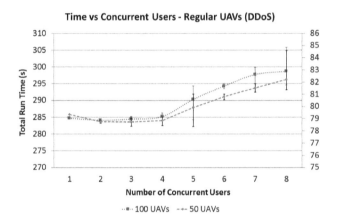

Figure 5. Run time variation of swarm simulation with increasing number of concurrent users in server mode operation for single frequency band

performance evaluation [21], the number of users was varied from 1 to 6. Here, we varied the number of concurrent users from 1 to 8 and extended the analysis to Jamming attack as well which is covered in the next subsection.

Fig. 5 show the evaluation results for Case II_a (50 UAV hosts) and II_b (100 UAV hosts) using the DDoS attack scenario. As mentioned earlier, these scenarios do not use any malicious node and all UAVs communicate at a single frequency of $5GHz$. This simulation was intended to analyze the performance variation for multiple concurrent users, simulating a swarm using single frequency in absence of an attack. Please note that the two separate vertical axes show the variation of total run time for the two Cases, II_a and II_b. The error bars show the maximum and minimum time while the points depict the average time.

Number of Concurrent Users - DDoS Security Simulation. The final analysis for DDoS attack targets the

performance evaluation of the testbed with multiple users using it concurrently in server mode operation. To this end, Cases III_a (5 attack nodes) and III_b (10 attack nodes) were used. Fig. 6 show the test results for this experiment. As mentioned earlier, the number of malicious hosts was changed for the two cases, keeping number of regular UAVs as 10. The number of malicious hosts used is much lesser than Case II because malicious nodes generate more traffic in the network and are responsible for increasing the execution time, as found in the two initial experiments discussed in this section. Just like the last analysis, the two separate vertical axes show the variation of total run time for two different numbers of attack nodes. The error bars show the maximum and minimum time while the points represent the average time.

5.3. Jamming Attack - Multiple Targets

This subsection covers the results of all simulations for Jamming attack. The various simulations were done for the 3 cases mentioned in subsection 5.1, namely Case I, II and III. We have analyzed the effect of increasing number of UAVs, attack nodes, concurrent users in Server Mode operation and use of GUI.

Number of UAVs. We have used Case I_a (number of UAVs varied) for this experiment. Please note that this simulation might seem similar to DDoS attack UAV-only simulation but it should be noted that here, multiple frequencies are being used for communication rather than single. As mentioned before, the frequency range for UAV-UAV communication lies between $5 - 15GHz$ for this case. This has been done in order to make sure that all frequencies are jammed in the attack area. It is clear from Fig. 7 that the testbed run-time varies exponentially with the increasing number of UAV nodes.

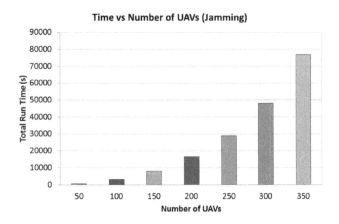

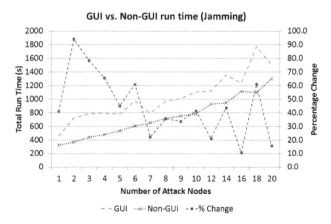

Figure 7. Run time variation with increasing number of UAVs for Jamming attack (different frequencies of hosts)

Figure 9. Run time variation with GUI and Non-GUI options, and the percentage change in two options for Jamming attack

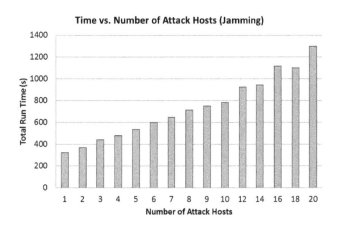

Figure 8. Run time variation with increasing number of Attack Nodes for Jamming attack (different frequencies of hosts as well as attack nodes)

Clearly, the run time is directly proportional to the power of each 50 nodes and thus, is easily predictable for higher number of UAVs and therefore, large number of UAVs can be used for swarm simulations scenarios where multiple frequency channels are used. It should also be noted that the exponent might be higher than that found in single frequency swarm simulation (DDoS UAV-only simulation) and thus, it can be seen that even for 350 nodes, the run time reaches almost 24 hours compared to one hour run time in case of single frequency communication.

Number of Attack Nodes. For the second analysis of Jamming attack, Case I_b (varying number of attack nodes) has been used. Fig. 8 shows the performance for simulations with increasing number of attack nodes while the number of UAVs is 10. It should be noted that the attack simulation trend for Jamming attack is non-linear polynomial instead of exponential as it was in case of DDoS attack. Since the trend is not

exponential, large number of attack nodes may be used for security simulations of Jamming attack. Comparing the two attack scenarios, it can be noted that the run time for Jamming attack for Case I_b is roughly half of DDoS attack simulation run time. Once again, keeping in mind the number of attack nodes required for a successful attack, large values (more than 100) can not and need not be used.

Graphical User Interface. We used Case I_b (varying number of attack nodes) to measure the speed of simulation for GUI and non-GUI options. Fig. 9 show the results obtained for GUI and non-GUI options on the server as a blue dashed line and a black dotted line respectively. The red dashed line (showing a heartbeat trend) represents the percentage difference between the two modes with respect to the lower value (non-GUI option).

The percentage change between GUI and non-GUI options for a Jamming attack is quite random and higher for lower number of attack nodes. Mostly, it is between $10-70\%$. This trend is exactly opposite of the trend shown by DDoS attack simulation. It is quite clear that the performance is affected very badly by use of GUI for a Jamming attack. The significant changes in performance for Jamming attack in current and previous case can be attributed to its anatomy and implementation and will be discussed in subsection 5.4.

Number of Concurrent Users - Multiple frequency swarm simulation. This performance test was performed varying number of concurrent users using the simulation testbed on a single server. The number of concurrent users was varied from 1 to 8, and simulation run time for Cases II_a (50 UAV hosts) and II_b (100 UAV hosts) were evaluated. Fig. 10 show the evaluation results using the Jamming attack scenario. These scenarios do not use any malicious node and all different UAVs communicate at different frequencies in the range

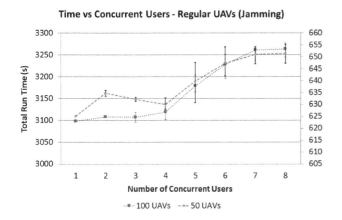

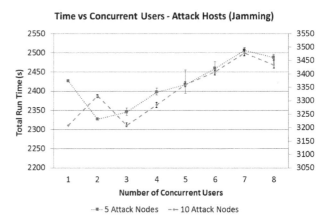

Figure 10. Run time variation of swarm simulation with increasing number of concurrent users in server mode for variable frequency bands (Jamming attack scenario)

Figure 11. Run time variation of security simulation with increasing number of concurrent users in server mode operation for Jamming attack

of $5-15GHz$. This simulation was intended to analyze the performance variation for multiple concurrent users, simulating a swarm using multiple frequencies in absence of an attack. Please note that the two separate vertical axes show the variation of total run time for the two Cases, II_a and II_b, lower numbers, obviously, depicting Case II_a. The error bars show the maximum and minimum time while the points depict the average time.

It should be noted that both Cases II_a and II_b follow similar trend after certain number of users and the run time seems to be becoming invariable. Another important aspect to note is the average percentage variation in both cases is less than 5% between time taken for single and 8 concurrent users.

Number of Concurrent Users – Jamming attack security simulation. The final performance test for Jamming attack involves the performance analysis for the Jamming attack simulation with multiple users using the testbed concurrently. Run time values for Cases III_a (5 attack nodes) and III_b (10 attack nodes) were evaluated. Fig. 11 show the performance test results for these cases. The two separate vertical axes in this evaluation also show the variation of total run time for two different numbers of attack nodes. The error bars show the maximum and minimum time while the points represent the average time.

It can be again noted that the plot for both cases follow similar trend once number of concurrent users increase to 3. Similar to the previous evaluation of swarm simulation, the overall percentage variation between maximum and minimum run times for each case is less than 10%. Maximum for both cases was at 7 concurrent users while minimum for Case III_a at 2 and for Case III_b at 3 concurrent users.

5.4. Analysis

The various test categories for number of UAVs, number of attacks hosts, GUI/non-GUI option and number of concurrent users were performed, which give us valuable insights in terms of the operational capability of the testbed. Although some simulation times are quite high in case of swarm simulations of large number of nodes, the performance is reasonable for security simulations. Some important points which can be noted from the analysis are as follows:

- Total run time varies exponentially with the number of attack nodes in security simulations as well as number of UAVs in swarm simulation. Despite of the trend, it should be noted that the variation in attack nodes are only by 2 while in case of UAVs, the number is varied by 50. This gives a clear indication of how many attack nodes and UAV hosts can be deployed in simulation scenarios.

- Using the GUI for any security simulation has little impact on performance for DDoS attacks but the variations are higher for Jamming attack. The Jamming attack requires more processing in terms of creation of channels for different frequencies and transmission of packets but this can not be attributed to slower execution times due to GUI. Therefore, it can be concluded that use of network animation while using several communication channels requires more processing in the base simulation framework of OMNeT++.

- Performance analysis for multiple users using the testbed in server mode reveals that performance gets affected with increasing number of users but average variation reduces as number of concurrent users increases after 4.

- The average simulation time saturates after certain number of users and shows a trend of becoming invariable with respect to number of concurrent users. The variation in minimum and maximum shows that total system performance is not affected much.

- It can be also noted that the simulation times for concurrent user analysis follows the same trend for same cases evaluated for different number of UAVs or attack nodes after number of concurrent users increases more than 4. This implies that immaterial of the value, the trend would be similar and thus, the run time can be estimated for higher number of users using the obtained trend.

- The attack simulation run time for 20 attack nodes for both types of attacks took approximately half an hour. Practically, the number of attack host in order to launch such attacks are much less. For example, we need 4 attack nodes for a GPS spoofing attack [27] and thus, the simulation capability is quite extensive.

- Since the variation in run time for concurrent users is not exponential, the simulation testbed seems quite capable of handling more than 20 users concurrently on a regular server. The evaluation was done for up to 8 users and showed no alarming trend.

- Although most of the trends are similar for both classes of attacks, the times are much higher for Jamming than DDoS attack simulation. The primary reason behind this is the anatomy of the simulation for these two attacks. Successful execution of a Single Target DDoS attack requires one node to be stopped from communicating while in Jamming, all frequencies, and thus, all nodes are required to be stopped from communicating. The underlying simulation engine of OMNeT++ simulates single object for a single channel (single frequency) and packets are transmitted in that channel within the same object (channel). When several channels are used, each channel is a different object and an inter-process communication takes place between two object for packet transmission and thus, causes increase in total run time for simulations related to Jamming attack.

6. Conclusion

Two classes of attacks - single and multiple target - were simulated in the in-house developed simulation testbed (UAVSim) for UAVNets. Simulation run time analysis for UAVSim were presented to demonstrate its use in generic computing environment. Various simulations indicate that the performance of the testbed

is reasonable and allows users to adjust various options according to the requirement. Along with attack simulation involving single and multiple targets, the testbed was proved to be capable of simulating large UAV swarm networks. Single target - single frequency network as well as multiple target - multiple frequency system (where different frequencies are used for communication between different UAVs) simulation was demonstrated. Overall use of UAVSim can also be extended in other domains of interest involving UAV networks.

Performance for server based concurrent multi-user operation was also tested for different scenarios using different number of simultaneous users were evaluated and the testbed was found to perform very well. Although a maximum of 8 concurrent users-scenario was tested, it is understood that for a reasonable amount of concurrent users, the testbed will perform without much delay in simulation. Interactive GUI, additional result analysis module, model browsing capability (from other model development software), enhanced high speed mode of operation, support of concurrent users, etc. are some of the features which makes this software simulation testbed an ideal simulation environment for UAV simulations in generic computing environment.

References

[1] IANS, "Mumbai pizza delivery drones raise security buzz." http://economictimes.indiatimes.com/industry/services/hotels-/-restaurants/mumbai-pizza-delivery-drones-raise-security-buzz/articleshow/35486155.cms, 2014. [Online; Last accessed: 30-July-2014].

[2] D. Aamoth, "Delivering Domino's Pizza by Unmanned Helicopter: What Could Possibly Go Wrong?." http://techland.time.com/2013/06/03/delivering-dominos-pizza-by-unmanned-helicopter-/-what-could-possibly-go-wrong/, 2013. [Online; Last accessed: 30-July-2014].

[3] A. Chang, "With Prime Air, Amazon plans to deliver purchases via drones." http://articles.latimes.com/2013/dec/02/business/la-fi-tn-amazon-prime-air-20131202, 2013. [Online; Last accessed: 30-July-2014].

[4] M. Bryson, A. Reid, C. Hung, F. Ramos, and S. Sukkarieh, "Cost-Effective Mapping Using Unmanned Aerial Vehicles in Ecology Monitoring Applications," in *Experimental Robotics* (O. Khatib, V. Kumar, and G. Sukhatme, eds.), vol. 79 of *Springer Tracts in Advanced Robotics*, pp. 509–523, Springer Berlin Heidelberg, 2014.

[5] FAA, "Integration of Civil Unmanned Aircraft Systems (UAS) in the National Airspace System (NAS) Roadmap, First Edition." http://www.faa.gov/about/initiatives/uas/media/uas_roadmap_2013.pdf, 2013. [Online; Published November 2013].

[6] A. Javaid, W. Sun, V. Devabhaktuni, and M. Alam, "Cyber security threat analysis and modeling of

an unmanned aerial vehicle system," in *2012 IEEE Conference on Technologies for Homeland Security (HST)*, pp. 585–590, Nov 2012.

[7] A. Kim, B. Wampler, J. Goppert, I. Hwang, and H. Aldridge, "Cyber attack vulnerabilities analysis for unmanned aerial vehicles," *The American Institute of Aeronautics and Astronautics: Reston, VA, USA*, 2012.

[8] P. Lu and Q. Geng, "Real-time simulation system for UAV based on Matlab/Simulink," in *2011 IEEE 2nd International Conference on Computing, Control and Industrial Engineering (CCIE)*, vol. 1, pp. 399–404, Aug 2011.

[9] J. Zhang, Q. Geng, and Q. Fei, "UAV flight control system modeling and simulation based on FlightGear," in *International Conference on Automatic Control and Artificial Intelligence (ACAI 2012)*, pp. 2231–2234, March 2012.

[10] J. Goppert, A. Shull, N. Sathyamoorthy, W. Liu, I. Hwang, and H. Aldridge, "Software/Hardware-in-the-Loop Analysis of Cyberattacks on Unmanned Aerial Systems," *Journal of Aerospace Information Systems*, vol. 11, no. 5, pp. 337–343, 2014.

[11] Y. Qiang, X. Bin, Z. Yao, Y. Yanping, L. Haotao, and Z. Wei, "Visual simulation system for quadrotor unmanned aerial vehicles," in *2011 30th Chinese Control Conference (CCC)*, pp. 454–459, July 2011.

[12] T. X. Brown, S. Doshi, S. Jadhav, and J. Himmelstein, "Test Bed for a Wireless Network on Small UAVs," in *In Proceedings of AIAA 3rd Unmanned Unlimited Technical Conference*, pp. 20–23, 2004.

[13] J. Wu, W. Wang, J. Zhang, and B. Wang, "Research of a kind of new UAV training simulator based on equipment simulation," in *2011 International Conference on Electronic and Mechanical Engineering and Information Technology (EMEIT)*, vol. 9, pp. 4812–4815, Aug 2011.

[14] J. Yang and H. Li, "UAV Hardware-in-loop Simulation System Based on Right-angle Robot," in *2012 4th International Conference on Intelligent Human-Machine Systems and Cybernetics (IHMSC)*, vol. 1, pp. 58–61, Aug 2012.

[15] J. Corner and G. Lamont, "Parallel simulation of UAV swarm scenarios," in *Proceedings of the 2004 Winter Simulation Conference*, vol. 1, pp. –363, Dec 2004.

[16] S. Hamilton, T. Schmoyer, and J. Drew Hamilton, "Validating a network simulation testbed for army UAVs," in *2007 Winter Simulation Conference*, pp. 1300–1305, Dec 2007.

[17] S. Chaumette, R. Laplace, C. Mazel, and R. Mirault, "SCUAL, swarm of communicating uavs at LaBRI: An open UAVNet testbed," in *Wireless Personal Multimedia Communications (WPMC), 2011 14th International Symposium on*, pp. 1–5, IEEE, 2011.

[18] M. Ashtiani and M. A. Azgomi, "A distributed simulation framework for modeling cyber attacks and the evaluation of security measures," *Simulation*, p. 0037549714540221, 2014.

[19] E. Pereira, K. Hedrick, and R. Sengupta, "The C3UV Testbed for Collaborative Control and Information Acquisition Using UAVs," in *2013 American Control Conference (ACC)*, pp. 1466–1471, IEEE, 2013.

[20] A. Y. Javaid, W. Sun, and M. Alam, "UAVSim: A simulation testbed for unmanned aerial vehicle network cyber security analysis," in *2013 IEEE Globecom Workshops (GC Wkshps)*, pp. 1432–1436, Dec 2013.

[21] A. Javaid, W. Sun, and M. Alam, "UAVNet Simulation in UAVSim: A Performance Evaluation and Enhancement," in *9th International Conference on Testbeds and Research Infrastructures for the Development of Networks & Communities (TRIDENTCOM 2014)*, May 2014.

[22] A. Varga *et al.*, "The OMNeT++ discrete event simulation system," in *Proceedings of the European Simulation Multiconference (ESM-2001)*, vol. 9, p. 185, sn, 2001.

[23] K. M. Elleithy, D. Blagovic, W. Cheng, and P. Sideleau, "Denial of Service Attack Techniques: Analysis, Implementation and Comparison," vol. 3, pp. 66–71, 2006.

[24] T. Karygiannis and L. Owens, "Wireless network security," *NIST special publication*, vol. 800, p. 48, 2002.

[25] K. Pelechrinis, M. Iliofotou, and S. V. Krishnamurthy, "Denial of service attacks in wireless networks: The case of jammers," *IEEE Communications Surveys & Tutorials*, vol. 13, no. 2, pp. 245–257, 2011.

[26] P. Katopodis, G. Katsis, O. Walker, M. Tummala, and J. Michael, "A Hybrid, Large-scale Wireless Sensor Network for Missile Defense," in *IEEE International Conference on System of Systems Engineering, 2007. SoSE '07.*, pp. 1–5, April 2007.

[27] N. O. Tippenhauer, C. Pöpper, K. B. Rasmussen, and S. Capkun, "On the Requirements for Successful GPS Spoofing Attacks," in *Proceedings of the 18th ACM Conference on Computer and Communications Security, CCS '11*, (New York, NY, USA), pp. 75–86, ACM, 2011.

9

Advancements of Outlier Detection: A Survey

Ji Zhang

Department of Mathematics and Computing
University of Southern Queensland, Australia

Abstract

Outlier detection is an important research problem in data mining that aims to discover useful abnormal and irregular patterns hidden in large datasets. In this paper, we present a survey of outlier detection techniques to reflect the recent advancements in this field. The survey will not only cover the traditional outlier detection methods for static and low dimensional datasets but also review the more recent developments that deal with more complex outlier detection problems for dynamic/streaming and high-dimensional datasets.

Keywords: Data Mining, Outlier Detection, High-dimensional Datasets

1. Introduction

Outlier detection is an important research problem in data mining that aims to find objects that are considerably dissimilar, exceptional and inconsistent with respect to the majority data in an input database [50]. Outlier detection, also known as anomaly detection in some literatures, has become the enabling underlying technology for a wide range of practical applications in industry, business, security and engineering, etc. For example, outlier detection can help identify suspicious fraudulent transaction for credit card companies. It can also be utilized to identify abnormal brain signals that may indicate the early development of brain cancers. Due to its inherent importance in various areas, considerable research efforts in outlier detection have been conducted in the past decade. A number of outlier detection techniques have been proposed that use different mechanisms and algorithms. This paper presents a comprehensive review on the major state-of-the-art outlier detection methods. We will cover different major categories of outlier detection approaches and critically evaluate their respective advantages and disadvantages.

In principle, an outlier detection technique can be considered as a mapping function f that can be expressed as $f(p) \to q$, where $q \in \Re^+$. Giving a data point p in the given dataset, a corresponding outlier-ness score is generated by applying the mapping function f to quantitatively reflect the strength of outlier-ness of p. Based on the mapping function f, there are typically two major tasks for outlier detection problem to accomplish, which leads to two corresponding problem formulations. From the given dataset that is under study, one may want to find the top k outliers that have the highest outlier-ness scores or all the outliers whose outlier-ness score exceeding a user specified threshold.

The exact techniques or algorithms used in different outlier methods may vary significantly, which are largely dependent on the characteristic of the datasets to be dealt with. The datasets could be static with a small number of attributes where outlier detection is relatively easy. Nevertheless, the datasets could also be dynamic, such as data streams, and at the same time have a large number of attributes. Dealing with this kind of datasets is more complex by nature and requires special attentions to the detection performance (including speed and accuracy) of the methods to be developed.

Given the abundance of research literatures in the field of outlier detection, the scope of this survey will be clearly specified first in order to facilitate a systematic survey of the existing outlier detection methods. After that, we will start the survey with a review of the conventional outlier detection techniques that are primarily suitable for relatively low-dimensional

---

*Corresponding author. Email: Ji.Zhang@usq.edu.au

static data, followed by some of the major recent advancements in outlier detection for high-dimensional static data and data streams.

2. Scope of This Survey

Before the review of outlier detection methods is presented, it is necessary for us to first explicitly specify the scope of this survey. There have been a lot of research work in detecting different kinds of outliers from various types of data where the techniques outlier detection methods utilize differ considerably. Most of the existing outlier detection methods detect the so-called *point outliers from vector-like data sets*. This is the focus of this review as well as of this thesis. Another common category of outliers that has been investigated is called *collective outliers*. Besides the vector-like data, outliers can also be detected from other types of data such as sequences, trajectories and graphs, etc. In the reminder of this subsection, we will discuss briefly different types of outliers.

First, outliers can be classified as point outliers and collective outliers based on the number of data instances involved in the concept of outliers.

- **Point outliers.** In a given set of data instances, an individual outlying instance is termed as a point outlier. This is the simplest type of outliers and is the focus of majority of existing outlier detection schemes [28]. A data point is detected as a point outlier because it displays outlier-ness at its own right, rather than together with other data points. In most cases, data are represented in vectors as in the relational databases. Each tuple contains a specific number of attributes. The principled method for detecting point outliers from vector-type data sets is to quantify, through some outlier-ness metrics, the extent to which each single data is deviated from the other data in the data set.

- **Collective outliers.** A collective outlier represents a collection of data instances that is outlying with respect to the entire data set. The individual data instance in a collective outlier may not be outlier by itself, but the joint occurrence as a collection is anomalous [28]. Usually, the data instances in a collective outlier are related to each other. A typical type of collective outliers are sequence outliers, where the data are in the format of an ordered sequence.

Outliers can also be categorized into vector outliers, sequence outliers, trajectory outliers and graph outliers, etc, depending on the types of data from where outliers can be detected.

- **Vector outliers.** Vector outliers are detected from vector-like representation of data such as the

relational databases. The data are presented in tuples and each tuple has a set of associated attributes. The data set can contain only numeric attributes, or categorical attributes or both. Based on the number of attributes, the data set can be broadly classified as low-dimensional data and high-dimensional data, even though there is not a clear cutoff between these two types of data sets. As relational databases still represent the mainstream approaches for data storage, therefore, vector outliers are the most common type of outliers we are dealing with.

- **Sequence outliers.** In many applications, data are presented as a sequence. A good example of a sequence database is the computer system call log where the computer commands executed, in a certain order, are stored. A sequence of commands in this log may look like the following sequence: *http-web, buffer-overflow, http-web, http-web, smtp-mail, ftp, http-web, ssh*. Outlying sequence of commands may indicate a malicious behavior that potentially compromises system security. In order to detect abnormal command sequences, normal command sequences are maintained and those sequences that do not match any normal sequences are labeled sequence outliers. Sequence outliers are a form of collective outlier.

- **Trajectory outliers.** Recent improvements in satellites and tracking facilities have made it possible to collect a huge amount of trajectory data of moving objects. Examples include vehicle positioning data, hurricane tracking data, and animal movement data [65]. Unlike a vector or a sequence, a trajectory is typically represented by a set of key features for its movement, including the coordinates of the starting and ending points; the average, minimum, and maximum values of the directional vector; and the average, minimum, and maximum velocities. Based on this representation, a weighted-sum distance function can be defined to compute the difference of trajectory based on the key features for the trajectory [60]. A more recent work proposed a partition-and-detect framework for detecting trajectory outliers [65]. The idea of this method is that it partitions the whole trajectory into line segments and tries to detect outlying line segments, rather than the whole trajectory. Trajectory outliers can be point outliers if we consider each single trajectory as the basic data unit in the outlier detection. However, if the moving objects in the trajectory are considered, then an abnormal sequence of such moving objects (constituting the sub-trajectory) is a collective outlier.

- **Graph outliers.** Graph outliers represent those graph entities that are abnormal when compared with their peers. The graph entities that can become outliers include nodes, edges and subgraphs. For example, Sun *et al.* investigate the detection of anomalous nodes in a bipartite graph [84][85]. Autopart detects outlier edges in a general graph [27]. Noble *et al.* study anomaly detection on a general graph with labeled nodes and try to identify abnormal substructure in the graph [72]. Graph outliers can be either point outliers (*e.g.*, node and edge outliers) or collective outliers (*e.g.*, sub-graph outliers).

Unless otherwise stated, all the outlier detection methods discussed in this review refer to those methods for detecting point outliers from vector-like data sets.

3. Related Work

4. Outlier Detection Methods for Low Dimensional Data

The earlier research work in outlier detection mainly deals with static datasets with relatively low dimensions. Literature on these work can be broadly classified into four major categories based on the techniques they used, *i.e.*, statistical methods, distance-based methods, density-based methods and clustering-based methods.

4.1. Statistical Detection Methods

Statistical outlier detection methods [23, 47] rely on the statistical approaches that assume a distribution or probability model to fit the given dataset. Under the distribution assumed to fit the dataset, the outliers are those points that do not agree with or conform to the underlying model of the data.

The statistical outlier detection methods can be broadly classified into two categories, *i.e.*, the parametric methods and the non-parametric methods. The major differences between these two classes of methods lie in that the parametric methods assume the underlying distribution of the given data and estimate the parameters of the distribution model from the given data [34] while the non-parametric methods do not assume any knowledge of distribution characteristics [31].

Statistical outlier detection methods (parametric and non-parametric) typically take two stages for detecting outliers, *i.e.*, the training stage and test stage.

- **Training stage.** The training stage mainly involves fitting a statistical model or building data profiles based on the given data. Statistical techniques can be performed in a supervised, semi-supervised, and unsupervised manner. Supervised techniques estimate the probability density for normal instances and outliers. Semi-supervised techniques estimate the probability density for either normal instances, or outliers, depending on the availability of labels. Unsupervised techniques determine a statistical model or profile which fits all or the majority of the instances in the given data set;

- **Test stage.** Once the probabilistic model or profile is constructed, the next step is to determine if a given data instance is an outlier with respect to the model/profile or not. This involves computing the posterior probability of the test instance to be generated by the constructed model or the deviation from the constructed data profile. For example, we can find the distance of the data instance from the estimated mean and declare any point above a threshold to be an outlier [42].

Parametric Methods. Parametric statistical outlier detection methods explicitly assume the probabilistic or distribution model(s) for the given data set. Model parameters can be estimated using the training data based upon the distribution assumption. The major parametric outlier detection methods include Gaussian model-based and regression model-based methods.

A. Gaussian Models

Detecting outliers based on Gaussian distribution models have been intensively studied. The training stage typically performs estimation of the mean and variance (or standard deviation) of the Gaussian distribution using Maximum Likelihood Estimates (MLE). To ensure that the distribution assumed by human users is the optimal or close-to-optima underlying distribution the data fit, statistical discordany tests are normally conducted in the test stage [23][16][18]. So far, over one hundred discordancy/outlier tests have been developed for different circumstances, depending on the parameter of dataset (such as the assumed data distribution) and parameter of distribution (such as mean and variance), and the expected number of outliers [50][58]. The rationale is that some small portion of points that have small probability of occurrence in the population are identified as outliers. The commonly used outlier tests for normal distributions are the *mean-variance test* and *box-plot test* [66][49][83][44]. In the mean-variance test for a Gaussian distribution $N(\mu, \sigma^2)$, where the population has a mean μ and variance σ, outliers can be considered to be points that lie 3 or more standard deviations (*i.e.*, $\geq 3\sigma$) away from the mean [41]. This test is general and can be applied to some other commonly used distributions such as Student t-distribution and Poisson distribution, which feature a fatter tail and a longer right tail than a normal distribution, respectively. The box-plot test draws on the box plot to graphically depict the distribution of data using five major attributes, *i.e.*, smallest non-outlier observation (min), lower quartile

(Q1), median, upper quartile (Q3), and largest non-outlier observation (max). The quantity Q3-Q1 is called the *Inter Quartile Range (IQR)*. IQR provides a means to indicate the boundary beyond which the data will be labeled as outliers; a data instance will be labeled as an outlier if it is located 1.5*IQR times lower than Q1 or 1.5*IQR times higher than Q3.

In some cases, a mixture of probabilistic models may be used if a single model is not sufficient for the purpose of data modeling. If labeled data are available, two separate models can be constructed, one for the normal data and another for the outliers. The membership probability of the new instances can be quantified and they are labeled as outliers if their membership probability of outlier probability model is higher than that of the model of the normal data. The mixture of probabilistic models can also be applied to unlabeled data, that is, the whole training data are modeled using a mixture of models. A test instance is considered to be an outlier if it is found that it does not belong to any of the constructed models.

B. Regression Models

If the probabilistic model is unknown regression can be employed for model construction. The regression analysis aims to find a dependence of one/more random variable(s) $\mathcal{Y}$ on another one/more variable(s) $\mathcal{X}$. This involves examining the conditional probability distribution $\mathcal{Y}|\mathcal{X}$. Outlier detection using regression techniques are intensively applied to time-series data [4][2][39][1][64]. The training stage involves constructing a regression model that fits the data. The regression model can either be a linear or non-linear model, depending on the choice from users. The test stage tests the regression model by evaluating each data instance against the model. More specifically, such test involves comparing the actual instance value and its projected value produced by the regression model. A data point is labeled as an outlier if a remarkable deviation occurs between the actual value and its expected value produced by the regression model.

Basically speaking, there are two ways to use the data in the dataset for building the regression model for outlier detection, namely the *reverse search* and *direct search* methods. The reverse search method constructs the regression model by using all data available and then the data with the greatest error are considered as outliers and excluded from the model. The direct search approach constructs a model based on a portion of data and then adds new data points incrementally when the preliminary model construction has been finished. Then, the model is extended by adding most fitting data, which are those objects in the rest of the population that have the least deviations from the model constructed thus far. The data added to the model in the last round, considered to be the least fitting data, are regarded to be outliers.

Non-parametric Methods. The outlier detection techniques in this category do not make any assumptions about the statistical distribution of the data. The most popular approaches for outlier detection in this category are histograms and Kernel density function methods.

A. Histograms

The most popular non-parametric statistical technique is to use histograms to maintain a profile of data. Histogram techniques by nature are based on the frequency or counting of data.

The histogram based outlier detection approach is typically applied when the data has a single feature. Mathematically, a histogram for a feature of data consists of a number of disjoint bins (or buckets) and the data are mapped into one (and only one) bin. Represented graphically by the histogram graph, the height of bins corresponds to the number of observations that fall into the bins. Thus, if we let n be the total number of instances, k be the total number of bins and m_i be the number of data point in the ith bin ($1 \leq i \leq k$), the histogram satisfies the following condition $n = \sum_{i=1}^{k} m_i$. The training stage involves building histograms based on the different values taken by that feature in the training data.

The histogram techniques typically define a measure between a new test instance and the histogram based profile to determine if it is an outlier or not. The measure is defined based on how the histogram is constructed in the first place. Specifically, there are three possible ways for building a histogram:

1. The histogram can be constructed only based on normal data. In this case, the histogram only represents the profile for normal data. The test stage evaluates whether the feature value in the test instance falls in any of the populated bins of the constructed histogram. If not, the test instance is labeled as an outlier [5] [54][48];

2. The histogram can be constructed only based on outliers. As such, the histogram captures the profile for outliers. A test instance that falls into one of the populated bins is labeled as an outlier [32]. Such techniques are particularly popular in intrusion detection community [34][38] [30] and fraud detection [40];

3. The histogram can be constructed based on a mixture of normal data and outliers. This is the typical case where histogram is constructed. Since normal data typically dominate the whole data set, thus the histogram represents an approximated profile of normal data. The sparsity

of a bin in the histogram can be defined as the ratio of frequency of this bin against the average frequency of all the bins in the histogram. A bin is considered as sparse if such ratio is lower than a user-specified threshold. All the data instance falling into the sparse bins are labeled as outliers.

The first and second ways for constructing histogram, as presented above, rely on the availability of labeled instances, while the third one does not.

For multivariate data, a common approach is to construct feature-wise histograms. In the test stage, the probability for each feature value of the test data is calculated and then aggregated to generate the so-called *outlier score*. A low probability value corresponds a higher outlier score of that test instance. The aggregation of per-feature likelihoods for calculating outlier score is typically done using the following equation:

$$Outlier\_Score = \sum_{f \in F} w_f \cdot (1 - p_f)/|F|$$

where w_f denotes the weight assigned for feature f, p_f denotes the probability for the value of feature f and F denotes the set of features of the dataset. Such histogram-based aggregation techniques have been used in intrusion detection in system call data [35], fraud detection [40], damage detection in structures [67] [70] [71], network intrusion detection [90] [91], web-based attack detection [63], Packet Header Anomaly Detection (PHAD), Application Layer Anomaly Detection (ALAD) [69], NIDES (by SRI International) [5] [12] [79]. Also, a substantial amount of research has been done in the field of outlier detection for sequential data (primarily to detect intrusions in computer system call data) using histogram based techniques. These techniques are fundamentally similar to the instance based histogram approaches as described above but are applied to sequential data to detect collective outliers.

Histogram based detection methods are simple to implement and hence are quite popular in domain such as intrusion detection. But one key shortcoming of such techniques for multivariate data is that they are not able to capture the interactions between different attributes. An outlier might have attribute values that are individually very frequent, but their combination is very rare. This shortcoming will become more salient when dimensionality of data is high. A feature-wise histogram technique will not be able to detect such kinds of outliers. Another challenge for such techniques is that users need to determine an optimal size of the bins to construct the histogram.

B. Kernel Functions

Another popular non-parametric approach for outlier detection is the parzen windows estimation due to Parzen [76]. This involves using Kernel functions to approximate the actual density distribution. A new instance which lies in the low probability area of this density is declared to be an outlier.

Formally, if $x_1, x_2, ..., x_N$ are IID (independently and identically distributed) samples of a random variable x, then the Kernel density approximation of its *probability density function (pdf)* is

$$f_h(x) = \frac{1}{Nh} \sum_{i=1}^{N} K\left(\frac{x - x_i}{h}\right)$$

where K is Kernel function and h is the bandwidth (smoothing parameter). Quite often, K is taken to be a standard Gaussian function with mean $\mu = 0$ and variance $\sigma^2 = 1$:

$$K(x) = \frac{1}{\sqrt{2\pi}} e^{-\frac{1}{2}x^2}$$

Novelty detection using Kernel function is presented by [17] for detecting novelties in oil flow data. A test instance is declared to be novel if it belongs to the low density area of the learnt density function. Similar application of parzen windows is proposed for network intrusion detection [29] and for mammographic image analysis [86]. A semi-supervised probabilistic approach is proposed to detect novelties [31]. Kernel functions are used to estimate the probability distribution function (pdf) for the normal instances. Recently, Kernel functions are used in outlier detection in sensor networks [80][25].

Kernel density estimation of pdf is applicable to both univariate and multivariate data. However, the pdf estimation for multivariate data is much more computationally expensive than the univariate data. This renders the Kernel density estimation methods rather inefficient in outlier detection for high-dimensional data.

Advantages and Disadvantages of Statistical Methods. Statistical outlier detection methods feature some advantages. They are mathematically justified and if a probabilistic model is given, the methods are very efficient and it is possible to reveal the meaning of the outliers found [75]. In addition, the model constructed, often presented in a compact form, makes it possible to detect outliers without storing the original datasets that are usually of large sizes.

However, the statistical outlier detection methods, particularly the parametric methods, suffer from some key drawbacks. First, they are typically not applied in a multi-dimensional scenario because most distribution models typically apply to the univariate feature space. Thus, they are unsuitable even for moderate multi-dimensional data sets. This greatly limits their applicability as in most practical applications the data is multiple or even high dimensional. In addition, a lack of the prior knowledge regarding the underlying distribution of the dataset makes the distribution-based methods difficult to use in practical applications. A single distribution may not model the entire data because the data may originate from multiple distributions. Finally, the quality of results cannot be guaranteed because they are largely dependent on the distribution chosen to fit the data. It is not guaranteed that the data being examined fit the assumed distribution if there is no estimate of the distribution density based on the empirical data. Constructing such tests for hypothesis verification in complex combinations of distributions is a nontrivial task whatsoever. Even if the model is properly chosen, finding the values of parameters requires complex procedures. From above discussion, we can see the statistical methods are rather limited to large real-world databases which typically have many different fields and it is not easy to characterize the multivariate distribution of exemplars.

For non-parametric statistical methods, such as histogram and Kernal function methods, they do not have the problem of distribution assumption that the parametric methods suffer and they both can deal with data streams containing continuously arriving data. However, they are not appropriate for handling high-dimensional data. Histogram methods are effective for a single feature analysis, but they lose much of their effectiveness for multi or high-dimensional data because they lack the ability to analyze multiple feature simultaneously. This prevents them from detecting subspace outliers. Kernel function methods are appropriate only for relatively low dimensional data as well. When the dimensionality of data is high, the density estimation using Kernel functions becomes rather computationally expensive, making it inappropriate for handling high-dimensional data streams.

4.2. Distance-based Methods

There have already been a number of different ways for defining outliers from the perspective of distance-related metrics. Most existing metrics used for distance-based outlier detection techniques are defined based upon the concepts of *local neighborhood* or *k* nearest neighbors (*k*NN) of the data points. The notion of distance-based outliers does not assume any underlying data distributions and generalizes many concepts from distribution-based methods. Moreover, distance-based methods scale better to multi-dimensional space and can be computed much more efficiently than the statistical-based methods.

In distance-based methods, distance between data points is needed to be computed. We can use any of the L_p metrics like the Manhattan distance or Euclidean distance metrics for measuring the distance between a pair of points. Alternately, for some other application domains with presence of categorical data (*e.g.*, text documents), non-metric distance functions can also be used, making the distance-based definition of outliers very general. Data normalization is normally carried out in order to normalize the different scales of data features before outlier detection is performed.

A. Local Neighborhood Methods

The first notion of distance-based outliers, called $DB(k, \lambda)$-Outlier, is due to Knorr and Ng [58]. It is defined as follows. A point p in a data set is a $DB(k, \lambda)$-Outlier, with respect to the parameters k and λ, if no more than k points in the data set are at a distance λ or less (*i.e.*, λ–neighborhood) from p. This definition of outliers is intuitively simple and straightforward. The major disadvantage of this method, however, is its sensitivity to the parameter λ that is difficult to specify a priori. As we know, when the data dimensionality increases, it becomes increasingly difficult to specify an appropriate circular local neighborhood (delimited by λ) for outlier-ness evaluation of each point since most of the points are likely to lie in a thin shell about any point [19]. Thus, a too small λ will cause the algorithm to detect all points as outliers, whereas no point will be detected as outliers if a too large λ is picked up. In other words, one needs to choose an appropriate λ with a very high degree of accuracy in order to find a modest number of points that can then be defined as outliers.

To facilitate the choice of parameter values, this first local neighborhood distance-based outlier definition is extended and the so-called $DB(pct, d_{min})$-Outlier is proposed which defines an object in a dataset as a $DB(pct, d_{min})$-Outlier if at least $pct\%$ of the objects in the datasets have the distance larger than d_{min} from this object [59][60]. Similar to $DB(k, \lambda)$-Outlier, this method essentially delimits the local neighborhood of data points using the parameter d_{min} and measures the outlierness of a data point based on the percentage, instead of the absolute number, of data points falling into this specified local neighborhood. As pointed out in [56] and [57], $DB(pct, d_{min})$ is quite general and is able to unify the exisiting statisical detection methods using discordancy tests for outlier detection. For exmaple, $DB(pct, d_{min})$ unifies the definition of outliers using a normal distribution-based discordancy test with $pct = 0.9988$ and $d_{min} = 0.13$. The specification of pct is obviously more intuitive and easier than the specification of k in $DB(k, \lambda)$-Outliers [59]. However, $DB(pct, d_{min})$-Outlier suffers a similar problem as $DB(pct, d_{min})$-Outlier in specifying the local neighborhood parameter d_{min}.

To efficiently calculate the number (or percentage) of data points falling into the local neighborhood of each point, three classes of algorithms have been presented, *i.e.*, the nested-loop, index-based and cell-based algorithms. For easy of presentation, these three algorithms are discussed for detecting $DB(k, \lambda)$-Outlier.

The *nested-loop algorithm* uses two nested loops to compute $DB(k, \lambda)$-Outlier. The outer loop considers each point in the dataset while the inner loop computes for each point in the outer loop the number (or percentage) of points in the dataset falling into the specified λ-neighborhood. This algorithm has the advantage that it does not require the indexing structure be constructed at all that may be rather expensive at most of the time, though it has a quadratic complexity with respect to the number of points in the dataset.

The *index-based algorithm* involves calculating the number of points belonging to the λ-neighborhood of each data by intensively using a pre-constructed multi-dimensional index structure such as R^*-tree [22] to facilitate kNN search. The complexity of the algorithm is approximately logarithmic with respect to the number of the data points in the dataset. However, the construction of index structures is sometimes very expensive and the quality of the index structure constructed is not easy to guarantee.

In the *cell-based algorithm*, the data space is partitioned into cells and all the data points are mapped into cells. By means of the cell size that is known a priori, estimates of pair-wise distance of data points are developed, whereby heuristics (pruning properties) are presented to achieve fast outlier detection. It is shown that three passes over the dataset are sufficient for constructing the desired partition. More precisely, the d–dimensional space is partitioned into cells with side length of $\frac{\lambda}{2\sqrt{d}}$. Thus, the distance between points in any 2 neighboring cells is guaranteed to be at most λ. As a result, if for a cell the total number of points in the cell and its neighbors is greater than k, then none of the points in the cell can be outliers. This property is used to eliminate the vast majority of points that cannot be outliers. Also, points belonging to cells that are more than 3 cells apart are more than a distance λ apart. As a result, if the number of points contained in all cells that are at most 3 cells away from the a given cell is less than k, then all points in the cell are definitely outliers. Finally, for those points that belong to a cell that cannot be categorized as either containing only outliers or only non-outliers, only points from neighboring cells that are at most 3 cells away need to be considered in order to determine whether or not they are outliers. Based on the above properties, the authors propose a three-pass algorithm for computing outliers in large databases. The time complexity of this cell-based algorithm is $O(c^d + N)$, where c is a number that is inversely proportional to

λ. This complexity is linear with dataset size N but exponential with the number of dimensions d. As a result, due to the exponential growth in the number of cells as the number of dimensions is increased, the cell-based algorithm starts to perform poorly than the nested loop for datasets with dimensions of 4 or higher.

In [36], a similar definition of outlier is proposed. It calculates the number of points falling into the w-radius of each data point and labels those points as outliers that have low neighborhood density. We consider this definition of outliers as the same as that for $DB(k, \lambda)$-Outlier, differing only that this method does not present the threshold k explicitly in the definition. As the computation of the local density for each point is expensive, [36] proposes a clustering method for an efficient estimation. The basic idea of such approximation is to use the size of a cluster to approximate the local density of all the data in this cluster. It uses the fix-width clustering [36] for density estimation due to its good efficiency in dealing with large data sets.

B. kNN-distance Methods

There have also been a few distance-based outlier detection methods utilizing the k nearest neighbors (kNN) in measuring the outlier-ness of data points in the dataset. The first proposal uses the distance to the kth nearest neighbors of every point, denoted as D^k, to rank points so that outliers can be more efficiently discovered and ranked [81]. Based on the notion of D^k, the following definition for D_n^k-Outlier is given: Given k and n, a point is an outlier if the distance to its kth nearest neighbor of the point is smaller than the corresponding value for no more than $n - 1$ other points. Essentially, this definition of outliers considers the top n objects having the highest D^k values in the dataset as outliers.

Similar to the computation of $DB(k, \lambda)$-Outlier, three different algorithms, *i.e.*, the nested-loop algorithm, the index-based algorithm, and the partition-based algorithm, are proposed to compute D^k for each data point efficiently.

The *nested-loop algorithm* for computing outliers simply computes, for each input point p, D^k, the distance of between p and its kth nearest neighbor. It then sorts the data and selects the top n points with the maximum D^k values. In order to compute D^k for points, the algorithm scans the database for each point p. For a point p, a list of its k nearest points is maintained, and for each point q from the database which is considered, a check is made to see if the distance between p and q is smaller than the distance of the kth nearest neighbor found so far. If so, q is included in the list of the k nearest neighbors for p. The moment that the list contains more than k neighbors, then the point that is furthest away from p is deleted from the list. In this algorithm, since only one point is processed at a time, the database would need to be scanned N times, where N is the number of points in the database. The computational complexity is in the order of $O(N^2)$, which is rather expensive for large datasets. However, since we are only interested in the top n outliers, we can apply the following pruning optimization to early-stop the computation of D^k for a point p. Assume that during each step of the algorithm, we store the top n outliers computed thus far. Let D^n_{min} be the minimum among these top n outliers. If during the computation of for a new point p, we find that the value for D^k computed so far has fallen below D^n_{min}, we are guaranteed that point p cannot be an outlier. Therefore, it can be safely discarded. This is because D^k monotonically decreases as we examine more points. Therefore, p is guaranteed not to be one of the top n outliers.

The *index-based algorithm* draws on index structure such as R*-tree [22] to speed up the computation. If we have all the points stored in a spatial index like R*-tree, the following pruning optimization can be applied to reduce the number of distance computations. Suppose that we have computed for point p by processing a portion of the input points. The value that we have is clearly an upper bound for the actual D^k of p. If the minimum distance between p and the *Minimum Bounding Rectangles* (MBR) of a node in the R*-tree exceeds the value that we have anytime in the algorithm, then we can claim that none of the points in the sub-tree rooted under the node will be among the k nearest neighbors of p. This optimization enables us to prune entire sub-trees that do not contain relevant points to the kNN search for p.

The major idea underlying the *partition-based algorithm* is to first partition the data space, and then prune partitions as soon as it can be determined that they cannot contain outliers. Partition-based algorithm is subject to the pre-processing step in which data space is split into cells and data partitions, together with the Minimum Bounding Rectangles of data partitions, are generated. Since n will typically be very small, this additional preprocessing step performed at the granularity of partitions rather than points is worthwhile as it can eliminate a significant number of points as outlier candidates. This partition-based algorithm takes the following four steps:

1. First, a clustering algorithm, such as BIRCH, is used to cluster the data and treat each cluster as a separate partition;

2. For each partition P, the lower and upper bounds (denoted as $P.lower$ and $P.upper$, respectively) on D^k for points in the partition are computed. For every point $p \in P$, we have $P.lower \leq D^k(p) \leq P.upper$;

3. The *candidate partitions*, the partitions containing points which are candidates for outliers, are identified. Suppose we could compute $minDkDist$, the lower bound on D^k for the n outliers we have detected so far. Then, if $P.upper < minDkDist$, none of the points in P can possibly be outliers and are safely pruned. Thus, only partitions P for which $P.upper \geq minDkDist$ are chosen as candidate partitions;

4. Finally, the outliers are computed from among the points in the candidate partitions obtained in Step 3. For each candidate partition P, let $P.neighbors$ denote the neighboring partitions of P, which are all the partitions within distance $P.upper$ from P. Points belonging to neighboring partitions of P are the only points that need to be examined when computing D^k for each point in P.

The D^k_n-Outlier is further extended by considering for each point the sum of its k nearest neighbors [10]. This extension is motivated by the fact that the definition of D^k merely considers the distance between an object with its kth nearest neighbor, entirely ignoring the distances between this object and its another $k - 1$ nearest neighbors. This drawback may make D^k fail to give an accurate measurement of outlier-ness of data points in some cases. For a better understanding, we present an example, as shown in Figure 1, in which the same D^k value is assigned to points p_1 and p_2, two points with apparently rather different outlier-ness. The $k - 1$ nearest neighbors for p_2 are populated much more densely around it than those of p_1, thus the outlier-ness of p_2 is obviously lower than p_1. Obviously, D^k is not robust enough in this example to accurately reveal the outlier-ness of data points. By summing up the distances between the object with all of its k nearest neighbors, we will be able to have a more accurate measurement of outlier-ness of the object, though this will require more computational effort in summing up the distances. This method is also used in [36] for anomaly detection.

The idea of kNN-based distance metric can be extended to consider the k nearest dense regions. The recent methods are the Largest_cluster method [61][98] and Grid-ODF [89], as discussed below.

Khoshgoftaar *et al.* propose a distance-based method for labeling wireless network traffic records in the data stream used as either normal or intrusive [61][98]. Let d be the largest distance of an instance to the centriod of the largest cluster. Any instance or cluster that has a distance greater than αd ($\alpha \geq 1$) to the largest cluster is defined as an attack. This method is referred to as the Largest_Cluster method. It can also be used to detect outliers. It takes the following several steps for outlier detection:

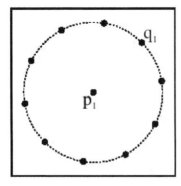 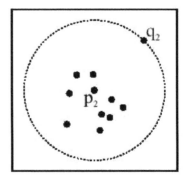

Figure 1. Points with the same D^k value but different outlier–ness

1. Find the largest cluster, *i.e.* the cluster with largest number of instances, and label it as normal. Let c_0 be the centriod of this cluster;

2. Sort the remaining clusters in ascending order based on the distance from their cluster centroid to c_0;

3. Label all the instances that have a distance to c_0 greater than αd, where α is a human-specified parameter;

4. Label all the other instances as normal.

When used in dealing with projected anomalies detection for high-dimensional data streams, this method suffers the following limitations:

- First and most importantly, this method does not take into account the nature of outliers in high-dimensional data sets and is unable to explore subspaces to detect projected outliers;

- k-means clustering is used in this method as the backbone enabling technique for detecting intrusions. This poses difficulty for this method to deal with data streams. k-means clustering requires iterative optimization of clustering centroids to gradually achieve better clustering results. This optimization process involves multiple data scans, which is infeasible in the context of data streams;

- A strong assumption is made in this method that all the normal data will appear in a single cluster (*i.e.*, the largest cluster), which is not properly substantiated in the paper. This assumption may be too rigid in some applications. It is possible that the normal data are distributed in two or more clusters that correspond to a few varying normal behaviors. For a simple instance, the network traffic volume is usually high during

the daytime and becomes low late in the night. Thus, network traffic volume may display several clusters to represent behaviors exhibiting at different time of the day. In such case, the largest cluster is apparently not where all the normal cases are only residing;

- In this method, one needs to specify the parameter α. The method is rather sensitive to this parameter whose best value is not obvious whatsoever. First, the distance scale between data will be rather different in various subspaces; the distance between any pair of data is naturally increased when it is evaluated in a subspace with higher dimension, compared to in a lower-dimensional subspace. Therefore, specifying an ad-hoc α value for each subspace evaluated is rather tedious and difficult. Second, α is also heavily affected by the number of clusters the clustering method produces, *i.e.*, k. Intuitively, when the number of clusters k is small, D will become relatively large, then α should be set relatively small accordingly, and vice versa.

Recently, an extension of the notion of kNN, called Grid-ODF, from the k nearest objects to the k nearest dense regions is proposed [89]. This method employed the sum of the distances between each data point and its k nearest dense regions to rank data points. This enables the algorithm to measure the outlier-ness of data points from a more global perspective. Grid-ODF takes into account the mechanisms used in detecting both global and local outliers. In the local perspective, human examine the point's immediate neighborhood and consider it as an outlier if its neighborhood density is low. The global observation considers the dense regions where the data points are densely populated in the data space. Specifically, the neighboring density of the point serves as a good indicator of its outlying degree from the local perspective. In the left sub-figure of Figure 2, two square boxes of equal size are used to delimit the neighborhood of points p_1 and p_2. Because the neighboring density of p_1 is less than that of p_2, so the outlying degree of p_1 is larger than p_2. On the other hand, the distance between the point and the dense regions reflects the similarity between this point and the dense regions. Intuitively, the larger such distance is, the more remarkably p is deviated from the main population of the data points and therefore the higher outlying degree it has, otherwise it is not. In the right sub-figure of 2, we can see a dense region and two outlying points, p_1 and p_2. Because the distance between p_1 and the dense region is larger than that between p_2 and the dense region, so the outlying degree of p_1 is larger than p_2.

Based on the above observations, a new measurement of outlying factor of data points, called *Outlying Degree Factor* (ODF), is proposed to measure the outlier-ness of points from both the global and local perspectives. The ODF of a point p is defined as follows:

$$ODF(p) = \frac{k\_DF(p)}{NDF(p)}$$

where $k\_DF(p)$ denotes the average distance between p and its k nearest dense cells and $NDF(p)$ denotes number of points falling into the cell to which p belongs.

In order to implement the computation of ODF of points efficiently, grid structure is used to partition the data space. The main idea of grid-based data space partition is to super-impose a multi-dimensional cube in the data space, with equal-volumed cells. It is characterized by the following advantages. First, $NDF(p)$ can be obtained instantly by simply counting the number of points falling into the cell to which p belongs, without the involvement of any indexing techniques. Secondly, the dense regions can be efficiently identified, thus the computation of $k\_DF(p)$ can be very fast. Finally, based on the density of grid cells, we will be able to select the top n outliers only from a specified number of points viewed as *outlier candidates*, rather than the whole dataset, and the final top n outliers are selected from these outlier candidates based on the ranking of their ODF values.

The number of outlier candidates is typically 9 or 10 times as large as the number of final outliers to be found (*i.e.*, top n) in order to provide a sufficiently large pool for outlier selection. Let us suppose that the size of outlier candidates is $m * n$, where the m is a positive number provided by users. To generate $m * n$ outlier candidates, all the cells containing points are sorted in ascending order based on their densities, and then the points in the first t cells in the sorting list that satisfy the following inequality are selected as the $m * n$ outlier

candidates:

$$\sum_{i=1}^{t-1} Den(C_i) \leq m * n \leq \sum_{i=1}^{t} Den(C_i)$$

The kNN-distance methods, which define the top n objects having the highest values of the corresponding outlier-ness metrics as outliers, are advantageous over the local neighborhood methods in that they order the data points based on their relative ranking, rather than on the distance cutoff. Since the value of n, the top outlier users are interested in, can be very small and is relatively independent of the underlying data set, it will be easier for the users to specify compared to the distance threshold λ.

C. Advantages and Disadvantages of Distance-based Methods

The major advantage of distance-based algorithms is that, unlike distribution-based methods, distance-based methods are non-parametric and do not rely on any assumed distribution to fit the data. The distance-based definitions of outliers are fairly straightforward and easy to understand and implement.

Their major drawback is that most of them are not effective in high-dimensional space due to the curse of dimensionality, though one is able to mechanically extend the distance metric, such as Euclidean distance, for high-dimensional data. The high-dimensional data in real applications are very noisy, and the abnormal deviations may be embedded in some lower-dimensional subspaces that cannot be observed in the full data space. Their definitions of a local neighborhood, irrespective of the circular neighborhood or the k nearest neighbors, do not make much sense in high-dimensional space. Since each point tends to be equi-distant with each other as number of dimensions goes up, the degree of outlier-ness of each points are approximately identical and significant phenomenon of deviation or abnormality cannot be observed. Thus, none of the data points can be viewed outliers if the concepts of proximity are used to define outliers. In addition, neighborhood and kNN search in high-dimensional space is a non-trivial and expensive task. Straightforward algorithms, such as those based on nested loops, typically require $O(N^2)$ distance computations. This quadratic scaling means that it will be very difficult to mine outliers as we tackle increasingly larger data sets. This is a major problem for many real databases where there are often millions of records. Thus, these approaches lack a good scalability for large data set. Finally, the existing distance-based methods are not able to deal with data streams due to the difficulty in maintaining a data distribution in the local neighborhood or finding the kNN for the data in the stream.

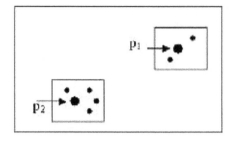 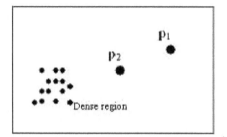

Figure 2. Local and global perspectives of outlier–ness of p_1 and p_2

4.3. Density–based Methods

Density-based methods use more complex mechanisms to model the outlier-ness of data points than distance-based methods. It usually involves investigating not only the local density of the point being studied but also the local densities of its nearest neighbors. Thus, the outlier-ness metric of a data point is relative in the sense that it is normally a ratio of density of this point against the the averaged densities of its nearest neighbors. Density-based methods feature a stronger modeling capability of outliers but require more expensive computation at the same time. What will be discussed in this subsection are the major density-based methods called LOF method, COF method, INFLO method and MDEF method.

A. LOF Method

The first major density-based formulation scheme of outlier has been proposed in [21], which is more robust than the distance-based outlier detection methods. An example is given in [21] (refer to figure 3), showing the advantage of a density-based method over the distance-based methods such as $DB(k, \lambda)$-Outlier. The dataset contains an outlier o, and C_1 and C_2 are two clusters with very different densities. The $DB(k, \lambda)$-Outlier method cannot distinguish o from the rest of the data set no matter what values the parameters k and λ take. This is because the density of o's neighborhood is very much closer to the that of the points in cluster C_1. However, the density-based method, proposed in [21], can handle it successfully.

This density-based formulation quantifies the outlying degree of points using *Local Outlier Factor (LOF)*. Given parameter $MinPts$, LOF of a point p is defined as

$$LOF_{MinPts}(p) = \frac{\sum_{o \in MinPts(p)} \frac{lrd_{MinPts}(o)}{lrd_{MinPts}(p)}}{|N_{MinPts}(p)|}$$

where $|N_{MinPts}(p)|$ denotes the number of objects falling into the $MinPts$-neighborhood of p and $lrd_{MinPts}(p)$ denotes the *local reachability density* of point p that is defined as the inverse of the average reachability distance based on the $MinPts$ nearest neighbors of p, i.e.,

$$lrd_{MinPts}(p) = 1 / \left(\frac{\sum_{o \in MinPts(p)} reach\_dist_{MinPts}(p, o)}{|N_{MinPts}(p)|} \right)$$

Further, the reachability distance of point p is defined as

$$reach\_dist_{MinPts}(p, o) = max(MinPts\_distance(o), dist(p, o))$$

Intuitively speaking, LOF of an object reflects the density contrast between its density and those of its neighborhood. The neighborhood is defined by the distance to the $MinPts^{th}$ nearest neighbor. The local outlier factor is a mean value of the ratio of the density distribution estimate in the neighborhood of the object analyzed to the distribution densities of its neighbors [21]. The lower the density of p and/or the higher the densities of p's neighbors, the larger the value of $LOF(p)$, which indicates that p has a higher degree of being an outlier. A similar outlier-ness metric to LOF, called OPTICS-OF, was proposed in [20].

Unfortunately, the LOF method requires the computation of LOF for all objects in the data set which is rather expensive because it requires a large number of kNN search. The high cost of computing LOF for each data point p is caused by two factors. First, we have to find the $MinPts^{th}$ nearest neighbor of p in order to specify its neighborhood. This resembles to computing D^k in detecting D_n^k-Outliers. Secondly, after the $MinPts^{th}$-neighborhood of p has been determined, we have to further find the $MinPts^{th}$-neighborhood for each data points falling into the $MinPts^{th}$-neighborhood of p. This amounts to $MinPts^{th}$ times in terms of computation efforts as computing D^k when we are detecting D_n^k-Outliers.

It is desired to constrain a search to only the top n outliers instead of computing the LOF of every object in the database. The efficiency of this algorithm is boosted by an efficient micro-cluster-based local outlier mining algorithm proposed in [52].

LOF ranks points by only considering the neighborhood density of the points, thus it may miss out the

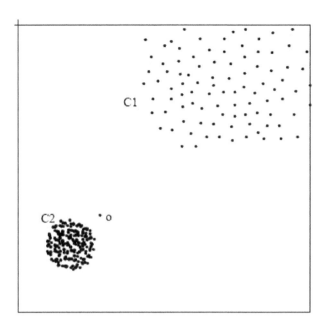

Figure 3. A sample dataset showing the advantage of LOF over $DB(k, \lambda)$-Outlier

potential outliers whose densities are close to those of their neighbors. Furthermore, the effectiveness of this algorithm using LOF is rather sensitive to the choice of *MinPts*, the parameter used to specify the local neighborhood.

B. COF Method

As LOF method suffers the drawback that it may miss those potential outliers whose local neighborhood density is very close to that of its neighbors. To address this problem, Tang *et al.* proposed a new *Connectivity-based Outlier Factor* (COF) scheme that improves the effectiveness of LOF scheme when a pattern itself has similar neighborhood density as an outlier [87]. In order to model the connectivity of a data point with respect to a group of its neighbors, a set-based nearest path (SBN-path) and further a set-based nearest trail (SBN-trail), originated from this data point, are defined. This SNB trail stating from a point is considered to be the pattern presented by the neighbors of this point. Based on SNB trail, the cost of this trail, a weighted sum of the cost of all its constituting edges, is computed. The final outlier-ness metric, COF, of a point p with respect to its k-neighborhood is defined as

$$COF_k(p) = \frac{|N_k(p)| * ac\_dist_{N_k(p)}(p)}{\sum_{o \in N_k(p)} ac\_dist_{N_k(o)}(o)}$$

where $ac\_dist_{N_k(p)}(p)$ is the average chaining distance from point p to the rest of its k nearest neighbors, which is the weighted sum of the cost of the SBN-trail starting from p.

It has been shown in [87] that COF method is able to detect outlier more effectively than LOF method for some cases. However, COF method requires more expensive computations than LOF and the time complexity is in the order of $O(N^2)$ for high-dimensional datasets.

C. INFLO Method

Even though LOF is able to accurately estimate outlier-ness of data points in most cases, it fails to do so in some complicated situations. For instance, when outliers are in the location where the density distributions in the neighborhood are significantly different, this may result in a wrong estimation. An example where LOF fails to have an accurate outlier-ness estimation for data points has been given in [53]. The example is presented in Figure 4. In this example, data p is in fact part of a sparse cluster $C2$ which is near the dense cluster $C1$. Compared to objects q and r, p obviously displays less outlier-ness. However, if LOF is used in this case, p could be mistakenly regarded to having stronger outlier-ness than q and r.

Authors in [53] pointed out that this problem of LOF is due to the inaccurate specification of the space where LOF is applied. To solve this problem of LOF, an improved method, called INFLO, is proposed [53]. The idea of INFLO is that both the nearest neighbors (NNs) and reverse nearest neighbors (RNNs) of a data point are taken into account in order to get a better estimation of the neighborhood's density distribution. The RNNs of an object p are those data points that have p as one of their k nearest neighbors. By considering the symmetric neighborhood relationship of both NN and RNN, the space of an object influenced by other objects is well determined. This space is called the *k-influence space* of a data point. The outlier-ness of a data point, called INFLuenced Outlierness (INFLO), is quantified. INFLO of a data point p is defined as

$$INFLO_k(p) = \frac{den_{avg}(IS_k(p))}{den(p)}$$

INFLO is by nature very similar to LOF. With respect to a data point p, they are both defined as the ratio of p's its density and the average density of its neighboring objects. However, INFLO uses only the data points in its k-influence space for calculating the density ratio. Using INFLO, the densities of its neighborhood will be reasonably estimated, and thus the outliers found will be more meaningful.

D. MDEF Method

In [77], a new density-based outlier definition, called Multi-granularity Deviation Factor (MEDF), is proposed. Intuitively, the MDEF at radius r for a point p_i is the relative deviation of its local neighborhood density from the average local neighborhood density in its r-neighborhood. Let $n(p_i, \alpha r)$ be the number of objects in the αr-neighborhood of p_i and $\hat{n}(p_i, r, \alpha)$ be the average, over all objects p in the r-neighborhood of p_i, of $n(p, \alpha r)$. In the example given by Figure 5, we have $n(p_i, \alpha r) = 1$, and $\hat{n}(p_i, r, \alpha) = (1 + 6 + 5 + 1)/4 = 3.25$.

MDEF of p_i, given r and α, is defined as

$$MDEF(p_i, r, \alpha) = 1 - \frac{n(p_i, \alpha r)}{\hat{n}(pi, r, \alpha)}$$

where $\alpha = \frac{1}{2}$. A number of different values are set for the sampling radius r and the minimum and the maximum values for r are denoted by r_{min} and r_{max}. A point is flagged as an outliers if for any $r \in [r_{min}, r_{max}]$, its MDEF is sufficient large.

E. Advantages and Disadvantages of Density-based Methods

The density-based outlier detection methods are generally more effective than the distance-based methods. However, in order to achieve the improved effectiveness, the density-based methods are more complicated and computationally expensive. For a data object, they have to not only explore its local density but also that of its neighbors. Expensive kNN search is expected for all the existing methods in this category. Due to the inherent complexity and non-updatability of their outlier-ness measurements used, LOF, COF, INFLO and MDEF cannot handle data streams efficiently.

4.4. Clustering–based Methods

The final category of outlier detection algorithm for relatively low dimensional static data is clustering-based. Many data-mining algorithms in literature find outliers as a by-product of clustering algorithms [6, 11, 13, 46, 101] themselves and define outliers as points that do not lie in or located far apart from any clusters. Thus, the clustering techniques implicitly define outliers as the background noise of clusters. So far, there are numerous studies on clustering, and some of them are equipped with some mechanisms to reduce the adverse effect of outliers, such as CLARANS [73], DBSCAN [37], BIRCH [101], WaveCluster [82]. More recently, we have seen quite a few clustering techniques tailored towards subspace clustering for high-dimensional data including CLIQUE [6] and HPStream [9].

Next, we will review several major categories of clustering methods, together with the analysis on their advantages and disadvantages and their applicability in dealing with outlier detection problem for high-dimensional data streams.

A. Partitioning Clustering Methods

The partitioning clustering methods perform clustering by partitioning the data set into a specific number of clusters. The number of clusters to be obtained, denoted by k, is specified by human users. They typically start with an initial partition of the dataset and then iteratively optimize the objective function until it reaches the optimal for the dataset. In the clustering process, center of the clusters (centroid-based methods) or the point which is located nearest to the cluster center (medoid-based methods) is used to represent a cluster. The representative partitioning clustering methods are PAM, CLARA, k-means and CLARANS.

PAM [62] uses a k-medoid method to identify the clusters. PAM selects k objects arbitrarily as medoids and swap with objects until all k objects qualify as medoids. PAM compares an object with entire dataset to find a medoid, thus it has a slow processing time with a complexity of $\mathcal{O}(k(N - k)^2)$, where N is number of data in the data set and k is the number of clusters.

CLARA [62] tries to improve the efficiency of PAM. It draws a sample from the dataset and applies PAM on the sample that is much smaller in size than the the whole dataset.

k-means [68] initially choose k data objects as seeds from the dataset. They can be chosen randomly or in a way such that the points are mutually farthest apart. Then, it examines each point in the dataset and assigns it to one of the clusters depending on the minimum distance. The centroid's position is recalculated and updated the moment a point is added to the cluster and this continues until all the points are grouped into the final clusters. The k-means algorithm is relatively

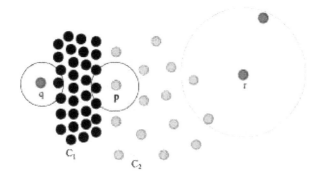

Figure 4. An example where LOF does not work

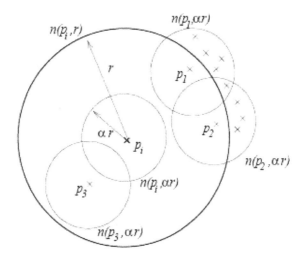

Figure 5. Definition of MDEF

scalable and efficient in processing large datasets because the computational complexity is $\mathcal{O}(nkt)$, where n is total number of points, k is the number of clusters and t is the number of iterations of clustering. However, because it uses a centroid to represent each cluster, k-means suffers the inability to correctly cluster with a large variation of size and arbitrary shapes, and it is also very sensitive to the noise and outliers of the dataset since a small number of such data will substantially effect the computation of mean value the moment a new object is clustered.

CLARANS [73] is an improved k-medoid method, which is based on randomized search. It begins with a random selection of k nodes, and in each of following steps, compares each node to a specific number of its neighbors in order to find a local minimum. When one local minimum is found, CLARANS continues to repeat this process for another minimum until a specific number of minima have been found. CLARANS has been experimentally shown to be more effective than both PAM and CLEAR. However, the computational complexity of CLARANS is close to quadratic w.r.t the number of points [88], and it is prohibitive for clustering large database. Furthermore, the quality of clustering result is dependent on the sampling method, and it is not stable and unique due to the characteristics of randomized search.

B. Hierarchical Clustering Methods

Hierarchical clustering methods essentially constructs a hierarchical decomposition of the whole dataset. It can be further divided into two categories based on how this dendrogram is operated to generate clusters, *i.e.*, *agglomerative methods* and *divisive methods*. An agglomerative method begins with each point as a distinct cluster and merges two closest clusters in each subsequent step until a stopping criterion is met. A divisive method, contrary to an agglomerative method, begins with all the point as a single cluster and splits it in each subsequent step until a stopping criterion is met. Agglomerative methods are seen more popular in practice. The representatives of hierarchical methods are MST clustering, CURE and CHAMELEON.

MST clustering [92] is a graph-based divisive clustering algorithm. Given n points, a MST is a set of edges that connects all the points and has a minimum total length. Deletion of edges with larger lengths will subsequently generate a specific number of clusters. The overhead for MST clustering is determined by the Euclidean MST construction, which is $\mathcal{O}(n^2)$ in time complexity, thus MST algorithm can be used for scalable clustering. However, MST algorithm can only work well on the clean dataset and are sensitive to outliers. The intervention of outliers, termed "chaining-effect" (that is, a line of outliers between two distinct clusters will make these two clusters be marked as one cluster due to its adverse effect), will seriously degrade the quality of the clustering results.

CURE [46] employs a novel hierarchical clustering algorithm in which each cluster is represented by a constant number of well-distributed points. A random sample drawn from the original dataset is first partitioned and each partition is partially clustered. The partial clusters are then clustered in a second pass to yield the desired clusters. The multiple representative points for each cluster are picked to be as disperse as possible and shrink towards the center using a pre-specified shrinking factor. At each step of the algorithm, the two clusters with the closest pair of representative (this pair of representative points are from different clusters) points are merged. Usage of multiple points representing a cluster enables CURE to well capture the shape of clusters and makes it suitable for clusters with non-spherical shapes and wide variance in size. The shrinking factor helps to dampen the adverse effect of outliers. Thus, CURE is more robust to outliers and identifies clusters having arbitrary shapes.

CHAMELEON [55] is a clustering technique trying to overcome the limitation of existing agglomerative hierarchical clustering algorithms that the clustering is irreversible. It operates on a sparse graph in which nodes represent data points and weighted edges represent similarities of among the data points. CHAMELEON first uses a graph partition algorithm to cluster the data points into a large number of relatively small sub-clusters. It then employs an agglomerative hierarchical clustering algorithm to genuine clusters by progressively merging these sub-clusters. The key feature of CHAMELEON lies in its mechanism determining the similarity between two sub-clusters in sub-cluster merging. Its hierarchical algorithm takes into consideration of both inter-connectivity and closeness of clusters. Therefore, CHAMELEON can dynamically adapt to the internal characteristics of the clusters being merged.

C. Density-based Clustering Methods

The density-based clustering algorithms consider normal clusters as dense regions of objects in the data space that are separated by regions of low density. Human normally identify a cluster because there is a relatively denser region compared to its sparse neighborhood. The representative density-based clustering algorithms are DBSCAN and DENCLUE.

The key idea of DBSCAN [37] is that for each point in a cluster, the neighborhood of a given radius has to contain at least a minimum number of points. DBSCAN introduces the notion of "density-reachable points" and based on which performs clustering. In DBSCAN, a cluster is a maximum set of density-reachable points w.r.t. parameters Eps and $MinPts$, where Eps is the given radius and $MinPts$ is the minimum number of points required to be in the Eps-neighborhood. Specifically, to discover clusters in the dataset, DBSCAN examines the Eps-neighborhood of each point in the dataset. If the Eps-neighborhood of a point p contains more than $MinPts$, a new cluster with p as the core object is generated. All the objects from within this Eps-neighborhood are then assigned to this cluster. All this newly entry points will also go through the same process to gradually grow this cluster. When there is no more core object can be found, another core object will be initiated and another cluster will grow. The whole clustering process terminates when there are no new points can be added to any clusters. As the clusters discovered are dependent on the specification of the parameters, DBSCAN relies on the user's ability to select a good set of parameters. DBSCAN outperforms CLARANS by a factor of more than 100 in terms of efficiency [37]. DBSCAN is also powerful in discovering of clusters with arbitrary shapes. The drawbacks DBSCAN suffers are: (1) It is subject to adverse effect resulting from "chaining-effect"; (2) The two parameters used in DBSCAN, i.e., Eps and $MinPts$, cannot be easily decided in advance and require a tedious process of parameter tuning.

DENCLUE [51] performs clustering based on density distribution functions, a set of mathematical functions used to model the influence of each point within its neighborhood. The overall density of the data space can be modeled as sum of influence function of all data points. The clusters can be determined by density attractors. Density attractors are the local maximum of the overall density function. DENCLUE has advantages that it can well deal with dataset with a large number of noises and it allows a compact description of clusters of arbitrary shape in high-dimensional datasets. To facilitate the computation of the density function, DENCLUE makes use of grid-like structure. Noted that even though it uses grids in clustering, DENCLUE is fundamentally different from grid-based clustering algorithm in that grid-based clustering algorithm uses grid for summarizing information about the data points in each grid cell, while DENCLUE uses such structure to effectively compute the sum of influence functions at each data point.

D. Grid-based Clustering Methods

Grid-based clustering methods perform clustering based on a grid-like data structure with the aim of enhancing the efficiency of clustering. It quantizes the space into a finite number of cells which form a grid structure on which all the operations for clustering are performed. The main advantage of the approaches in this category is their fast processing time which is typically only dependent on the number of cells in the quantized space, rather than the number of data objects. The representatives of grid-based clustering algorithms are STING, WaveCluster and DClust.

STING [88] divides the spatial area into rectangular grids, and builds a hierarchical rectangle grids structure. It scans the dataset and computes the necessary statistical information, such as mean, variance, minimum, maximum, and type of distribution, of each grid. The hierarchical grid structure can represent the statistical information with different resolutions at different levels. The statistical information in this hierarchical structure can be used to answer queries. The likelihood that a cell is relevant to the query at some confidence level is computed using the parameters of this cell. The likelihood can be defined as the proportion of objects in this cell that satisfy the query condition. After the confidence interval is obtained, the cells are labeled as relevant or irrelevant based on some confidence threshold. After examining the current layer, the clustering proceeds to the next layer and repeats the process. The algorithm will subsequently only examine the relevant cells instead of all the cells. This process terminates when all the layers have been examined. In this way, all the relevant regions (clusters) in terms of query are found and returned.

WaveCluster [82] is grid-based clustering algorithm based on wavelet transformation, a commonly used technique in signal processing. It transforms the multi-dimensional spatial data to the multi-dimensional signal, and it is able to identify dense regions in the transformed domain that are clusters to be found.

In DClust [95], the data space involved is partitioned into cells with equal size and data points are mapped into the grid structure. A number of representative points of the database are picked using the density criterion. A *Minimum Spanning Tree* (MST) of these representative points, denoted as R-MST, is built.

After the R-MST has been constructed, multi-resolution clustering can be easily achieved. Suppose a user wants to find k clusters. A graph search through the R-MST is initiated, starting from the largest cost edge, to the lowest cost edge. As an edge is traversed, it is marked as deleted from the R-MST. The number of partitions resulting from the deletion is computed. The process stops when the number of partitions reaches k. Any change in the value of k simply implies re-initiating the search-and-marked procedure on the R-MST. Once the R-MST has been divided into k partitions, we can now propagate this information to the original dataset so that each point in the dataset is assigned to one and only one partition/cluster. DClust is equipped with more robust outlier elimination mechanisms to identify and filter the outliers during the various stages of the clustering process. First, DClust uses a uniform random sampling approach to sample the large database. This is effective in ruling out the majority of outliers in the database. Hence, the sample database obtained will be reasonably clean; Second, DClust employs a grid structure to identify representative points. Grid cells whose density is less than the threshold are pruned. This pre-filtering step ensures that the R-MST constructed is an accurate reflection of the underlying cluster structure. Third, the clustering of representative points may cause a number of the outliers that are in close vicinity to form a cluster. The number of points in such outlier clusters will be much smaller than the number of points in the normal clusters. Thus, any small clusters of representative points will be treated as outlier clusters and eliminated. Finally, when the points in the dataset are labeled, some of these points may be quite far from any representative point. DClust will regard such points as outliers and filter them out in the final clustering results.

E. Advantages and Disadvantage of Clustering-based Methods

Detecting outliers by means of clustering analysis is quite intuitive and consistent with human perception of outliers. In addition, clustering is a well-established research area and there have been abundant clustering algorithms that users can choose from for performing clustering and then detecting outliers.

Nevertheless, many researchers argue that, strictly speaking, clustering algorithms should not be considered as outlier detection methods, because their objective is only to group the objects in dataset such that clustering functions can be optimized. The aim to eliminate outliers in dataset using clustering is only to dampen their adverse effect on the final clustering result. This is in contrast to the various definitions of outliers in outlier detection which are more objective and independent of how clusters in the input data set are identified. One of the major philosophies in designing new outlier detection approaches is to directly model outliers and detect them without going though clustering the data first. In addition, the notions of outliers in the context of clustering are essentially binary in nature, without any quantitative indication as to how outlying each object is. It is desired in many applications that the outlier-ness of the outliers can be quantified and ranked.

5. Outlier Detection Methods for High Dimensional Data

There are many applications in high-dimensional domains in which the data can contain dozens or even hundreds of dimensions. The outlier detection techniques we have reviewed in the preceding sections use various concepts of proximity in order to find the outliers based on their relationship to the other points in the data set. However, in high-dimensional space, the data are sparse and concepts using the notion of proximity fail to achieve most of their effectiveness. This is due to the curse of dimensionality that renders the high-dimensional data tend to be equi-distant to each other as dimensionality increases. They does not consider the outliers embedded in subspaces and are not equipped with the mechanism for detecting them.

5.1. Methods for Detecting Outliers in High-dimensional Data

To address the challenge associated with high data dimensionality, two major categories of research work have been conducted. The first category of methods project the high dimensional data to lower dimensional data. Dimensionality deduction techniques, such as *Principal Component Analysis*(PCA), *Independent Component Analysis* (ICA), *Singular Value Decomposition* (SVD), etc can be applied to the high-dimensional data before outlier detection is performed. Essentially, this category of methods perform feature selection and can be considered as the pre-processing work for outlier detection. The second category of approaches is more promising yet challenging. They try to re-design the mechanism to accurately capture the proximity relationship between data points in the high-dimensional space [14].

A. Sparse Cube Method. Aggarwal *et al.* conducted some pioneering work in high-dimensional outlier detection [15][14]. They proposed a new technique for outlier detection that finds outliers by observing the density distributions of projections from the data. This new definition considers a point to be an outlier if in some lower-dimensional projection it is located in a local region of abnormally low density. Therefore, the outliers in these lower-dimensional projections are detected by simply searching for these projections featuring lower density. To measure the sparsity of a lower-dimensional projection quantitatively, the

authors proposed the so-called *Sparsity Coefficient*. The computation of Sparsity Coefficient involves a grid discretization of the data space and making an assumption of normal distribution for the data in each cell of the hypercube. Each attribute of the data is divided into φ equi-depth ranges. In each range, there is a fraction $f = 1/\varphi$ of the data. Then, a k-dimensional cube is made of ranges from k different dimensions. Let N be the dataset size and $n(D)$ denote the number of objects in a k-dimensional cube D. Under the condition that attributes were statistically independent, the Sparsity Coefficient $S(D)$ of the cube D is defined as:

$$S(D) = \frac{n(D) - N * f^k}{\sqrt{N * f^k * (1 - f^k)}}$$

Since there are no closure properties for Sparsity Coefficient, thus no fast subspace pruning can be performed and the lower-dimensional projection search problem becomes a NP-hard problem. Therefore, the authors employ evolutionary algorithm in order to solve this problem efficiently. After lower-dimensional projections have been found, a post-processing phase is required to map these projections into the data points; all the sets of data points that contain in the abnormal projections reported by the algorithm.

B. Example-based Method. Recently, an approach using outlier examples provided by users are used to detect outliers in high-dimensional space [96][97]. It adopts an *"outlier examples → subspaces → outliers"* manner to detect outliers. Specifically, human users or domain experts first provide the systems with a few initial outlier examples. The algorithm finds the subspaces in which most of these outlier examples exhibit significant outlier-ness. Finally, other outliers are detected from these subspaces obtained in the previous step. This approach partitions the data space into equi-depth cells and employs the Sparsity Coefficient proposed in [14] to measure the outlier-ness of outlier examples in each subspace of the lattice. Since it is untenable to exhaustively search the space lattice, the author also proposed to use evolutionary algorithms for subspace search. The fitness of a subspace is the average Sparsity Coefficients of all cubes in that subspace to which the outlier examples belong. All the objects contained in the cubes which are sparser than or as sparse as cubes containing outlier examples in the subspace are detected as outliers.

However, this method is limited in that it is only able to find the outliers in the subspaces where most of the given user examples are outlying significantly. It cannot detect those outliers that are embedded in other subspaces. Its capability for effective outlier detection is largely depended on the number of given examples and, more importantly, how these given examples are similar to the majority of outliers in the dataset. Ideally, this set of user examples should be a good sample of all the outliers in the dataset. This method works poorly when the number of user examples is quite small and cannot provide enough clues as to where the majority of outliers in the dataset are. Providing such a good set of outlier examples is a difficult task whatsoever. The reasons are two-fold. First, it is not trivial to obtain a set of outlier examples for a high-dimensional data set. Due to a lack of visualization aid in high-dimensional data space, it is not obvious at all to find the initial outlier examples unless they are detected by some other techniques. Secondly and more importantly, even when a set of outliers have already been obtained, testing the representativeness of this outlier set is almost impossible. Given these two strong constraints, this approach becomes inadequate in detecting outliers in high-dimensional datasets. It will miss out those projected outliers that are not similar to those given outlier examples.

C. Outlier Detection in Subspaces. Since outlier-ness of data points mainly appear significant in some subspaces of moderate dimensionality in high-dimensional space and the quality of the outliers detected varies in different subspaces consisting of different combinations of dimension subsets. The authors in [24] employ evolutionary algorithm for feature selection (find optimal dimension subsets which represent the original dataset without losing information for unsupervised learning task of outlier detection as well as clustering). This approach is a wrapper algorithm in which the dimension subsets are selected such that the quality of outlier detected or the clusters generated can be optimized. The originality of this work is to combine the evolutionary algorithm with the data visualization technique utilizing parallel coordinates to present evolution results interactively and allow users to actively participate in evolutionary algorithm searching to achieve a fast convergence of the algorithm.

D. Subspace Outlier Detection for Categorical Data. Das *et al.* study the problem of detecting anomalous records in categorical data sets [33]. They draw on a probability approach for outlier detection. For each record in the data set, the probabilities for the occurrence of different subsets of attributes are investigated. A data record is labeled as an outlier if the occurrence probability for the values of some of its attribute subsets is quite low. Specifically, the probability for two subsets of attributes a_t and b_t to occur together in a record, denoted by $r(a_t, b_t)$, is quantified as:

$$r(a_t, b_t) = \frac{P(a_t, b_t)}{P(a_t)P(b_t)}$$

Due to the extremely large number of possible attribute subsets, only the attribute subsets with a length not exceeding than k are studied.

Because it always evaluates pairs of attribute subsets, each of which contain at least one attribute, therefore, this method will miss out the abnormality evaluation for 1-dimensional attribute subsets. In addition, due to the exponential growth of the number of attribute subsets w.r.t k, the value of k is set typically small in this method. Hence, this method can only cover attribute subsets not larger than $2k$ for a record (this method evaluates a pair of attribute subsets at a time). This limits the ability of this method for detecting records that have outlying attribute subsets larger than $2k$.

5.2. Outlying Subspace Detection for High-dimensional Data

All the outlier detection algorithms that we have discussed so far, regardless of in low or high dimensional scenario, invariably fall into the framework of detecting outliers in a specific data space, either in full space or subspace. We term these methods as *"space → outliers"* techniques. For instance, outliers are detected by first finding locally sparse subspaces [14], and the so-called Strongest/Weak Outliers are discovered by first finding the Strongest Outlying Spaces [59].

A new research problem called *outlying subspace detection* for multiple or high dimensional data has been identified recently in [94][99][93]. The major task of outlying subspace detection is to find those subspaces (subset of features) in which the data points of interest exhibit significant deviation from the rest of population. This problem can be formulated as follows: given a data point or object, find the subspaces in which this data is considerably dissimilar, exceptional or inconsistent with respect to the remaining points or objects. These points under study are called *query points*, which are usually the data that users are interested in or concerned with. As in [94][99], a distance threshold T is utilized to decide whether or not a data point deviates significantly from its neighboring points. A subspace s is called an outlying subspace of data point p if $OD_s(p) \geq T$, where OD is the outlier-ness measurement of p.

Finding the correct subspaces so that outliers can be detected is informative and useful in many practical applications. For example, in the case of designing a training program for an athlete, it is critical to identify the specific subspace(s) in which an athlete deviates from his or her teammates in the daily training performances. Knowing the specific weakness (subspace) allows a more targeted training program to be designed. In a medical system, it is useful for the Doctors to identify from voluminous medical data the subspaces in which a particular patient is found abnormal and therefore a corresponding medical treatment can be provided in a timely manner.

The unique feature of the problem of outlying subspace detection is that, instead of detecting outliers in specific subspaces as did in the classical outlier detection techniques, it involves searching from the space lattice for the associated subspaces whereby the given data points exhibit abnormal deviations. Therefore, the problem of outlying subspace detection is called an *"outlier → spaces"* problem so as to distinguish the classical outlier detection problem which is labeled as a *"space → outliers"* problem. It has been theoretically and experimentally shown that the conventional outlier detection methods, irrespectively dealing with low or high-dimensional data, cannot successfully cope with the problem of outlying subspace detection problem in [94]. The existing high-dimensional outlier detection techniques, *i.e.*, find outliers in given subspaces, are theoretically applicable to solve the outlying detection problem. To do this, we have to detect outliers in all subspaces and a search in all these subspaces is needed to find the set of outlying subspaces of p, which are those subspaces in which p is in their respective set of outliers. Obviously, the computational and space costs are both in an exponential order of d, where d is the number of dimensions of the data point. Such an exhaustive space searching is rather expensive in high-dimensional scenario. In addition, they usually only return the top n outliers in a given subspace, thus it is impossible to check whether or not p is an outlier in this subspace if p is not in this top n list. This analysis provides an insight into the inherent difficulty of using the existing high-dimensional outlier detection techniques to solve the new outlying subspace detection problem.

A. HighDoD

Zhang *et al.* proposed a novel dynamic subspace search algorithm, called HighDoD, to efficiently identify the outlying subspaces for the given query data points [94][99]. The outlying measure, *OD*, is based on the sum of distances between a data and its k nearest neighbors [10]. This measure is simple and independent of any underlying statistical and distribution characteristics of the data points. The following two heuristic pruning strategies employing upward-and downward closure property are proposed to aid in the search for outlying subspaces: *If a point p is not an outlier in a subspace s, then it cannot be an outlier in any subspace that is a subset of s. If a point p is an outlier in a subspace s, then it will be an outlier in any subspace that is a superset of s.* These two properties can be used to quickly detect the subspaces in which the point is not an outlier or the subspaces in which the point is an outlier. All these subspaces can be removed from further consideration in the later

stage of the search process. A fast dynamic subspace search algorithm with a sample-based learning process is proposed. The learning process aims to quantitize the prior probabilities for upward- and downward pruning in each layer of space lattice. The *Total Saving Factor* (TSF) of each layer of subspaces in the lattice, used to measure the potential advantage in saving computation, is dynamically updated and the search is performed in the layer of lattice that has the highest TSF value in each step of the algorithm.

However, HighDoD suffers the following major limitations. First, HighDoD relies heavily on the closure (monotonicity) property of the outlying measurement of data points, termed OD, to perform the fast bottom-up or top-down subspace pruning in the space lattice, which is the key technique HighDoD utilizes for speeding up subspace search. Under the definition of OD, a subspace will always be more likely to be an outlying subspace than its subset subspaces. This is because that OD of data points will be naturally increased when the dimensionality of the subspaces under study goes up. Nevertheless, this may not be a very accurate measurement. The definition of a data point's outlier-ness makes more sense if its measurement can be related to other points, meaning that the averaged level of the measurement for other points in the same subspace should be taken into account simultaneously in order to make the measurement statistically significant. Therefore, the design of a new search method is desired in this situation. Secondly, HighDoD labels each subspace in a binary manner, either an outlying subspace or a non-outlying one, and most subspaces are pruned away before their outlying measurements are virtually evaluated in HighDoD. Thus, it is not possible for HighDoD to return a ranked list of the detected outlying subspaces. Apparently, a ranked list will be more informative and useful than an unranked one in many cases. Finally, a human-user defined cutoff for deciding whether a subspace is outlying or not with respect to a query point is used. This parameter will define the "outlying front" (the boundary between the outlying subspaces and the non-outlying ones). Unfortunately, the value of this parameter cannot be easily specified due to the lack of prior knowledge concerning the underlying distribution of data point that maybe very complex in the high-dimensional spaces.

B. SOF Method

In [100], a novel technique based on genetic algorithm is proposed to solve the outlying subspace detection problem and well copes with the drawbacks of the existing methods. A new metric, called *Subspace Outlying Factor (SOF)*, is developed for measuring the outlying degree of each data point in different subspaces. Based on SOF, a new definition of outlying subspace, called SOF Outlying Subspaces, is proposed. Given an input dataset D, parameters n and k, a subspace s is a SOF Outlying Subspace for a given query data point p if there are no more than $n-1$ other subspaces s' such that $SOF(s',p) > SOF(s,p)$. The above definition is equivalent to say that the top n subspaces having the largest SOF values are considered to be outlying subspaces. The parameters used in defining SOF Outlying Subspaces are easy to be specified, and do not require any prior knowledge about the data distribution of the dataset. A genetic algorithm (GA) based method is proposed for outlying subspace detection. The upward and downward closure property is no longer required in the GA-based method, and the detected outlying subspaces can be ranked based on their fitness function values. The concepts of the lower and upper bounds of D^k, the distance between a given point and its kth nearest neighbor, are proposed. These bounds are used for a significant performance boost in the method by providing a quick approximation of the fitness of subspaces in the GA. A technique is also proposed to compute these bounds efficiently using the so-called kNN Look-up Table.

5.3. Clustering Algorithms for High–dimensional Data

We have witnessed some recent developments of clustering algorithms towards high-dimensional data. As clustering provides a possible, even though not the best, means to detect outliers, it is necessary for us to review these new developments. The representative methods for clustering high-dimensional data are CLIQUE and HPStream.

A. CLIQUE

CLIQUE [7] is a grid-based clustering method that discretizes the data space into non-overlapping rectangular units, which are obtained by partitioning every dimension into a specific number of intervals of equal length. A unit is dense if the fraction of total data points contained in this unit is greater than a threshold. Clusters are defined as unions of connected dense units within a subspace. CLIQUE first identifies a subspace that contains clusters. A bottom-up algorithm is used that exploits the monotonicity of the clustering criterion with respect to dimensionality: if a k-dimensional unit is dense, then so are its projections in $(k-1)$-dimensional space. A candidate generation procedure iteratively determines the candidate k-dimensional units C_k after determining the $(k-1)$-dimensional dense units D_{k-1}. A pass is made over the data to determine those candidates units that are dense D_k. A depth-first search algorithm is then used to identify clusters in the subspace: it starts with some unit u in D, assign it the first cluster label number, and find all the units it is connected

to. Then, if there are still units in D that have yet been visited, it finds one and repeats the procedure. CLIQUE is able to automatically finds dense clusters in subspaces of high-dimensional dataset. It can produce identical results irrespective of the order in which input data are presented and not presume any specific mathematical form of data distribution. However, the accuracy of this clustering method maybe degraded due to the simplicity of this method. The clusters obtained are all of the rectangular shapes, which is obviously not consistent with the shape of natural clusters. In addition, the subspaces obtained are dependent on the choice of the density threshold. CLIQUE uses a global density threshold (i.e., a parameter that is used for all the subspaces), thus it is difficult to specify its value especially in high-dimensional subspaces due to curse of dimensionality. Finally, the subspaces obtained are those where dense units exist, but this has nothing to do with the existence of outliers. As a result, CLIQUE is not suitable for detecting projected outliers.

B. HPStream

In order to find the clusters embedded in the subspaces of high-dimensional data space in data streams, a new clustering method, called HPStream, is proposed [9]. HPStream introduces the concept of *projected clustering* to data streams as significant and high-quality clusters only exist in some low-dimensional subspaces. The basic idea of HPStream is that it does not only find clusters but also updates the set of dimensions associated with each cluster where more compact clusters can be found. The total number of clusters obtained in HPStream is initially obtained through k–means clustering and the initial set of dimensions associated with each of these k clusters is the full set of dimensions of the data stream. As more streaming data arrive, the set of dimensions for each cluster evolves such that each cluster can become more compact with a smaller radius.

HPStream is innovative in finding clusters that are embedded in subspaces for high-dimensional data streams. However, the number of subspaces returned by HPStream is equal to the number of clusters obtained that is typically of a small value. Consequently, if HPStream is applied to detect projected outliers, then it will only be able to detect the outliers in those subspaces returned and miss out a significant potions of outliers existing in other subspaces that are not returned by HPStream. Of course, it is possible to increase the number of subspaces returned in order to improve the detection rate. However, the increase of subspaces will imply an increase of the number of clusters accordingly. An unreasonably large number of clusters is not consistent with the formation of natural clusters and will therefore affect the detection accuracy of projected outliers.

6. Outlier Detection Methods for Data Streams

The final major category of outlier detection methods we will discuss in this section are those outlier detection methods for handling data streams. We will first discuss Incremental LOF, and then the outlier detection methods for sensor networks that use Kernel density function. The incremental clustering methods that can handle continuously arriving data will also be covered at the end of this subsection.

A. Incremental LOF Method

Since LOF method is not able to handle data streams, thus an incremental LOF algorithm, appropriate for detecting outliers from dynamic databases where frequently data insertions and deletions occur, is proposed in [78]. The proposed incremental LOF algorithm provides an equivalent detection performance as the iterated static LOF algorithm (applied after insertion of each data record), while requiring significantly less computational time. In addition, the incremental LOF algorithm also dynamically updates the profiles of data points. This is an appealing property, since data profiles may change over time. It is shown that insertion of new data points as well as deletion of obsolete points influence only limited number of their nearest neighbors and thus insertion/deletion time complexity per data point does not depend on the total number of points N[78].

The advantage of Incremental LOF is that it can deal with data insertions and deletions efficiently. Nevertheless, Incremental LOF is not economic in space. The space complexity of this method is in the order of the data that have been inserted but have not been deleted. In other words, Incremental LOF has to maintain the whole length of data stream in order to deal with continuously arriving data because it does not utilize any compact data summary or synopsis. This is clearly not desired for data stream applications that are typically subject to explicit space constraint.

B. Outlier Detection Methods for Sensor Networks

There are a few recent anomaly detection methods for data streams. They mainly come from sensor networks domain such as [80] and [25]. However, the major effort taken in these works is the development of distributable outlier detection methods from distributed data streams and does not deal with the problem of outlier detection in subspaces of high-dimensional data space. Palpanas *et al.* proposed one of the first outlier detection methods for distributed data streams in the context of sensor networks [80]. The author classified the sensor nodes in the network as the low capacity and high capacity nodes, through which a multi-resolution structure of the sensor network is created. The high capacity nodes are nodes equipped with relatively strong computational strength that can detect local outliers. The Kernel density function is employed to model local data distribution in a single or multiple dimensions of space. A point is detected as an outlier if the number of values that have fallen into its neighborhood (delimited by a sphere of radius r) is less than an application-specific threshold. The number of values in the neighborhood can be computed by the Kernel density function. Similarly, the authors in [25] also emphasize the design of distributed outlier detection methods. Nevertheless, this work employs a number of different commonly used outlier-ness metric such as the distance to k^{th} nearest neighbor, average distance to the k nearest neighbors, the inverse of the number of neighbors within a specific distance. Nevertheless, these metrics are not applicable to data streams.

C. Incremental Clustering Methods

Most clustering algorithms we have discussed earlier in this section assume a complete and static dataset to operate. However, new data becomes continuously available in many applications such as the data streams. With the aforementioned classical clustering algorithms, reclustering from scratch to account for data updates is too costly and inefficient. It is highly desired that the data can be processed and clustered in an incremental fashion. The recent representative clustering algorithms having mechanisms to handle data updates are BIRCH*, STREAM and CluStream.

BIRCH* [45] is a framework for fast, scalable and incremental clustering algorithms. In the BIRCH* family of algorithms, objects are read from the databases sequentially and inserted into incrementally evolving clusters which are represented by generalized cluster features (CF*s), the condensed and summarized representation of clusters. A new objects reading from the databases is inserted into the closest cluster. BIRCH* organizes all clusters in an in-memory index, and height-balanced tree, called CF*-tree. For a new object, the search for an appropriate cluster requires time logarithmic in the number of the clusters to a linear scan. CF*s are efficient because: (1) they occupy much less space than the naive representation; (2) the calculation of inter-cluster and intra-cluster measurements using the CF* is much faster than calculations involving all objects in clusters. The purpose of the CF*-tree is to direct a new object to the cluster closest to it. The non-leaf and leaf entries function differently, non-leaf entries are used to guide new objects to appropriate leaf clusters, whereas leaf entries represent the dynamically evolving clusters. However, clustering of high-dimensional datasets has not been studied in BIRCH*. In addition, BIRCH* cannot perform well when the clusters are not spherical in shape due to the fact that it relies on spherical summarization to produce the clusters.

STREAM [74] considers the clustering of continuously arriving data, and provides a clustering algorithm superior to the commonly used k-means algorithm. STREAM assumes that the data actually arrives in chunks $X_1, X_2, \cdots, X_n$, each of which fits into main memory. The streaming algorithm is as follows. For each chunk i, STREAM first assigns weight to points in the chunks according to their respective appearance frequency in the chunks ensuring that each point appear only once. The STREAM clusters each chunk using procedure LOCALSEARCH. For each chunk, only k weighted cluster centers are retained and the whole chunk is discarded in order to free the memory for new chunks. Finally, LOCALSEARCH is applied to the weighted centers retained from $X_1, X_2, \cdots, X_n$, to obtain a set of (weighted) centers for the entire stream $X_1, X_2, \cdots, X_n$.

In order to find clusters in different time horizons (such as the last month, last year or last decade), a new clustering method for data stream, called CluStream, is proposed in [8]. This approach provides the user the flexibility to explore the nature of the evolution of the clusters over different time periods. In order to avoid bookkeeping the huge amount of information about the clustering results in different time horizons, CluStream divides the clustering process into an online micro-clustering component and an offine macro-clustering component. The micro-clustering phase mainly collects online the data statistics for clustering purpose. This process is not dependent on any user input such as the time horizon or the required granularity of the clustering process. The aim is to maintain statistics at a sufficiently high level of granularity so that it can be effectively used by the offline components of horizon-specific macro-clustering as well as evolution analysis. The micro-clusters generated by the algorithm serve as an intermediate statistical representation which can be maintained in an efficient way even for a data stream of large volume. The macro-clustering process does not work on the original data stream that may be very large in size. Instead, it uses the compactly stored summary statistics of the micro-clusters. Therefore, the micro-clustering phase is not subject to the one-pass constraint of data stream applications.

D. Advantages and Disadvantages of Outlier Detection for Data Streams

The methods discussed in this subsection can detect outliers from data streams. The incremental LOF method is able to deal with continuously arriving data, but it may face an explosion of space consumption. Moreover, the incremental LOF method is not able to find outliers in subspaces in an automatic manner. The outlier detection methods for sensor networks cannot find projected outliers either. Unlike the clustering methods that are only appropriate for static databases,

BIRCH*, STREAM and CluStream go one step further and are able to handle incrementally the continuously arriving data. Nevertheless, they are designed to use all the features of data in detecting outliers and are difficult to detect projected outliers.

6.1. Summary

This section presents a comprehensive survey on the major existing methods for detecting point outliers from vector-like data sets. Both the conventional outlier detection methods that are mainly appropriate for relatively low dimensional static databases and the more recent methods that are able to deal with high-dimensional projected outliers or data stream applications have been discussed. For a big picture of these methods, we present a summary in Table 6. In this table, we evaluate each method against two criteria, namely whether it can detect projected outliers in a high-dimensional data space and whether it can handle data streams. The symbols of tick and cross in the table indicate respectively whether or not the corresponding method satisfies the evaluation criteria. From this table, we can see that the conventional outlier detection methods cannot detect projected outliers embedded in different subspaces; they detect outliers only in the full data space or a given subspace. Amongst these methods that can detect projected outliers, only HPStream can meet both criteria. However, being a clustering method, HPStream cannot provide satisfactory support for projected outliers detection from high-dimensional data streams.

7. Conclusions

In this paper, a comprehensive survey is presented to review the existing methods for detecting point outliers from various kinds of vector-like datasets. The outlier detection techniques that are primarily suitable for relatively low-dimensional static data, which serve the technical foundation for many of the methods proposed later, are reviewed first. We have also reviewed some of recent advancements in outlier detection for dealing with more complex high-dimensional static data and data streams.

It is important to be aware of the limitation of this survey. As it has clearly stated in Section 2, we only focus on the point outlier detection methods from vector-like datasets due to the space limit. Also, outlier detection is a fast developing field of research and more new methods will quickly emerge in the foreseeable near future. Driven by their emergence, it is believed that outlier detection techniques will play an increasingly important role in various practical applications where they can be applied to.

Category	Method	High-D subspace outliers	Data Stream
Statistical detection methods	Gaussian models	X	X
	Regression models	X	√
	Histograms	X	√
	Kernel functions	X	√
Distance-based methods	DB(k, λ)-Outliers	X	X
	DB(pct, d_{min})-Outliers	X	X
	K^{th} NN method	X	X
	K^{th} NN sum method	X	X
	Grid-ODF	X	X
Density-based methods	LOF	X	X
	COF	X	X
	INFLO	X	X
	MDEF	X	X
	Incremental LOF	X	√
Clustering-based methods	PAM/CLARA	X	X
	k-means	X	X
	CLARANS	X	X
	MST clustering	X	X
	BIRCH	X	√
	CURE	X	X
	CHAMELEON	X	X
	DBSCAN	X	X
	DENCLUE	X	X
	STING	X	√
	WaveCluster	X	√
	DCLUST	X	√
	STREAM	X	√
	BIRCH*	X	√
	CluStream	X	√
High-D outlier detection methods	Sparse Coefficient method	√	X
	Example-based method	√	X
	HighDoD	√	X
	SOF method	√	X
High-D clustering-based methods	CLIQUE	√	X
	HPStream	√	√

Figure 6. A summary of major existing outlier detection methods

References

[1] C. C. Aggarwal. On Abnormality Detection in Spuriously Populated Data Streams. SIAM International Conference on Data Mining (SDM'05), Newport Beach, CA, 2005.

[2] B. Abraham and G. E. P. Box. Bayesian analysis of some outlier problems in time series. *Biometrika* 66, 2, 229-236, 1979.

[3] A. Arasu, B. Babcock, S. Babu, M. Datar, K. Ito, I. Nishizawa, J. Rosenstein an J. Widom. STREAM: The Stanford Stream Data Manager, *SIGMOD'03*, 2003.

[4] B. Abraham and A. Chuang. Outlier detection and time series modeling. *Technometrics* 31, 2, 241-248, 1989.

[5] D. Anderson, T. Frivold, A. Tamaru, and A. Valdes. Next-generation intrusion detection expert system (nides), software users manual, beta-update release. *Technical Report*, Computer Science Laboratory, SRI International, 1994.

[6] R. Agrawal, J. Gehrke, D. Gunopulos and P. Raghavan. Automatic subspace clustering of high dimensional data for data mining applications. In Proc. of *1998 ACM SIGMOD International Conference on Management of Data (SIGMOD'98)*, pp 94-105, 1998.

[7] R. Agrawal, J. Gehrke, D. Gunopulos, and P. Raghavan. Automatic Subspace Clustering of High Dimensional Data Mining Application. In proceeding of *ACM SIGMOD'99*, Philadelphia, PA, USA, 1999.

[8] C. C. Aggarwal, J. Han, J. Wang, P. S. Yu: A Framework for Clustering Evolving Data Streams. In Proc. of *29th Very Large Data Bases (VLDB'03)*,pp 81-92, Berlin, Germany, 2003.

[9] C. C. Aggarwal, J. Han, J. Wang, P. S. Yu. A Framework for Projected Clustering of High Dimensional Data Streams. In Proc. of *30th Very Large Data Bases (VLDB'04)*, pp 852-863, Toronto, Canada, 2004.

[10] F. Angiulli and C. Pizzuti. Fast Outlier Detection in High Dimensional Spaces. In Proc. of *6th European Conference on Principles and Practice of Knowledge Discovery in Databases (PKDD'02)*,Helsinki, Finland, pp 15-26, 2002.

[11] C. C. Aggarwal, C. M. Procopiuc, J. L. Wolf, P. S. Yu and J. S. Park. Fast algorithms for projected clustering. In Proc. of *1999 ACM SIGMOD International Conference on Management of Data (SIGMOD'99)*, pp 61-72, 1999.

[12] D. Anderson, A. Tamaru, and A. Valdes. Detecting unusual program behavior using the statistical components of NIDES. *Technical Report*, Computer Science Laboratory, SRI International, 1995.

[13] C. C. Aggarwal and P. Yu. Finding generalized projected clusters in high dimensional spaces. In Proc. of *2000 ACM SIGMOD International Conference on Management of Data (SIGMOD'00)*, pp 70-81, 2000.

[14] C. C. Aggarwal and P. S. Yu. Outlier Detection in High Dimensional Data. In Proc. of *2001 ACM SIGMOD International Conference on Management of Data (SIGMOD'01)*, Santa Barbara, California, USA, 2001.

[15] Charu C. Aggarwal and Philip S. Yu. 2005. An effective and efficient algorithm for high-dimensional outlier detection. *VLDB Journal*, 14: 211-221, Springer-Verlag Publisher.

[16] V. Barnett. The ordering of multivariate data (with discussion). *Journal of the Royal Statistical Society*. Series A 139, 318-354, 1976.

[17] C. Bishop. Novelty detection and neural network validation. In Proceedings of *IEEE Vision, Image and Signal Processing*, Vol. 141. 217-222, 1994.

[18] R. J. Beckman and R. D. Cook. Outliers. *Technometrics* 25, 2, 119-149, 1983.

[19] K. Beyer, J. Goldstein, R. Ramakrishnan and U. Shaft. When is nearest neighbors meaningful? In Proc. of *7th International Conference on Database Theory (ICDT'99)*, pp 217-235, Jerusalem, Israel, 1999.

[20] M. M. Breunig, H-P Kriegel, R. T. Ng and J. Sander. OPTICS-OF: Identifying Local Outliers. *PKDD'99*, 262-270, 1999.

[21] M. Breuning, H-P. Kriegel, R. Ng, and J. Sander. LOF: Identifying Density-Based Local Outliers. In Proc. of *2000 ACM SIGMOD International Conference on Management of Data (SIGMOD'00)*, Dallas, Texas, pp 93-104, 2000.

[22] N. Beckmann, H.-P. Kriegel, R. Schneider, and B. Seeger. The R^*-tree: an efficient and robust access method for points and rectangles. In Proc. of *1990 ACM SIGMOD International Conference on Management of Data (SIGMOD'90)*, pp 322-331, Atlantic City, NJ, 1990.

[23] V. Barnett and T. Lewis. *Outliers in Statistical Data*. John Wiley, 3rd edition, 1994.

[24] L. Boudjeloud and F. Poulet. Visual Interactive Evolutionary Algorithm for High Dimensional Data Clustering and Outlier Detection. In Proc. of *9th Pacific-Asia Conference on Advances in Knowledge Discovery and Data Mining (PAKDD'05)*, Hanoi, Vietnam, pp426-431, 2005.

[25] Branch, J. Szymanski, B. Giannella, C. Ran Wolff Kargupta, H. n-Network Outlier Detection in Wireless Sensor Networks. In Proc. of. *26th IEEE International Conference on Distributed Computing Systems (ICDCS)*, 2006.

[26] H. Cui. Online Outlier Detection Detection Over Data Streams. *Master thesis*, Simon Fraser University, 2002.

[27] D. Chakrabarti. Autopart: Parameter-free graph partitioning and outlier detection. In *PKDD'04*, pages 112-124, 2004.

[28] V. Chandola, A. Banerjee, and V. Kumar. Outlier Detection-A Survey, *Technical Report*, TR 07-017, Department of Computer Science and Engineering, University of Minnesota, 2007.

[29] C. Chow and D. Y. Yeung. Parzen-window network intrusion detectors. In Proceedings of *the 16th International Conference on Pattern Recognition*, Vol. 4, Washington, DC, USA, 40385, 2002.

[30] D. E. Denning. An intrusion detection model. *IEEE Transactions of Software Engineering* 13, 2, 222-232, 1987.

[31] M. Desforges, P. Jacob, and J. Cooper. Applications of probability density estimation to the detection of abnormal conditions in engineering. In Proceedings of *Institute of Mechanical Engineers*, Vol. 212. 687-703, 1998.

[32] D. Dasgupta and F. Nino. A comparison of negative and positive selection algorithms in novel pattern detection. In Proceedings of *the IEEE International Conference on Systems, Man, and Cybernetics*, Vol. 1. Nashville, TN, 125-130, 2000.

[33] K. Das and J. G. Schneider: Detecting anomalous records in categorical datasets. *KDD'07*, 220-229, 2007.

[34] E. Eskin. Anomaly detection over noisy data using learned probability distributions. In Proceedings of *the Seventeenth International Conference on Machine Learning (ICML)*. Morgan Kaufmann Publishers Inc., 2000.

[35] D. Endler. Intrusion detection: Applying machine learning to solaris audit data. In Proceedings of the 14th *Annual Computer Security Applications Conference*, 268, 1998.

[36] E. Eskin, A. Arnold, M. Prerau, L. Portnoy and S. Stolfo. A Geometric Framework for Unsupervised Anomaly Detection: Detecting Intrusions in Unlabeled Data. *Applications of Data Mining in Computer Security*, 2002.

[37] M. Ester, H-P Kriegel, J. Sander, and X.Xu. A Density-based Algorithm for Discovering Clusters in Large Spatial Databases with Noise. In proceedings of *2nd International Conference on Knowledge Discovery and Data Mining (KDD'96)*, Portland, Oregon, USA, 1996.

[38] E. Eskinand and S. Stolfo. Modeling system call for intrusion detection using dynamic window sizes. In Proceedings of *DARPA Information Survivability Conference and Exposition*, 2001.

[39] A. J. Fox. Outliers in time series. *Journal of the Royal Statistical Society*, Series B (Methodological) 34, 3, 350-363, 1972.

[40] T. Fawcett. and F. Provost. Activity monitoring: noticing interesting changes in behavior. In Proceedings of the 5th *ACM SIGKDD International Conference on Knowledge Discovery and Data Mining*, 53-62, 1999.

[41] D. Freedman, R. Pisani and R. Purves. *Statistics*, W. W. Norton, New York, 1978.

[42] F. Grubbs Procedures for detecting outlying observations in samples. *Technometrics* 11, 1, 1-21, 1969.

[43] D. E. Goldberg. *Genetic Algorithms in Search, Optimization, and Machine Learning*. Addison-Wesley, Reading, Massachusetts, 1989.

[44] S. Guttormsson, R. M. II, and M. El-Sharkawi. Elliptical novelty grouping for on-line short-turn detection of

excited running rotors. *IEEE Transactions on Energy Conversion* 14, 1, 1999.

[45] V. Ganti, R. Ramakrishnan, J. Gehrke, A. Powell, and J. French. Clustering Large Datasets in Arbitrary Metric Spaces. In Proc.s of *the 15th International Conference on Data Engineering (ICDE'99)*, Sydney, Australia, 1999.

[46] S. Guha, R. Rastogi, and K. Shim. CURE: An Efficient Clustering Algorithm for Large Databases. In Proceedings of *the 1998 ACM SIGMOD International Conference on Management of Data (SIGMOD'98)*, Seattle, WA, USA, 1998.

[47] D. Hawkins. *Identification of Outliers*. Chapman and Hall, London, 1980.

[48] P. Helman and J. Bhangoo. A statistically based system for prioritizing information exploration under uncertainty. In *IEEE International Conference on Systems, Man, and Cybernetics*, Vol. 27, 449-466, 1997.

[49] P. S. Horn, L. Feng, Y. Li, and A. J. Pesce. Effect of outliers and nonhealthy individuals on reference interval estimation. *Clinical Chemistry* 47, 12, 2137-2145, 2001.

[50] J. Han and M Kamber. *Data Mining: Concepts and Techniques*. Morgan Kaufman Publishers, 2000.

[51] A. Hinneburg, and D.A. Keim. An Efficient Approach to Cluster in Large Multimedia Databases with Noise. *KDD'98*, 1998.

[52] W. Jin, A. K. H. Tung and J. Han. Finding Top n Local Outliers in Large Database. In Proc. of *7th ACM International Conference on Knowledge Discovery and Data Mining (SIGKDD'01)*, San Francisco, CA, pp 293-298, 2001.

[53] W. Jin, A. K. H. Tung, J. Han and W. Wang: Ranking Outliers Using Symmetric Neighborhood Relationship. *PAKDD'06*, 577-593, 2006.

[54] H. S. Javitz and A. Valdes. The SRI IDES statistical anomaly detector. In Proceedings of *the 1991 IEEE Symposium on Research in Security and Privacy*, 1991.

[55] G. Karypis, E-H. Han, and V. Kumar. CHAMELEON: A Hierarchical Clustering Algorithm Using Dynamic Modeling. *IEEE Computer*, 32, Pages 68-75, 1999.

[56] E. M. Knorr and R. T. Ng. A unified approach for mining outliers. *CASCON'97*, 11, 1997.

[57] E. M. Knorr and R. T. Ng. A Unified Notion of Outliers: Properties and Computation. *KDD'97*, 219-222, 1997.

[58] E. M. Knorr and R. T. Ng. Algorithms for Mining Distance-based Outliers in Large Dataset. In Proc. of *24th International Conference on Very Large Data Bases (VLDB'98)*, New York, NY, pp 392-403, 1998.

[59] E. M. Knorr and R. T. Ng (1999). Finding Intentional Knowledge of Distance-based Outliers. In Proc. of *25th International Conference on Very Large Data Bases (VLDB'99)*, Edinburgh, Scotland, pp 211-222, 1999.

[60] E. M. Knorr, R. T. Ng and V. Tucakov. Distance-Based Outliers: Algorithms and Applications. *VLDB Journal*, 8(3-4): 237-253, 2000.

[61] T. M. Khoshgoftaar, S. V. Nath, and S. Zhong. Intrusion Detection in Wireless Networks using Clusterings Techniques with Expert Analysis. Proceedings of *the Fourth International Conference on Machine Leaning and Applications (ICMLA'05)*, Los Angeles, CA, USA, 2005.

[62] L. Kaufman and P.J. Rousseeuw. *Finding Groups in Data: an Introduction to Cluster Analysis*. John wiley&Sons, 1990.

[63] C. Kruegel, T. Toth, and E. Kirda. Service specific anomaly detection for network intrusion detection. In Proceedings of *the 2002 ACM Symposium on Applied computing*, 201-208, 2002.

[64] X. Li and J. Han: Mining Approximate Top-K Subspace Anomalies in Multi-Dimensional Time-Series Data. *VLDB*, 447-458, 2007.

[65] J. Lee, J. Han and X. Li. Trajectory Outlier Detection: A Partition-and-Detect Framework. *ICDE'08*, 140-149, 2008.

[66] J. Laurikkala, M. Juhola1, and E. Kentala. 2000. Informal identification of outliers in medical data. In *Fifth International Workshop on Intelligent Data Analysis in Medicine and Pharmacology*, 20-24, 2000.

[67] G. Manson. Identifying damage sensitive, environment insensitive features for damage detection. In Proceedings of *the IES Conference*. Swansea, UK, 2002.

[68] J. MacQueen. Some methods for classification and analysis of multivariate observations. In Proc. of *5th Berkeley Symp. Math. Statist, Prob.*, 1, pages 281-297, 1967.

[69] M. V. Mahoney and P. K. Chan. Learning nonstationary models of normal network traffic for detecting novel attacks. In Proceedings of *the 8th ACM SIGKDD International Conference on Knowledge Discovery and Data Mining*, 376-385, 2002.

[70] G. Manson, G. Pierce, and K. Worden. On the long-term stability of normal condition for damage detection in a composite panel. In Proceedings of *the 4th International Conference on Damage Assessment of Structures*, Cardiff, UK, 2001.

[71] G. Manson, S. G. Pierce, K. Worden, T. Monnier, P. Guy, and K. Atherton. Long-term stability of normal condition data for novelty detection. In Proceedings of *Smart Structures and Integrated Systems*, 323-334, 2000.

[72] C. C. Noble and D. J. Cook. Graph-based anomaly detection. In *KDD'03*, pages 631-636, 2003.

[73] R. Ng and J. Han. Efficient and Effective Clustering Methods for Spatial Data Mining. In proceedings of *the 20th VLDB Conference*, pages 144-155, 1994.

[74] L. O'Callaghan, N. Mishra, A. Meyerson, S. Guha, and R. Motwani. Streaming-Data Algorithms For High-Quality Clustering. In Proceedings of *the 18th International Conference on Data Engineering (ICDE'02)*, San Jose, California, USA, 2002.

[75] M. I. Petrovskiy. Outlier Detection Algorithms in Data Mining Systems. *Programming and Computer Software*, Vol. 29, No. 4, pp 228-237, 2003.

[76] E. Parzen. On the estimation of a probability density function and mode. *Annals of Mathematical Statistics* 33, 1065-1076, 1962.

[77] S. Papadimitriou, H. Kitagawa, P. B. Gibbons and C. Faloutsos: LOCI: Fast Outlier Detection Using the Local Correlation Integral. *ICDE'03*, 315, 2003.

[78] D. Pokrajac, A. Lazarevic, L. Latecki. Incremental Local Outlier Detection for Data Streams, *IEEE symposiums on computational Intelligence and Data Mining (CIDM'07)*, 504-515, Honolulu, Hawaii, USA, 2007.

[79] P. A. Porras and P. G. Neumann. EMERALD: Event monitoring enabling responses to anomalous live disturbances. In Proceedings of *20th NIST-NCSC National Information Systems Security Conference*, 353-365, 1997.

[80] T. Palpanas, D. Papadopoulos, V. Kalogeraki, D. Gunopulos. Distributed deviation detection in sensor networks. *SIGMOD Record* 32(4): 77-82, 2003.

[81] S. Ramaswamy, R. Rastogi, and S. Kyuseok. Efficient Algorithms for Mining Outliers from Large Data Sets. In Proc. of *2000 ACM SIGMOD International Conference on Management of Data (SIGMOD'00)*, Dallas, Texas, pp 427-438, 2000.

[82] G. Sheikholeslami, S. Chatterjee, and A. Zhang. WaveCluster: A Wavelet based Clustering Approach for Spatial Data in Very Large Database. *VLDB Journal*, vol.8 (3-4), pages 289-304, 1999.

[83] H. E. Solberg and A. Lahti. Detection of outliers in reference distributions: Performance of horn's algorithm. *Clinical Chemistry* 51, 12, 2326-2332, 2005.

[84] J. Sun, H. Qu, D. Chakrabarti and C. Faloutsos. Neighborhood Formation and Anomaly Detection in Bipartite Graphs. *ICDM'05*, 418-425, 2005.

[85] J. Sun, H. Qu, D. Chakrabarti and C. Faloutsos. Relevance search and anomaly detection in bipartite graphs. *SIGKDD Explorations* 7(2): 48-55, 2005.

[86] L. Tarassenko. Novelty detection for the identification of masses in mammograms. In Proceedings of *the 4th IEEE International Conference on Artificial Neural Networks*, Vol. 4. Cambridge, UK, 442-447, 1995.

[87] J. Tang, Z. Chen, A. Fu, and D. W. Cheung. Enhancing Effectiveness of Outlier Detections for Low Density Patterns. In Proc. of *6th Pacific-Asia Conference on Knowledge Discovery and Data Mining (PAKDD'02)*, Taipei, Taiwan, 2002.

[88] W. Wang, J. Yang, and R. Muntz. STING: A Statistical Information Grid Approach to Spatial Data Mining. In Proceedings of *23rd VLDB Conference*, pages 186-195, Athens, Green, 1997.

[89] W. Wang, J. Zhang and H. Wang. Grid-ODF: Detecting Outliers Effectively and Efficiently in Large Multidimensional Databases. In Proc. of *2005 International Conference on Computational Intelligence and Security (CIS'05)*, pp 765-770, Xi'an, China, 2005.

[90] K. Yamanishi and J. I. Takeuchi. Discovering outlier filtering rules from unlabeled data: combining a supervised learner with an unsupervised learner. In Proceedings of *the 7th ACM SIGKDD International Conference on Knowledge Discovery and Data Mining*, 389-394, 2001.

[91] K. Yamanishi, J. I. Takeuchi, G. Williams, and P. Milne. On-line unsupervised outlier detection using finite mixtures with discounting learning algorithms. *Data Mining and Knowledge Discovery* 8, 275-300, 2004.

[92] C.T. Zahn. Graph-theoretical Methods for Detecting and Describing Gestalt Clusters. *IEEE Transaction on Computing*, C-20, pages 68-86, 1971.

[93] J. Zhang, Q. Gao and H. Wang. Outlying Subspace Detection in High dimensional Space. *Encyclopedia of Database Technologies and Applications (2nd Edition)*, Idea Group Publisher, 2009.

[94] J. Zhang and H. Wang. Detecting Outlying Subspaces for High-dimensional Data: the New Task, Algorithms and Performance. *Knowledge and Information Systems: An International Journal (KAIS)*, Springer-Verlag Publisher, 2006.

[95] J. Zhang, W. Hsu and M. L. Lee. Clustering in Dynamic Spatial Databases. *Journal of Intelligent Information Systems (JIIS)* 24(1): 5-27, Kluwer Academic Publisher, 2005.

[96] C. Zhu, H. Kitagawa and C. Faloutsos. Example-Based Robust Outlier Detection in High Dimensional Datasets. In Proc. of *2005 IEEE International Conference on Data Management(ICDM'05)*, pp 829-832, 2005.

[97] C. Zhu, H. Kitagawa, S. Papadimitriou and C. Faloutsos. OBE: Outlier by Example. *PAKDD'04*, 222-234, 2004.

[98] S. Zhong, T. M. Khoshgoftaar, and S. V. Nath. A clustering approach to wireless network intrusion detection. In *ICTAI*, pages 190-196, 2005.

[99] J. Zhang, M. Lou, T. W. Ling and H. Wang. HOS-Miner: A System for Detecting Outlying Subspaces of High-dimensional Data. In Proc. of *30th International Conference on Very Large Data Bases (VLDB'04)*, demo, pages 1265-1268, Toronto, Canada, 2004.

[100] J. Zhang, Q. Gao and H. Wang. A Novel Method for Detecting Outlying Subspaces in High-dimensional Databases Using Genetic Algorithm. *2006 IEEE International Conference on Data Mining (ICDM'06)*, pages 731-740, Hong Kong, China, 2006.

[101] T. Zhang, R. Ramakrishnan, and M. Livny. BIRCH: An Efficient Data Clustering Method for Very Large Databases. In proceedings of *the 1996 ACM International Conference on Management of Data (SIGMOD'96)*, pages 103-114, Montreal, Canada, 1996.

[102] J. Zhang and H. Wang. Detecting Outlying Subspaces for High-dimensional Data: the New Task, Algorithms and Performance. *Knowledge and Information Systems (KAIS)*, 333-355, 2006.

[103] J. Zhang, Q. Gao, H. Wang, Q. Liu, K. Xu. Detecting Projected Outliers in High-Dimensional Data Streams. DEXA 2009: 629-644, 2009.

Non-stationary Parallel Multisplitting Two-Stage Iterative Methods with Self-Adaptive Weighting Schemes

GuoYan Meng[1], ChuanLong Wang[2,*], XiHong Yan[2], QingShan Zhao[1]

[1]Department of Compute Science, Xinzhou teacher University, Xinzhou 034000, Shanxi Province, P. R. China
[2]Department of mathematics, Taiyuan Normal University, Taiyuan 030012, Shanxi Province, P. R. China

Abstract

In this paper, we study the non-stationary parallel multisplitting two-stage iterative methods with self-adaptive weighting matrices for solving a linear system whose coefficient matrix is symmetric positive definite. Two choices of Self-adaptive weighting matrices are given, especially, the nonnegativity is eliminated. Moreover, we prove the convergence of the non-stationary parallel multisplitting two-stage iterative methods with self-adaptive weighting matrices. Finally, the numerical comparisons of several self-adaptive non-stationary parallel multisplitting two-stage iterative methods are shown.

Keywords: Self-adaptive weighting matrices, non-stationary, multisplitting, two-stage, linear systems

1. Introduction

To solve large sparse linear system of equations on multiprocessor systems,

$$Ax = b, \quad A = (a_{ij}) \in R^{n \times n} \text{ nonsingular and } b \in R^n \ . \quad (1)$$

O'Leary and White [14] first proposed parallel methods based on multisplitting of matrices in 1985, after this, combing with two-stage iterative methods (see [2, 4, 10]), the multisplitting two-stage iterative methods [15] were proposed, where several basic convergence results were found. The scheme was proposed as following

$$A = B_i - C_i, \quad B_i = M_i - N_i, \quad i = 1, 2, \cdots, m \ , \quad (2)$$

$$M_i x_i^{(k,l)} = N_i x_i^{(k,l-1)} + C_i x^{(k)} + b \ , \quad (3)$$

$$x^{(k+1)} = \sum_{i=1}^{m} E_i x_i^{(k,q(i,k))} \ , \quad (4)$$

where $E_i \geq 0$, diagonal, and $\sum_{i=1}^{m} E_i = I$. $(M_i, N_i, C_i, E_i)_{i=1}^{m}$ will be unchanged and independent of the iterative number k.

Later, many authors studied the methods for the case that A is an M-matrix, an H-matrix and a symmetric positive definite matrix. When A is an M-matrix or an H-matrix, many parallel multisplitting two-stage iterative methods (see [3, 5, 6, 12, 15, 17]) were presented, and the weighting matrices $E_i, i = 1, 2, \cdots, m$ were generalized (see [1, 11])

$$\sum_{i=1}^{m} E_i^{(k)} = I(\text{or} \neq I), \quad E_i^{(k)} \geq 0, \ k = 1, 2, \cdots, \quad (5)$$

and $E_i^{(k)}$ is diagonal, but these weighting matrices were preset as multi-parameter.

When A is a symmetric positive definite matrix, generally, which require the assumption that the weighting matrices are multiples of the identity matrix, that is $E_i = \alpha_i I, i = 1, 2, \cdots, m$ (see [8, 14]), but these results have little applicability for analysis of parallel processing. In order to improve the weighting matrices, White [19, 20] and Wen [18] presented the multisplitting which had a very special structure,

---

★This paper is an extended version of [22]. We have added a kind of self-adaptive weighting schemes in Algorithm 1, and also proven the convergence of Algorithm 1 in this condition. In addition, we have added the numerical example and completely recalculated the numerical examples with highly precision and higher size of coefficient matrix.
*Chuan-Long Wang. Email: clwang218@126.com

Chen [21] discussed asynchronous multisplitting, Cao [7] gave a nonstandard multisplitting, Migallón [13] proposed the non-stationary multisplittings, Wang and Bai [17] discussed the non-stationary two-stage multisplitting, but the non-stationary multisplitting usually had a block splitting for parallel processing. Furthermore, as we know, the weighting matrices have important role in parallel multisplitting methods, but the weighting matrices in all above-mentioned methods are determined previously, they are not known to be good or bad, this influences the efficiency of parallel methods. Fortunately, Wang [23] has presented modified parallel multisplitting iterative methods by optimizing the weighting matrices based on the sparsity of the coefficient matrix A. But none has ever studied that how to choose optimal weighting matrices for the parallel multisplitting two-stage iterative algorithms, we will discuss this problem in the paper.

Here, we still use the scalar weighting matrices

$$E_i^{(k)} = \alpha_i^{(k)} I, \ i = 1, 2, \cdots, m, \ k = 1, 2, \cdots . \tag{6}$$

in the parallel multisplitting two-stage iterative method, but $\alpha_i^{(k)}$ $(i = 1, 2, \cdots, m, \ k = 1, 2, \cdots)$ are chosen by finding the optimal point in the hyperplane H_k, where

$$H_k = \left\{ x | x = \sum_{i=1}^{m} \alpha_i^{(k)} x_i^{(k)}, \ \sum_{i=1}^{m} \alpha_i^{(k)} = 1 \right\}, \ k = 1, 2, \cdots . \tag{7}$$

Thus, $\alpha_i^{(k)} (i = 1, 2, \cdots, m, \ k = 1, 2, \cdots)$ are the optimal parameters in k-th iteration. In other words, the point $x^{(k)} = \sum_{i=1}^{m} \alpha_i^{(k)} x_i^{(k)}$ generated by the optimal weighting matrices (6) may be the optimal point to the solution of linear systems (1) in H_k. Thus, we search the optimal weighting matrices without nonnegative condition. In fact, numerical examples (will be seen in section 4) show that the methods with the weighting matrices (6) are effective.

The paper is organized as follows. In Section 1, we give some notations and preliminaries. In Section 2, the non-stationary parallel multisplitting two-stage iterative methods with self-adaptive weighting schemes are put forward. In Section 3, the convergence of the new method is established. We provide numerical results in Section 4.

Here are some essential notations and preliminaries. $R^{n \times n}$ is used to denote the $n \times n$ real matrix set, the matrix A^T denotes the transpose of A. Similarly the transpose of a vector x is denoted by x^T. A matrix $A \in R^{n \times n}$ is called symmetric positive definite(or semidefinite), if it is symmetric and for all $x \in R^n, x \neq 0$, it holds that $x^T A x > 0$(or $x^T A x \geq 0$). $A = M - N$ is called a splitting of the matrix A if $M \in R^{n \times n}$ is nonsingular; this splitting is called a convergent splitting

if $\rho(M^{-1}N) < 1$; a P-regular splitting of the symmetric positive definite matrix A if $M^T + N$ is positive definite, a symmetric positive definite splitting if N is symmetric positive semi-definite (see [6, 16]).

2. Algorithms

In this section, we give the non-stationary parallel multisplitting two-stage iterative methods with self-adaptive weighting schemes.

Let

$$E_i^{(k)} = \alpha_i^{(k)} I, \ i = 1, 2, \cdots, m, \ \sum_{i=1}^{m} \alpha_i^{(k)} = 1, \ k = 1, 2, \cdots .$$
$$\tag{8}$$

It is denoted $\alpha^{(k)} = (\alpha_1^{(k)}, \alpha_2^{(k)}, \cdots, \alpha_m^{(k)})^T$.

Algorithm 1. (SMTS) The non-stationary parallel multisplitting two-stage iterative methods with self-adaptive weighting schemes

Step 0. Given the precision $\epsilon > 0$, the initial point $x^{(0)}$ and set $k := 0$; For $k = 0, 1, \cdots$, until convergence.

Step 1. For all processors

$$x_i^{(k,0)} = x^{(k)} \ ,$$

Step 2. For processor i, for $l = 0, 1, \cdots, q(i, k) - 1$

$$M_i x_i^{(k,l+1)} = N_i x^{(k,l)} + C_i x^{(k)} + b, \ i = 1, 2, \cdots, m \ . \tag{9}$$

Step 3. Computing $\alpha_i^{(k)} (i = 1, 2, \cdots, m)$ by the following quadratic programming models.

(a) Let $x = \sum_{i=1}^{m} \alpha_i x_i^{(k,q(i,k))}$,

$$\min_{\alpha} \frac{1}{2} x^T A x - x^T b$$
$$s.t. \sum_{i=1}^{m} \alpha_i = 1 \ . \tag{10}$$

(b) Let $r_i^{(k,q(i,k))} = A x_i^{(k,q(i,k))} - b, \ r = \sum_{i=1}^{m} \alpha_i r_i^{(k,q(i,k))},$

$$\min_{\alpha} r^T r$$
$$s.t. \sum_{i=1}^{m} \alpha_i = 1 \ . \tag{11}$$

Step 4.

$$x^{(k+1)} = \sum_{i=1}^{m} \alpha_i^{(k)} x_i^{(k,q(i,k))} \ . \tag{12}$$

Step 5. If $\|A x^{(k+1)} - b\| < \varepsilon$, stop; Otherwise, set $k := k+1$; Go to Step 1.

By introducing matrices

$$G(i,k) = \sum_{l=0}^{q(i,k)-1} (M_i^{-1} N_i)^l M_i^{-1} , \qquad (13)$$

$$H(i,k) = (M_i^{-1} N_i)^{q(i,k)} + \sum_{l=0}^{q(i,k)-1} (M_i^{-1} N_i)^l M_i^{-1} C_i . \quad (14)$$

We can rewrite the **SMTS** as the following iteration

$$x^{(k+1)} = \sum_{i=1}^{m} E_i^{(k)} (H(i,k) x^{(k)} + G(i,k) b) = H(k) x^{(k)} + G(k) b , \qquad (15)$$

where

$$H(k) = \sum_{i=1}^{m} E_i^{(k)} H(i,k), \quad G(k) = \sum_{i=1}^{m} E_i^{(k)} G(i,k) . \quad (16)$$

It follows from straightforward derivation that

$$H(i,k) = I - G(i,k) A, \quad i = 1, 2, \cdots, m, \quad k = 0, 1, \ldots , \qquad (17)$$

and the iteration matrix

$$H(k) = I - G(k) A, \quad k = 0, 1, 2, \cdots . \qquad (18)$$

For the quadratic programming, we have following results (see [9]).

Let

$$X(k) = (x_1^{(k,q(1,k))}, \cdots, x_m^{(k,q(m,k))}),$$

$$\alpha = (\alpha_1, \cdots, \alpha_m)^T, \quad e = (1, \cdots, 1)^T .$$

Theorem 2.0.1. Let $\{x_1^{(k,q(1,k))}, \cdots, x_m^{(k,q(m,k))}\}$ be linear independent, the solution of the quadratic programming (10) is as following

$$\alpha = (X(k)^T A X(k))^{-1} (X(k)^T b + \mu e) , \qquad (19)$$

where $\mu = \frac{1 - e^T (X(k)^T A X(k))^{-1} X(k)^T b}{e^T (X(k)^T A X(k))^{-1} e}$.

Theorem 2.0.2. Let $\{r_1^{(k,q(1,k))}, \cdots, r_m^{(k,q(m,k))}\}$ be linear independent, the solution of the quadratic programming (11) is as following

$$\alpha = (R(k)^T R(k))^{-1} (R(k)^T b + \mu e) \qquad (20)$$

where $\mu = \frac{1 - e^T (R(k)^T R(k))^{-1} R(k)^T b}{e^T (R(k)^T R(k))^{-1} e}$.

3. Convergence Analysis

In this section, we study the convergence theories for algorithm 1 with self-adaptive weighting matrices.

Lemma 3.0.3. [11] Assume that A is a symmetric positive definite matrix, let $A = M - N$ be P-regular splitting. Then there exists a positive number r such that

$$\|A^{\frac{1}{2}} (M^{-1} N) A^{-\frac{1}{2}}\|_2 \leq r < 1 . \qquad (21)$$

Lemma 3.0.4. [8] Assume that A is a symmetric positive definite matrix, let $A = F - G$ is a P-regular splitting. Given $m \geq 1$, there exists a unique splitting $A = P - Q$ such that $(F^{-1} G)^m = P^{-1} Q$ and $A = P - Q$ is also a P-regular splitting.

Lemma 3.0.5. Assume that A is a symmetric positive definite matrix. Let $A = B_i - C_i, i = 1, 2, \cdots, m$ be symmetric positive definite splittings, and $B_i = M_i - N_i$ be P-regular splittings. If there exists a positive integer q such that the non-stationary iteration number

$$q(i,k) \leq q, \quad k = 1, 2, \cdots .$$

Then there exists a positive number r such that

$$\|A^{\frac{1}{2}} H(i,k) A^{-\frac{1}{2}}\|_2 \leq r < 1, \quad i = 1, 2, \cdots, m, \quad k = 1, 2, \cdots . \qquad (22)$$

Proof. We compute $G(i,k)$ directly

$$\begin{aligned} G(i,k) &= \sum_{l=0}^{q(i,k)-1} (M_i^{-1} N_i)^l M_i^{-1} \qquad (23) \\ &= (I - (M_i^{-1} N_i)^{q(i,k)})(I - M_i^{-1} N_i)^{-1} M_i^{-1} \\ &= (I - (M_i^{-1} N_i)^{q(i,k)}) B_i^{-1} . \qquad (24) \end{aligned}$$

From Lemma 3.0.4, there exists a unique P-regular splitting

$$B_i = P_i(k) - Q_i(k), \quad i = 1, 2, \cdots, m, \quad k = 1, 2, \cdots,$$

such that $P_i^{-1}(k) Q_i(k) = (M_i^{-1} N_i)^{q(i,k)}$. Hence, it is derived that

$$G(i,k) = (I - P_i^{-1}(k) Q_i(k)) B_i^{-1} = P_i^{-1}(k),$$
$$i = 1, 2, \cdots, m, \quad k = 1, 2, \cdots .$$

and thereby,

$$\begin{aligned} H(i,k) &= I - P_i^{-1}(k) A = P_i^{-1}(k)(P_i(k) - A) \\ &= P_i^{-1}(k)(B_i + Q_i(k) - (B_i - C_i)) \\ &= P_i^{-1}(k)(Q_i(k) + C_i), \qquad (25) \\ &\quad i = 1, 2, \cdots, m, \quad k = 1, 2, \cdots . \end{aligned}$$

From the assumptions of Lemma 3.0.5, the splitting

$$A = P_i(k) - (Q_i(k) + C_i), \quad i = 1, 2, \cdots, m, \quad k = 1, 2, \cdots , \quad (26)$$

are P-regular splittings. Thus, there exist the positive numbers $r(i,k), i = 1, 2, \cdots, m, k = 1, 2, \cdots$, such that

$$\|A^{\frac{1}{2}} H(i,k) A^{-\frac{1}{2}}\|_2 \leq r(i,k) < 1,$$

$$i = 1, 2, \cdots, m, \ k = 1, 2, \cdots .$$

Because of the $q(i, k) \leq q$, $q(i, k) = 1, 2, \cdots, q$ has q different values. Thus, the splittings (26) have at most q different splittings, so are the positive numbers $r(i, k), i = 1, 2, \cdots, m, \ k = 1, 2, \cdots$. Hence, there exists a positive number r such that (22) holds. $\square$

Theorem 3.0.6. Assume that A is a symmetric positive definite matrix. Let $A = B_i - C_i, i = 1, 2, \cdots, m$ be symmetric positive definite splitting, and $B_i = M_i - N_i$ be P-regular splittings. Suppose that weighting matrices $E_i^{(k)} = \alpha_i^{(k)} I, \ k = 1, 2, \cdots$ are given by (10). If there exists a positive integer q such that the non-stationary iteration number $q(i, k) \leq q$. Then $\{x^{(k)}\}$ generated by algorithm 1 converges to the unique solution of the linear system of equations (1).

Proof. Let x^* be the unique solution of linear system of equations (1), and let $\varepsilon^{(k)} = x^{(k)} - x^*, \ k = 1, 2, \cdots$. From the algorithm 1, we have

$$\varepsilon^{(k+1)} = H(k) \varepsilon^{(k)}, \ k = 1, 2, \cdots , \qquad (27)$$

where

$$H(k) = \sum_{i=1}^{m} \alpha_i^{(k)} \left((M_i^{-1} N_i)^{q(i,k)} + \sum_{l=0}^{q(i,k)-1} (M_i^{-1} N_i)^l M_i^{-1} C_i \right),$$

$$\qquad (28)$$

$$k = 1, 2, \cdots .$$

On the other hand, model (10) is equivalent to the following quadratic programming model,

$$\min_{\alpha} \frac{1}{2} (x - x^*)^T A (x - x^*)$$

$$s.t. \sum_{i=1}^{m} \alpha_i = 1. \qquad (29)$$

From (29), we have

$$\varepsilon^{(k+1)^T} A \varepsilon^{(k+1)} \ \leq \ \tilde{\varepsilon}_i^{(k+1)^T} A \tilde{\varepsilon}_i^{(k+1)}, \qquad (30)$$

$$i = 1, 2, \cdots, m, \ k = 1, 2, \cdots ,$$

where

$$\tilde{\varepsilon}_i^{(k+1)} = H(i, k) \varepsilon^{(k)}, \qquad (31)$$

$$i = 1, 2, \cdots, m, \ k = 1, 2, \cdots .$$

(27) and (28) combine (30) and (31), for $k = 1, 2, \cdots$, it holds that

$$\begin{aligned}
\|A^{\frac{1}{2}} \varepsilon^{(k+1)}\|_2 &= \|A^{\frac{1}{2}} H(k) \varepsilon^{(k)}\|_2 \leq \|A^{\frac{1}{2}} H(i, k) \varepsilon^{(k)}\|_2 \\
&= \|A^{\frac{1}{2}} H(i, k) A^{-\frac{1}{2}} A^{\frac{1}{2}} \varepsilon^{(k)}\|_2 \\
&\leq \|A^{\frac{1}{2}} H(i, k) A^{-\frac{1}{2}}\|_2 \|A^{\frac{1}{2}} \varepsilon^{(k)}\|_2 \\
&\leq \cdots \\
&\leq \Pi_{k=0}^{\infty} \|A^{\frac{1}{2}} (H(i, k)) A^{-\frac{1}{2}}\|_2 \|A^{\frac{1}{2}} \varepsilon^{(0)}\|_2,
\end{aligned}$$

$$i = 1, 2, \cdots, m .$$

From Lemma 3.0.5, we have

$$\|A^{\frac{1}{2}} H(i, k) A^{-\frac{1}{2}}\|_2 \leq r < 1, \ i = 1, 2, \cdots, m .$$

Thus,

$$\lim_{k \to \infty} \varepsilon^{(k+1)^T} A \varepsilon^{(k+1)} = 0 ,$$

which is equivalent to $\lim_{k \to \infty} \varepsilon^{(k+1)} = 0$. $\square$

Lemma 3.0.7. Assume that A is a nonsingular matrix, let $A = M - N$ be a convergent splitting. If the matrix $A^T M + M^T A - A^T A$ is symmetric positive definite, then

$$\|(A^T A)^{\frac{1}{2}} (M^{-1} N)((A^T A)^{-\frac{1}{2}})\|_2 < 1 .$$

Proof. At first, the matrix $A^T A - (M^{-1} N)^T A^T A (M^{-1} N)$ follows from direct operation that

$$\begin{aligned}
A^T A \ - \ & (M^{-1} N)^T A^T A (M^{-1} N) \\
= \ & A^T A - (I - A^T M^{-T}) A^T A (I - M^{-1} A) \\
= \ & A^T M^{-T} A^T A + A^T A M^{-1} A - A^T M^{-T} A^T A M^{-1} A \\
= \ & A^T M^{-T} (A^T M + M^T A - A^T A) M^{-1} A .
\end{aligned}$$

Hence, the matrix $A^T A - (M^{-1} N)^T A^T A M^{-1} N$ is symmetric positive definite if and only if the matrix $A^T M + M^T A - A^T A$ is symmetric positive definite. On the other hand, the matrix $A^T A - (M^{-1} N)^T A^T A M^{-1} N$ is symmetric positive definite if and only if $\|(A^T A)^{\frac{1}{2}} (M^{-1} N)(A^T A)^{-\frac{1}{2}}\|_2 < 1$. $\square$

Lemma 3.0.8. Assume that A is a nonsingular matrix. Let $A = B_i - C_i, i = 1, 2, \cdots, m$ be convergent splittings, and let $B_i = M_i - N_i, \ i = 1, 2, \cdots, m$ be also convergent splittings. Suppose the induced splitting

$$B_i = P_i(k) - Q_i(k), \ i = 1, 2, \cdots, m, \ k = 1, 2, \cdots ,$$

such that

$$P_i^{-1}(k) Q_i(k) = (M_i^{-1} N_i)^{q(i,k)}, i = 1, 2, \cdots, m, \ k = 1, 2, \cdots ,$$

and

$$A^T P_i(k) + P_i(k)^T A - A^T A, i = 1, 2, \cdots, m, \ k = 1, 2, \cdots .$$

are symmetric positive definite. If there exists a positive integer q such that the non-stationary iteration number $q(i, k) \leq q, \ i = 1, 2, \cdots, m, \ k = 1, 2, \cdots .$ Then

$$\|(A^T A)^{\frac{1}{2}} H(i, k)((A^T A)^{-\frac{1}{2}})\|_2 < r(i, k) \leq r < 1, \qquad (32)$$

$$i = 1, 2, \cdots, m, \ k = 1, 2, \cdots ,$$

Proof. We apply Lemma 3.0.7 to the splitting

$$A = P_i(k) - (Q_i(k) + C_i), \ i = 1, 2, \cdots, m, \ k = 1, 2, \cdots ,$$

the (32) is obtained directly. $\square$

Theorem 3.0.9. Assume that A is a nonsingular matrix. Let $A = B_i - C_i$, $i = 1, 2, \cdots, m$ be convergent splittings, and let $B_i = M_i - N_i$, $i = 1, 2, \cdots, m$, $k = 1, 2, \cdots$ be also convergent splittings. Suppose that weighting matrices $E_i^{(k)} = \alpha_i^{(k)} I$, $i = 1, 2, \cdots, m$, $k = 1, 2, \cdots$ are given by (11). If the induced splitting $B_i = P_i(k) - Q_i(k)$, $i = 1, 2, \cdots, m$, $k = 1, 2, \cdots$ such that

$$P_i^{-1}(k)Q_i(k) = (M_i^{-1}N_i)^{(q(i,k))}, i = 1, 2, \cdots, m, k = 1, 2, \cdots,$$

and

$$A^T P_i(k) + P_i(k)^T A - A^T A, i = 1, 2, \cdots, m, k = 1, 2, \cdots,$$

are symmetric positive definite, then $\{x^{(k)}\}$ generated by algorithm 1 converges to the unique solution of the linear system of equations (1).

Proof. The model (11) is equivalent to the following quadratic programming model

$$\min_{\alpha}(x - x^*)^T A^T A(x - x^*)$$

$$s.t. \sum_{i=1}^{m} \alpha_i = 1 . \tag{33}$$

Thus, similar to Theorem 3.0.6, for $i = 1, 2, \cdots, m$, $k = 1, 2, \cdots$, it is derived that

$$\|(A^T A)^{\frac{1}{2}} \varepsilon^{(k+1)}\|_2 = \|(A^T A)^{\frac{1}{2}} H(k)\varepsilon^{(k)}\|_2$$

$$\leq \|(A^T A)^{\frac{1}{2}} H(i, k)\varepsilon^{(k)}\|_2$$

$$= \|(A^T A)^{\frac{1}{2}} H(i, k)(A^T A)^{-\frac{1}{2}}(A^T A)^{\frac{1}{2}} \varepsilon^{(k)}\|_2$$

$$\leq \|(A^T A)^{\frac{1}{2}} H(i, k)(A^T A)^{-\frac{1}{2}}\|_2 \|(A^T A)^{\frac{1}{2}} \varepsilon^{(k)}\|_2$$

$$\leq \cdots$$

$$\leq \Pi_{k=0}^{\infty} \|(A^T A)^{\frac{1}{2}} H(i, k)(A^T A)^{-\frac{1}{2}}\|_2 \|(A^T A)^{\frac{1}{2}} \varepsilon^{(0)}\|_2 ,$$

$$i = 1, 2, \cdots, m.$$

From Lemma 3.0.8 we have

$$\|(A^T A)^{\frac{1}{2}} H(i, k)(A^T A)^{-\frac{1}{2}}\|_2 \leq r(i, k) \leq r < 1$$

$$i = 1, 2, \cdots, m, k = 1, 2, \cdots .$$

Thus,

$$\lim_{k \to \infty} \varepsilon^{(k+1)^T} (A^T A)\varepsilon^{(k+1)} = 0 .$$

so is the sequence $\{\varepsilon^{(k)}\}$. Hence, we have proved this theorem. □

Remark 3.0.10. The choice the optimization model of weighting matrices in k-th iteration can be various. Here, we only consider two schemes of optimizing weighting matrices for a linear system. In order to obtain self-adaptive weighting matrices, we need to solve the quadratic programming, but it may decrease the iterations largely because of the inequality implied in Theorem 3.0.6 and Theorem 3.0.9. Furthermore, we can parallel compute α as (19) and (20).

4. Numerical Experiments

In this section, we give some preliminary computational results. We implement our Algorithm 1 with three splittings (Gauss-Seidel splitting, Relaxation splitting and upper Gauss-Seidel splitting) to solve the linear system (1).

The test PDE problem we are considering in this paper is

$$-\Delta u \equiv -\left(\frac{\partial^2 u}{\partial x^2} + \frac{\partial^2 u}{\partial y^2}\right) = f(x, y) \tag{34}$$

with $(x, y) \in \Omega$, where $\Omega = (0, 1) \times (0, 1)$ is a square region. In all cases, the initial vector $x^{(0)}$ is set to zero and the stopping criterion for Algorithm 1 is

$$\frac{\|b - Ax^{(k)}\|_2}{\|b\|_2} \leq 10^{-6} .$$

where $\| \cdot \|_2$ refers to L_2-norm. In the following Tables, IT stands for the number of iterations satisfying the stopping criterion mentioned above, CPU stands for the parallel execution time of Algorithm 1. All timing results are reported in seconds. For the test problems, only the matrix A, which is constructed from finite difference discretization of the given PDE (34), is of importance, so the right-hand side vector b is created artificially. Hence, the right-hand side function $f(x, y)$ in Examples 1 and 2 is not relevant.

Example 1 This example considers equation $Ax = b$ obtained from nine-point finite difference discretization of the given PDE (34). So the coefficient matrix

$$A = \begin{pmatrix} D_p & G_p & & & \\ G_p & D_p & G_p & & \\ & \ddots & \ddots & \ddots & \\ & & G_p & D_p & G_p \\ & & & G_p & D_p \end{pmatrix}_{q \times q},$$

where

$$D_p = \begin{pmatrix} 20 & -4 & & & \\ -4 & 20 & -4 & & \\ & \ddots & \ddots & \ddots & \\ & & -4 & 20 & -4 \\ & & & -4 & 20 \end{pmatrix}_{p \times p},$$

$$G_p = \begin{pmatrix} -4 & -1 & & & \\ -1 & -4 & -1 & & \\ & \ddots & \ddots & \ddots & \\ & & -1 & -4 & -1 \\ & & & -1 & -4 \end{pmatrix}.$$

and the right-hand side vector b is chosen so that $b = (1, 2, 3, \cdots, n)^T$.

Example 2 This example considers equation $Ax = b$ from five-point finite difference discretization of the given PDE (34). So the matrix A is constructed as in Example 1, but D_p and G_p are different from Example 1,

that is $D_p = \begin{pmatrix} 4 & -1 & & & \\ -1 & 4 & -1 & & \\ & \ddots & \ddots & \ddots & \\ & & -1 & 4 & -1 \\ & & & -1 & 4 \end{pmatrix}_{p \times p}$ and $G_p = -I$,

and the right-hand side vector is chosen so that $b = (1, 1, \cdots, 1)^T$.

In all our numerical experiments, three splittings of the matrix A are proposed as following. Let

$$A = B_i - C_i, \quad i = 1, 2, 3$$

with $B_i = \begin{pmatrix} D_{ip} & G_{ip} & & & \\ G_{ip} & D_{ip} & G_{ip} & & \\ & \ddots & \ddots & \ddots & \\ & & G_{ip} & D_{ip} & G_{ip} \\ & & & G_{ip} & D_{ip} \end{pmatrix}.$$

Especially in Examples 1, we chose

$$D_{1p} = \begin{pmatrix} 24 & -4 & & & \\ -4 & 24 & -4 & & \\ & \ddots & \ddots & \ddots & \\ & & -4 & 24 & -4 \\ & & & -4 & 24 \end{pmatrix},$$

$$G_{1p} = \begin{pmatrix} -2 & -1 & & & \\ -1 & -2 & -1 & & \\ & \ddots & \ddots & \ddots & \\ & & -1 & -2 & -1 \\ & & & -1 & -2 \end{pmatrix},$$

$$D_{2p} = \begin{pmatrix} 22 & -4 & & & \\ -4 & 22 & -4 & & \\ & \ddots & \ddots & \ddots & \\ & & -4 & 22 & -4 \\ & & & -4 & 22 \end{pmatrix},$$

$$G_{2p} = \begin{pmatrix} -3 & -1 & & & \\ -1 & -3 & -1 & & \\ & \ddots & \ddots & \ddots & \\ & & -1 & -3 & -1 \\ & & & -1 & -3 \end{pmatrix},$$

$$D_{3p} = \begin{pmatrix} 26 & -3 & & & \\ -3 & 26 & -3 & & \\ & \ddots & \ddots & \ddots & \\ & & -3 & 26 & -3 \\ & & & -3 & 26 \end{pmatrix},$$

$$G_{3p} = \begin{pmatrix} -4 & & & \\ & -4 & & \\ & & \ddots & \\ & & & -4 \end{pmatrix}.$$

and in Examples 2, we chose

$$D_{1p} = \begin{pmatrix} 10 & -1 & & & \\ -1 & 10 & -1 & & \\ & \ddots & \ddots & \ddots & \\ & & -1 & 10 & -1 \\ & & & -1 & 10 \end{pmatrix},$$

$$G_{1p} = \begin{pmatrix} -3 & & & \\ & -3 & & \\ & & \ddots & \\ & & & -3 \\ & & & & -3 \end{pmatrix},$$

$$D_{2p} = \begin{pmatrix} 8 & -2 & & & \\ -2 & 8 & -2 & & \\ & \ddots & \ddots & \ddots & \\ & & -2 & 8 & -2 \\ & & & -2 & 8 \end{pmatrix},$$

$$G_{2p} = \begin{pmatrix} -2 & & & \\ & -2 & & \\ & & \ddots & \ddots & \ddots \\ & & & -2 & \\ & & & & -2 \end{pmatrix},$$

$$D_{3p} = \begin{pmatrix} 12 & -2 & & & \\ -2 & 12 & -2 & & \\ & \ddots & \ddots & \ddots & \\ & & -2 & 12 & -2 \\ & & & -2 & 12 \end{pmatrix},$$

$$G_{3p} = \begin{pmatrix} -2 & -1 & & & \\ -1 & -2 & -1 & & \\ & \ddots & \ddots & \ddots & \\ & & -1 & -2 & -1 \\ & & & -1 & -2 \end{pmatrix}.$$

Let

$$B_i = D_i - L_i - L_i^T, i = 1, 2, 3 , \tag{35}$$

where $D_i = diag(D_{i,p}, \cdots, D_{i,p}), i = 1, 2, 3.$ and corresponding to the D_i block, L_i is strictly block lower triangular matrix. M_i and N_i of Algorithm 1 are determined by the following three splitting methods.

The Gauss-Seidel splitting method

$$M_1 = D_1 - L_1, \quad N_1 = L_1^T ; \tag{36}$$

Table 1. Comparison of computational results for Example 1

p		SMTS with (11)	SMTS with (10)	old Alg with (i)	old Alg with (ii)	old Alg with (iii)
20	IT	49	53	538	382	284
	CPU	0.541037	0.537481	3.817864	3.772102	2.799893
30	IT	99	111	1152	819	636
	CPU	2.417201	2.775316	27.843393	19.953243	15.499852
40	IT	186	193	2003	1424	1003
	CPU	9.365693	9.571924	100.408128	70.903005	50.194703
50	IT	294	274	3089	2196	1902
	CPU	23.093113	21.421537	258.096561	173.782059	151.376797
60	IT	424	377	4412	3136	2233
	CPU	50.778948	44.560516	531.147234	396.874754	267.580311
70	IT	582	391	5970	4244	3448
	CPU	103.469121	68.018041	1067.300887	748.579123	609.459062
80	IT	769	463	7765	5520	4126
	CPU	308.959919	149.931800	2107.084033	1363.542538	1039.778497

Table 2. Comparison of computational results for Example 2

p		SMTS with (11)	SMTS with (10)	old Alg with (i)	old Alg with (ii)	old Alg with (iii)
20	IT	14	20	162	136	112
	CPU	0.132377	0.194257	1.552457	1.284103	1.080881
40	IT	40	44	529	440	424
	CPU	1.929001	2.143014	25.336636	20.901150	20.130860
60	IT	89	67	1130	940	851
	CPU	10.163563	7.775480	129.333241	107.352299	105.052453
80	IT	161	110	1967	1637	1671
	CPU	40.565789	26.947623	496.462925	389.863508	398.499981
100	IT	251	175	3039	2531	2469
	CPU	105.032853	73.645276	1177.090124	1046.563264	1021.525050
120	IT	363	244	4346	3621	3467
	CPU	224.098729	151.474390	2630.780502	2307.731365	2101.069915

The SOR splitting method

$$M_2 = \frac{1}{\omega}(D_2 - \omega L_2), \quad N_2 = \frac{1}{\omega}((1 - \omega)D_2 + \omega L_2^T) \ ; \ (37)$$

The upper Gauss-Seidel splitting method

$$M_3 = D_3 - L_3^T, \quad N_3 = L_3 \ . \tag{38}$$

In addition, the weighting matrices $E_i^{(k)} = \alpha_i^{(k)} I$, $i = 1, 2, 3$, $k = 1, 2, \cdots$.

In order to compare old algorithm with the fixed weighting matrices, we propose the fixed weighting matrices as following,

(i) $E_i = \alpha_i I$, $i = 1, 2, 3$, with $\alpha_1 = 0.2$, $\alpha_2 = 0.2$ $\alpha_3 = 0.6$;

(ii) $E_i = \alpha_i I$, $i = 1, 2, 3$, with $\alpha_1 = 0.4$, $\alpha_2 = 0.3$, $\alpha_3 = 0.3$;

(iii) $E_1 = diag(\alpha_1 I_p, \alpha_2 I_p, \cdots, \alpha_q I_p)$,
$E_2 = diag(\beta_1 I_p, \beta_2 I_p, \cdots, \beta_q I_p)$,
$E_3 = diag(\gamma_1 I_p, \gamma_2 I_p, \cdots, \gamma_q I_p)$,
where α_i and $\beta_i (i = 1, 2 \cdots q)$ are generated **randomly** in $(0,1)$, and $\gamma_i = 1 - \alpha_i - \beta_i$.

In all our tests we take $p = q$, $\omega = 1.5$, $q(i, k) = 5$. Numerical results for Example 1 and Example 2 are listed in Tables 1 and Tables 1, respectively.

In Example 2, the coefficient matrix A itself contains more zero entries than the matrix of Example 1. So we choose larger p. From Table 1 and Table 2 we see that the iteration counts and the CPU times of SMTS with (11) grow rapidly than SMTS with (10) with problem size, but they are much less than the usual old algorithm with fixed weighting matrices. The reason is that the nonnegativity of weighting

matrices are deleted, the range for finding the optimal weighting matrices is extended. The iteration counts and the CPU times of the old algorithm with (iii) is not stable because of randomly, so we have chosen lesser iteration number than the old algorithms with (i) and (ii). Numerical experiments have been presented showing the effectiveness of the self-adaptive strategy for weighting matrices.

Acknowledgement This work is supported by NSF of China (11071184) and NSF of Shanxi Province (201001006, 2012011015-6).

References

[1] Bai Z. Z. and Sun J. C. and Wang D. R. (1996) A unified framework for the construction of various matrix multisplitting iterative methods for large sparse system of linear equations. *Comput. Math. Appl.* **32**(12):51–76. doi:10.1016/S0898-1221(96)00207-6.

[2] Bai Z. Z. (1997) A class of two-stage iterative methods for systems of weakly nonlinear equations. *Numer. Alg.* **14**(4): 295–319. doi:10.1023/A:1019125332723.

[3] Bai Z. Z. (1997) Parallel multisplitting two-stage iterative methods for large sparse systems of weakly nonlinear equations. *Numer. Alg.* **15**(3–4):(347–372). doi:10.1023/A:1019110324062.

[4] Bai Z. Z. (1998) The convergence of the two-stage iterative method for hermitian positive definite linear systems. *Appl. Math. Lett.* **11**(2): 1–5. doi:10.1016/S0893-9659(98)00001-9.

[5] Bai Z. Z. (1999) Convergence analysis of two-stage multisplitting method. *Calcolo* **36**(63–74).

[6] Bai Z. Z. and Wang C. L. (2003) Convergence theorems for parallel multisplitting two-stage iterative methods for mildly nonlinear systems. *Lin .Alg. Appl.* **362**: 237–250. doi:10.1016/S0024-3795(02)00518-9.

[7] Cao Z. H. and Liu Z. Y. (1998) Symmetric multisplitting of a symmetric positive definite matrix. *Lin .Alg. Appl.* **285**(1): 309–319. doi:10.1016/S0024-3795(98)10151-9.

[8] Castel M. J. and Migallón V. and Penadés J. (1998) Convergence of non-stationary parallel multisplitting methods

for Hermitian positive definite matrices. *Math. Comput.* **67**(221): 209–220. doi:10.1090/S0025-5718(98)00893-X.

[9] FLETCHER R. (2000) Practical Methods of Optimization. *Americ: John Wiley and Sons Inc.* 2th ed.

[10] FROMMER A. and SZYLD D. B. (1992) *H*-splitting and two-stage iterative methods. *Numer. Math.* **63**(1): 345–356. doi:10.1007/BF01385865.

[11] FROMMER A. and SZYLD D. B. (1999) Weighted max norms, splitttngs,and overlapping additive Schwarz iterations. *Numer. Math.* **83**(2): 259–278. doi:10.1007/s002110050449.

[12] GU T. X. and LIU X. P. and SHEN L. J. (2000) Relaxed parallel two-stage multisplitting methods. *Int. J. Computer Math.* **75**(3): 351–363. doi:10.1080/00207160008804990.

[13] MIGALLÓN V. and PENADÉS J. and SZYLD D. B. (2001) Nonstationary multisplittings with general weighting matrices. *SIAM J. Matrix Anal. Appl.* **22**(4): 1089–1094. doi:10.1137/S0895479800367038.

[14] O'LEARY D. P. and WHITE R. E. (1985) Multi-splittings of matrices and parallel solutions of linear systems. *SIAM J. Alg. Disc. Meth.* **6**(4): 630–640. doi:10.1137/0606062.

[15] SZYLD D. B. and JONES M. T. (1992) Two-stage and multisplitting methods for the parallel solution of linear systems. *SIAM J. Matrix Anal. Appl.* **13**(2): 671–679. doi:10.1137/0613042.

[16] VARGA R. S. (2000) Matrix Iterative Analysis. *Germany:Springer Berlin Heidelberg.* 2th ed.

[17] WANG C. L. and BAI Z. Z. (2002) On the convergence of nonstationary multisplitting two-stage iteration methods for Hermitian positive definite linear systems. *J. Comput. Appl. Math.* **138**(2): 287–296. doi:10.1016/S0377-0427(01)00376-4.

[18] WEN R. P. and WANG C. L. and MENG G. Y. (2007) Convergence theorems for block splitting iterative methods for linear systems. *J. Comput.Appl. Math.* **202**(2): 540–547. doi:10.1016/j.cam.2006.03.006.

[19] WHITE R. E. (1989) Multisplitting with different weighting schemes. *SIAM J. Matrix Anal. Appl.* **10**(4): 481–493. doi:10.1137/0610034.

[20] WHITE R. E. (1990) Multisplitting of a symmetric positive definite matrix. *SIAM J. Matrix Anal. Appl.* **11**(1): 69–82. doi:10.1137/0611004

[21] CHEN F. (2010) Asynchronous multisplitting iteration with different weighting schemes. *Applied Mathematics and Computation.* **216**(6): 1771-1776. doi:10.1016/j.amc.2009.12.027.

[22] MENG G. Y. and WANG C. L. Yan X. H. (2012) Self-Adaptive Non-stationary Parallel Multisplitting Two-Stage Iterative Methods for Linear Systems. *In Proceedings of Third International conference ICDKE2012, Wuyishan, Fujian, China* (Germany:Springer Berlin Heidelberg), LNCS 7696, 38–47. doi:10.1007/978-3-642-34679-8-4.

[23] WANG C. L. and MENG G. Y. and YONG X. R. (2011) Modified parallel multisplitting iterative methods for non-Hermitian positive definite systems. *Adv. Comput. Math.* 1–14. doi: 10.1007/s10444-011-9262-8.

Towards Privacy-Preserving Web Metering Via User-Centric Hardware

Fahad Alarfi[*,1], Maribel Fernández[1]

[1]King's College London, Department of Informatics, Strand, London WC2R 2LS, UK.

Abstract

Privacy is a major issue today as more and more users are connecting and participating in the Internet. This paper discusses privacy issues associated with web metering schemes and explores the dilemma of convincing interested parties of the merits of web metering results with sufficient detail, and still preserving users' privacy. We analyse different categories of web metering schemes using an established privacy guideline and show how web metering can conflict with privacy. We propose a web metering scheme utilising user-centric hardware to provide web metering evidence in an enhanced privacy-preserving manner.

Keywords: Web Metering, Privacy-Preserving Technologies, Privacy Protection, Cryptographic Hardware, Smart Cards And Privacy, Network & Distributed Systems Security

1. Introduction

1.1. Web Metering Problem

Consider a service provider, which in the context of this paper will simply be a *webserver*, and a *user*, who is a person using a platform to access the webserver through an open network. The *web metering problem* is the problem of counting the number of visits done by such user to the webserver, additionally capturing data about these visits. Automated visits done by machines are outside the scope of this research partly because the research is mainly motivated by "Online Advertising". A *web metering scheme* produces the number of visits and supporting evidence to interested enquirers. The web metering scheme can be run by an *Audit Agency* or a less trusted third party *Metering Provider*. Figure 1 shows the four entities and their connections.

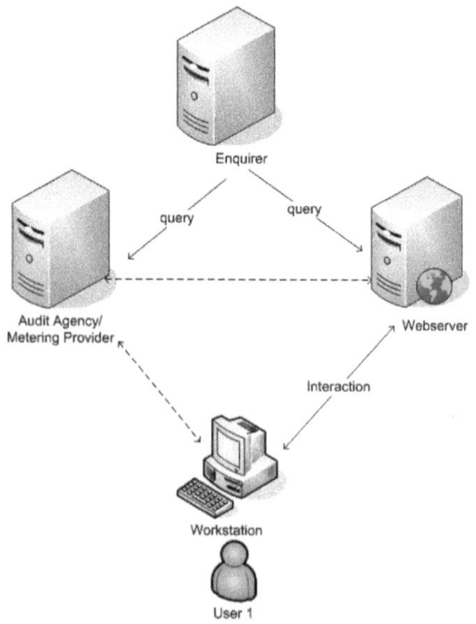

We classify web metering schemes as user-centric, webserver-centric or third-party-centric, depending on the entity controlling the scheme or having a major role in setting up the scheme.

Besides Online Advertising applications, the webserver might want to improve its content organisation to confidently allocate (or prioritise the display of) its content according to the web metering results. We

Figure 1. Web Metering Entities

consider a hostile environment where the adversary is motivated to fake users' visits or can invade users' privacy. The adversary can be a corrupt webserver or an outside attacker.

Privacy is the right of individuals to control or influence what information related to them may be

---

[*]Corresponding author. Email: fahad.alarifi@kcl.ac.uk

collected and stored and by whom; and to whom that information may be disclosed [28]. Another stronger notion is unlinkability. Unlinkability of two or more items of interest requires that these items of interest are no more and no less related after the adversary observation [37]. In the web metering context, unlinkability of two or more user's visits requires that the observer cannot determine if the visits are related or not.

There are trade-offs between designing secure web metering schemes and preserving users' privacy. The schemes become more difficult to design when the main interacting party is not interested to participate and operations need to be carried out transparently. To satisfy such *transparency* property[1] [31], the scheme needs to execute inside or behind another existing action or property in the web interaction so it does not require a new explicit action from the user. Such user non-cooperation or simply disinterest further enforces low cost solutions that can provide web metering results without the user involvement or breach of his privacy rights. Also, determining the qualities of captured visits (e.g. time of each visit or age of the user) can be a requirement for some web metering applications and such granularity of data can help in disputes resolution regarding web metering evidence. However, it is a trade-off; the more non-repudiated information collected about the web interaction, the greater likelihood of invading users' privacy.

Following Dolev-Yao threat model [20], an adversary has also control over the communication channels and can obtain data sent through the channels, and send data to entities impersonating another entity. Consequently, an adversary could impersonate a valid user and could receive replies from the Audit Agency that contain private data about the user. The adversary could also capture (and possibly correlate) private data sent from the user. In addition, a corrupt webserver could store non-private data that could be correlated at a later stage and invade the users' privacy. Also, a corrupt webserver could ask for more information from the user and receive private data.

Despite the desired web metering granularity and the existence of adversary attacks, the concept of using privacy as an economic rationality [21] can be applied in the context of web metering. That is, web metering evidence can be generated by trading services to the user in exchange for information. This approach inherently has a limited scope because it assumes users wish to participate in the web metering scheme and therefore, it has questionable efficiency. However, when such benefits outweigh the risks, users tend to

accept and adjust to metered interactions [24]; such an approach requires a web metering privacy policy for webservers to be able to fairly trade their services. Getting user information could be designed in new products so the activity happens transparently to the user. Furthermore, balancing the users' privacy right with the conflicting [9] webservers' and interested parties' freedom of expression right, complicates this interdisciplinary problem for privacy-preserving web metering schemes. Without closer analysis and specific metrics, service providers could pragmatically argue that information about visits to the webservers can be published as an exercise of their right of freedom of expression and for public interest.

Paper Contribution. The contributions of this paper are as follows. We propose a new web metering scheme that uses a hardware device at the user side to provide web metering evidence in a privacy-preserving manner. To the best of our knowledge, the proposed scheme is the first generic hardware-based user-centric web metering scheme. We show that the proposed scheme has the required security properties and enhances the privacy of users. In addition, we show that, aside the presence of the hardware component, the scheme can be implemented in a way that makes web metering transparent to the user. We also use privacy measurements to analyse and compare different categories of web metering schemes, showing the benefits of the proposed scheme. This paper is an extended version of [1].

1.2. Related Work

User-centric schemes. User-centric web metering schemes can use *digital signatures* and hash chaining to construct non-repudiation evidences of visits as proposed by Harn and Lin [25]. To exempt the user from producing a costly signature for each visit, a hash chain is proposed. That is, the webserver uses the received signature and the hash values as evidence for the number of visits. However, the received signature can be linked to the user's identity, which is a privacy problem.

To avoid the apparent privacy problem with digital signatures, *Secret Sharing schemes* were proposed by Naor and Pinkas [35] and used in many works e.g. by Masucci [5, 6] and others [32, 42]. As evidence of the visits, the webserver here needs to receive a specific number of shares from users to be able to compute a required result using a Secret Sharing scheme e.g. Shamir Secret Sharing [40]. However, the user has to be authenticated (which is another privacy problem) so that the webserver cannot impersonate him and have the required shares. Also, if the Metering Provider is generating and sending the shares, it has

---

[1]www.sites.google.com/site/yuriyarbitman/Home/on-metering-schemes

to be trusted not to collude with the webserver to link user identity with visits after the user-webserver interaction. Similarly, an adversary can observe and correlate user authentication data with the visits. The users' identities have also to be revealed to the Audit Agency to resolve disputes about collected shares by the webserver which can potentially be linked to the visits. With our assumptions, the proposed scheme addresses these issues.

Webserver-centric schemes. A webserver-centric *voucher* scheme uses e-coupons [29] as an attempt to map traditional advertisements models into the electronic ones. Such schemes can be used when a corrupt webserver is motivated to deflate number of visits [30]; however, the user has to be authenticated when forwarding the e-coupon to the issuing party to stop the webserver from forwarding the e-coupons itself. Also, a questionable Metering Provider can potentially use received e-coupons and authentication data and collude with the webserver to link the information to visits. Or an adversary can observe and correlate authentication data and e-coupons with the visits. Improvements have been proposed in [18] to address these issues in environments where the adversary is motivated to deflate number of visits. However, we only consider hit inflation attacks in this paper.

Another webserver-centric *processing-based* scheme was proposed by Chen and Mao [12] which uses computational complexity problems like prime factorisation. These computational problems attempt to force the webserver to use users resources in order to solve them and consequently provide web metering evidence via the produced result. However, besides using users' resources, an adversary can fake users' visits and possibly figure out computing resources at the user side e.g. by analysing the speed of the computations.

The use of a physical web metering *hardware* box attached to the webserver was proposed in [4]. In such webserver-centric scheme, the webserver connects to an audited hardware box which intercepts users requests and stores a log. Randomly, the box also produces a Message Authentication Code (MAC) on a user request which is then redirected to the Audit Agency as an additional verification step. The Audit Agency verifies the MAC code and the request and if valid, the received request is redirected back to the webserver. User impersonation is still a successful attack here in which the webserver can inflate the number of visits. The proposed scheme increases the cost of running a successful impersonation attack so that it is not feasible for a corrupt webserver.

Third-Party-Centric Schemes. A third-party-centric web metering scheme was proposed in [2] which tracks the user using an *HTTP proxy*. The intercepting HTTP proxy adds a JavaScript code to returned HTML pages to track users' actions e.g. mouse movements. Consequently, all visits have to go through the proxy, which does not preserve users' privacy.

Another third-party-centric scheme is *Google Analytics (GA)* [23] which can provide more granular information than the number of visits. During users' visits, referenced web metering code is loaded into the webserver script domain. The code is executed under the webserver control, setting a *webserver-owned* cookie [38] to track returning users to the webserver and not Google-Analytics.com. Then, the code extracts the user's assigned identifier in a cookie (set earlier or updated by the running script) and captures some user's data, all to be sent back to Google-Analytics.com. Despite the privacy improvement of webserver-owned cookie of not figuring out users visiting different webservers incorporating GA script, returning users will still be identified to the webserver and Google-Analytics.com. Besides the cookie issue, the referenced code captures private data about the user, e.g. user's Internet Protocol (IP) address to provide geographic results. In the proposed scheme, we ensure that such private information is preserved.

1.3. Paper Organisation

The remainder of this paper is organised as follows. Section 2 proposes a generic web metering scheme and provides an analysis covering assumptions, goals and practical aspects. Section 3 describes techniques to implement the generic scheme. Section 4 analyses the proposed scheme from security and privacy perspectives. Section 5 provides a proof-of-concept implementation analysis. Section 6 concludes the paper.

2. Web Metering Via User–Centric Hardware

2.1. High Level Description Of Proposed Scheme

Inspired by the webserver-centric hardware-based web metering scheme in [4] and the use of secure user-centric hardware-based broadcasting technique (e.g. pay television) in [14], we propose here a new web metering scheme that relies on a hardware device at the user side.

Definition 1. A **secure device** is an abstraction for an integrated circuit that can securely store a secret value. To access that secret value, a processor is needed which can be inside that hardware device or inside an attached computing platform. The device has to be equipped with a technique (e.g. *zeroization*) so that the secret key cannot be extracted.

The device contains a secured secret key used for authentication, and has the ability to store another signature secret key, inside or outside the device. Examples of such hardware devices are a smart card or an enhanced version e.g. a Trusted Platform Module (TPM) [27]. In addition to TPMs, two factor authentication is an authentication mechanism in which the user uses two different credentials e.g. a password and a hardware token. Banks two factor authentication token[2] is a non-transparent example of a hardware device distributed to the user for a secure webserver visit. The adversary could still *purchase* hardware devices for "fake" users' identities. The cost should typically be higher than the gained benefits, as in [22]. Our generic web metering scheme operates in an environment which consists of a webserver, a user, who owns a hardware device, and an Audit Agency. The three parties follow the protocol specified below. First, we define hardware authentication which will be used as a step in the generic scheme as follows.

Definition 2. Hardware authentication is a unilateral authentication [16] in which the Audit Agency is assured of the claimed communicating user's identity.

The following is a generic protocol for the proposed web metering scheme.

1. **User → Webserver** : Access request
2. **Webserver → User** : Certificate request
3. **User → Audit Agency** : Hardware certificate
4. **User ↔ Audit Agency** : Hardware authentication
5. **User → Audit Agency** : New key
6. **Audit Agency → User** : Certificate for new key
7. **User ↔ Webserver** : Certificate & signature
8. **Webserver ↔ Audit Agency** : Verification key & evidence

The protocol for the proposed web metering scheme consists of eight steps, as follows.

1. User sends an access request to webserver.
2. Webserver replies with a certificate request.
3. User sends the certificate for the secret key, inside the hardware device, to Audit Agency.
4. Audit Agency checks the received hardware device certificate. If the certificate is valid, Audit Agency authenticates the communicating user by checking whether he can access the corresponding hardware device.

[2]www.hsbc.co.uk/1/2/customer-support/online-banking-security/secure-key

5. If authentication succeeds, the user generates a signature key pair and sends public part of it to Audit Agency.
6. Audit Agency signs the received public key and sends a certificate back to user.
7. User signs webserver URL and time using the new signature key and sends the signed URL and certificate received in step 6 (or a form of it) to webserver.
8. Webserver checks the certificate and the signature, and stores them as evidence.

In step 1, the user sends an access request to an object in the webserver. In step 2, the webserver checks whether the user has submitted a valid (attestation) certificate. If not, the webserver requests a certificate signed by the Audit Agency. In step 3, the user checks if he holds a valid certificate. If so, step 7 is instead executed. Otherwise, the user sends to the Audit Agency, the certificate for the secret key embedded in the hardware device. For example, the *hardware certificate* can be a signature by a *hardware authority agency* (e.g. Intel) for a public key, where its corresponding private part is embedded in the hardware device. In step 4, the Audit Agency checks the validity of the received certificate (e.g. not revoked). If the certificate is valid, the Audit Agency checks whether the user holds the corresponding secret key in relation to the certificate. For this step, the user is asked to encrypt fresh nonces using the embedded secret key. In step 5, if the user is authenticated, he generates a new signature key pair and sends the public part of it (verification key) to the Audit Agency. This step can be executed for x number of key pairs. In step 6, the Audit Agency signs the received verification key (blindly if privacy is required) using its signature key and sends the produced signature (requested certificate) to the user. In step 7, the user forwards the received certificate in step 6 to the visited webserver or convinces the webserver that he has obtained a certificate. The user also sends his verification key to the webserver if it is not included in the submitted certificate. The user also signs a webserver identifier (e.g. URL) and possibly other information (e.g. time) and sends the signature to the webserver as evidence of the visit. For efficiency reasons, the webserver could periodically publish reference numbers or pseudonyms that can be linked to the webserver and time instead of concatenating URL and time for the *evidential signature*. In step 8, the webserver checks that the certificate was somehow *signed* using Audit Agency verification key. The webserver also checks (possibly using a privacy-preserving protocol) that the received signature was signed by the user's new signature key. If both checks succeed, the webserver stores the certificate and signature as web metering evidence.

The following is an example of the proposed scheme. Assume some webservers can only be accessed if the users own web metering hardware devices. Once the user accesses a webserver at some point in time, the user gets redirected to an Audit Agency. The Audit Agency first ensures that the user owns a valid hardware device. Then, the user generates and sends a session key and asks for a certificate. The Audit Agency produces the certificate using different possible schemes and sends it to the user. The user gets redirected back to the webserver when he submits the certificate and a new signature message.

2.2. Security And Privacy Assumptions And Goals

We assume that number of corrupt users is small as done in [3]. In particular, the webserver cannot convince significant number of users to collude with it, to create fake web metering evidence. The rationale behind this assumption here is that the number of users captured by web metering evidence should typically be large and unlikely for the webserver to be able to cost-effectively motivate a considerable number of users into colluding. Also, colluding participants have to risk losing the functionality of their hardware devices once tagged as rogue.

User-centric hardware-based web metering schemes have a potential to overcome user impersonation attacks and can be designed to preserve users' privacy. This can be achieved by involving the Audit Agency in the user setup or increasing the cost of webserver faking visits, as followed in the lightweight security approach in [22]. The hardware introduction is used here to increase the cost for a corrupt webserver to fake visits by requiring it to own a hardware device for each fake user. At the same time, the scheme has to ensure that it is impossible for a corrupt webserver with one authentic hardware device to be able to generate an unlimited number of evidences e.g. using a periodic hardware authentication with a limit of issued certificates. Therefore, we need a hardware device at the user side containing a secret key. Also, the secret keys certificates and public cryptographic values have to be available to the Audit Agency as they are required in step 3. In steps 3 and 7, the user is assumed to be securely redirected and may not necessarily be aware of this ongoing web metering operation, if a privacy-preserving scheme is being used in a *transparent* mode.

A summary of the assumptions we followed in this paper are as follows.

1. Number of corrupt users is far less than the total number of metered users.

2. User owns a secure hardware device (as in Definition 1).

3. The Audit Agency can obtain a list of valid devices certificates (e.g. from Intel) and recognise revoked or expired ones. Alternatively, users could be incentivised to register their authentic hardware devices for privacy-preserving browsing.

4. The web metering environment is where the user's privacy is a concern.

5. There is limited value of the online content (affecting the cost for webserver owning hardware devices).

In the rest of this section we further describe attacks that can happen during a hostile web metering operation and then highlight the required security goals to counter such attacks. We derive the following security attacks from the adversary capabilities described in Dolev-Yao threat model [20]: replay, impersonation and man in the middle attacks.

Replay Attack. A replay attack occurs when an adversary captures data sent from the user to the Metering Provider, the Audit Agency or the webserver and sends the data again. Similarly, an adversary captures data sent from the webserver to the Metering Provider or the Audit Agency and sends the data again.

If a replay attack is not detected, the visits number may be increased.

Impersonation Attack. An adversary in an impersonation attack (which is more powerful than the replay attack scenario where attack effect is limited to captured data), creates fake data and sends it to the Metering Provider or the Audit Agency impersonating a valid webserver or user. Or an adversary creates a fake request to a webserver impersonating a valid user.

If an impersonation attack is not detected, the visits number may be increased or the evidence data may have invalid properties.

Man In The Middle Attack. Man in the middle attack occurs when an adversary receives data from the user or the webserver not intended to him and modifies it before forwarding it to the intended party.

If such attacks are not detected, the visits number may be increased or the data have invalid properties.

Besides the three communication attacks, there is also a threat that a corrupt webserver may not follow the required **web metering operations**. A corrupt webserver is inherently motivated to change the number of visits. Also, a corrupt webserver can be motivated to change some metering operations without changing number of visits. For example, a corrupt webserver intentionally changes a webpage identifier, which is going to be recorded in the web metering evidence, to a different webpage that charges higher fees for advertisements.

To counter such attacks, there have to be data integrity of the web metering results and secure web metering operation (we define these concepts below). Data integrity is a property that counters threats to the validity of data [16]. Once this property is satisfied, it provides protection against unauthorised modification or destruction of data. Data Integrity in web metering refers to the integrity of stored and transferred evidences and data, as follows.

Evidential Integrity Goal. Evidential Integrity assures that evidences are kept as they were originally produced and stored. That is, once evidences are generated, they have to maintain their exact state and not change (maintaining evidences state includes intentional and accidental changes). Also, this integrity includes stored data that requires post processing work to produce the final web metering evidence.

Evidential integrity requirement is needed as a countermeasure to the web metering operation and stored web metering result attacks.

Communication Integrity Goal. Communication Integrity is concerned with integrity of the communication channels used for transferring web metering data. Transferred web metering data refers to pieces of data transmitted between users, webservers and Audit Agency that can be used to constitute the web metering evidence. Communicating this data has to be done in a way that if the data is changed en route, the change is going to be detected.

Communication integrity requirement is needed as a countermeasure to man in the middle communication channels attack.

The following security requirement is needed as a countermeasure to communication channels (replay and impersonation) and web metering operation attacks.

Security Goal. A web metering scheme is secure if its web metering operations are executed as expected per its specifications and can not be affected by an adversary, to eventually provide consistent results and evidence.

High Level Analysis Of Proposed Scheme. To ensure that an impersonation attack is not feasible, step 3 has been included as only users who have valid hardware devices will be authenticated (because the key cannot be extracted from the hardware device). As a result, an impersonation attack for imaginary set of users will require an adversary to own a hardware device for each impersonated user, which is not feasible. To ensure that man in the middle attack is not

successful, step 4 has been included as an adversary listening to communications will not be able to satisfy the required authentication. Similarly, an adversary interfering the certificate in step 6 will not posses the corresponding secret part. To ensure that replay attack is not successful, time should be included in the signed messages. We provide a more detailed analysis of security properties of the scheme in section 4.1.

To preserve user's privacy, in step 6, the Audit Agency has to blindly sign the new user's key and send the blind signature (i.e. certificate) to the user. Owing to the blind signature production, the Audit Agency does not know the user's key. In step 7, the user submits a form of the received signature or proves to the webserver that she possesses an Audit Agency signature on the new web metering signature key. The webserver would store the signatures as evidence for number of visits that are done by users carrying authentic hardware devices. We provide a more detailed analysis of privacy properties of the scheme in section 4.2.

2.3. Practical Aspects

The use of hardware devices is common today. Commercial hardware tokens[3] can be used in the proposed scheme as long as they hold a *zeroizable* secret for authentication. Also, a relevant application that uses hardware decoders but not for web metering purposes, is pay television. Here, the user has to have hardware decoders to get multimedia content sent by a broadcasting server. Only authorised users' decoders can decrypt the broadcast content, using the embedded decryption keys. The server encrypts the broadcast content, which will be decrypted using the corresponding decryption key, inside the hardware decoder. The technique can also have other security properties like a tracing capability to detect rogue decoders that share the decryption keys [14].

In case the user is not motivated to explicitly participate in the web metering scheme but still have an applicable hardware device, the scheme can still be run transparently to the user, where a program (or a script[4]) anonymously attests the user. One current application requiring TPMs are digital wallets [13]. In a typical application, the user uses a virtual wallet program on a device to make a transaction. Potential motivations for such a wallet over credit cards could be finding better deals or further authenticating communicating users with customised information set in the wallet. Recent commercial devices (e.g. iPhone 6[5]) use Near-Field Communication (NFC) technology

---

[3]www.safenet-inc.com/uploadedfiles/about_safenet/resource_library/ resource_items/product_briefs_-_edp/safenet_product_brief_ikey_4000.pdf
[4]www.cometway.com
[5]www.apple.com/iphone-6/

for such digital wallet. NFC devices can use TPMs [26]. On the other hand, an organisation might want to restrict accesses to their local network once users have certain hardware devices in a fashion similar to Virtual Private Network (VPN) connections. For example, the distributed hardware devices can provide the required connectivity and privacy-preserving web metering results. On a larger scale another non-transparent scenario could be to distribute free zeroizable devices to users (e.g. USB storage sticks).

There is also a trend of developing extra hardware devices (rather than traditional Personal Computers or mobile phones) for various desirable functions e.g. Google Glass[6]. Along the main functions like cameras or games, accessing certain webservers can be an additional function using a privacy-preserving web metering scheme.

3. Techniques To Implement Proposed Scheme

In this section, we start by describing mechanisms to implement each step in the proposed generic scheme.

Steps 1 and 2 can be implemented using standard mechanisms for issuing requests e.g. HTTP requests.

Steps 3 and 4 address the identification and authentication of the hardware device. As mentioned in section 2.1, a TPM can be used as a web metering hardware device for the required hardware authentication step. A *trusted computing* platform is a device which has an embedded TPM, which has Endorsement Key (EK) and a certificate on the public part of it to prove the platform is genuine. We can follow with such hardware device the lightweight security approach, where it is still possible for an adversary to construct fake web metering evidence but its cost does not offset the earned benefit.

Steps 5, 6 and 7 are included in the proposed scheme to take into account the privacy requirements. Step 8 is optional depending on whether the webserver needs to contact the Audit Agency for certificates or evidence redemption.

In the rest of this section, we describe existing protocols and schemes that can be used to implement steps 5, 6 and 7 in the web metering scheme in section 2.1. Using them, we give a technique to implement the scheme, satisfying both the security and users' privacy requirements. Table 1 outlines the mechanisms, satisfying the different requirements.

Security Without Privacy. To implement step 5, the following secure but *non-privacy-preserving* key transport protocol [7] can be used, where the Audit Agency and the user have public key pairs.

---

[6]www.google.com/glass/

1. **User → Audit Agency** : $ENC_{AuditAgency}[identifier, key, count]$

2. **Audit Agency → User** : $ENC_{key}[count, nonce]$

3. **User → Audit Agency** : $SIG_{User}[AuditAgency, H(nonce, count, key)]$

In the first message, the user encrypts, using a standard encryption scheme, his identifier, a new signature key and its count number using Audit Agency public key and sends it to the Audit Agency. The Audit Agency decrypts the message and encrypts the received count number and a nonce using the new key and sends it to the user. The user decrypts the message to reveal the nonce to hash it with the count number and the new key. Then, the user signs, using a standard signature scheme, the hash code along the Audit Agency identifier with his public key and sends the signature to the Audit Agency. This signed message can be used to link the new signature key to the user's identity. For step 6, once the new signature key is securely sent to Audit Agency, the Audit Agency signs it with its private key and sends the signature to the user as a certificate. For step 7, the user produces evidential signatures using the new signature key and uses the Audit Agency certificate to confirm its validity.

Security And Privacy. By contrast, to provide a privacy-preserving web metering scheme, the user has to commit to a new signature value for step 5 in the generic scheme, for example using Pedersen commitment scheme [36]. For the next step, an Audit Agency has to blindly sign the committed value (once the user is authenticated) and allow the user to prove its possession, without revealing it. For step 7, the user uses the new signature value, without linking it to the former authenticated credential.

A general view of the privacy-preserving technique required in step 5 can be two interacting entities in which one can prove to the other that it holds a secret without revealing it. New secrets can be generated with the help of a trusted third party while the former secret is "buried away" in another value. For example, using *Schnorr* zero-knowledge protocol [39], a secret s can be embedded in a smart card and used for signing such that $y = g^s \bmod p$ where g is a group generator and p and q are two large prime numbers such that q is a divisor of $p-1$ (y, g, p and q are public values). A commitment scheme can be used in constructing a zero-knowledge protocol. In the web metering context, the user can convince the Audit Agency that the interacted messages are correctly formed using zero-knowledge proof of knowledge of a discrete logarithm. The following are the corresponding steps for Schnorr protocol, that can be used for step 5.

Table 1. Mechanisms Comparison

Security Without Privacy	Security And Privacy
Key transport protocol	Zero-knowledge protocol
Audit Agency signature	Audit Agency blind signature
User non-repudiation signature	User evidential signature of knowledge

1. **User → Audit Agency** : User chooses a random r and sends $a = g^r mod$ p

2. **Audit Agency → User** : Audit Agency sends a challenge c

3. **User → Audit Agency** : User sends $b = r + c * s$ mod q

In step 1, the user chooses a random value r where $1 \leq r \leq q - 1$ and sends the commitment $a = g^r mod$ p. In step 2, the Audit Agency sends a challenge c where $1 \leq c \leq 2^t$ for some security parameter t. In step 3, the user sends to Audit Agency $b = r + c * s \, mod$ q. The Audit Agency checks whether $a * y^c = g^b$. If the check is correct, the Audit Agency is convinced that the user knows the secret s. The result of this check can be used as a proof that the user knows s without revealing it. For implementing step 6 in the generic scheme, the Audit Agency has to document the result as a "redeemable" privacy-preserving certificate. Then, for step 7, the zero-knowledge protocol has to run again between the user and the webserver.

In the following sections, we first show a secure technique that can be used to implement the generic scheme (without preserving the users' privacy). We then show a technique that can be used to implement the generic scheme, satisfying both the security and users' privacy requirements.

3.1. Privacy Certification Authority

In this part, we show a secure technique that can be used to implement the generic scheme (without preserving the users' privacy). The first attestation method published by Trusted Computing Group (TCG)[7] was to use Privacy Certification Authority (CA). Privacy CA has the same role as the Audit Agency. The following are seven steps using this attestation method for web metering purposes. In step 1, the user sends an access request to the webserver. In step 2, the webserver sends a request for attestation to the user, to enable the webserver to reply to the user request and reliably record web metering evidence. In step 3, the user submits his hardware certificate (for EK) to Privacy CA when the Privacy CA cross checks it with published certificates. All hardware certificates are initially published by the hardware authority agency

and their status can be updated by the Privacy CA. In step 4, Privacy CA validates the used EK with respect to the submitted hardware certificate. In step 5, if the checks are correct, the user generates Attestation Identity Key (AIK) and sends the public part of it to the Privacy CA. In step 6, Privacy CA signs the received AIK and sends a signed AIK certificate. In step 7, the user uses AIK private key for producing evidential signatures to the webserver and the received AIK certificate to authenticate himself. Figure 2 shows the message flow for the described steps.

The privacy problem with using Privacy CA method is as follows. The webserver would send to Privacy CA (for example, on conflict resolution) the received signatures along the corresponding AIK certificate. Privacy CA can link the self-issued AIK certificate along the non-repudiation signatures to the corresponding used TPM's EK.

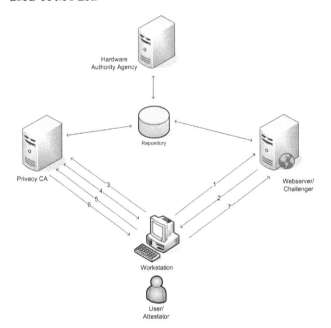

Figure 2. Attestation Using Privacy Certification Authority (CA)

3.2. Direct Anonymous Attestation Protocol

Direct Anonymous Attestation (DAA) protocol [8] can fortunately provide the needed public commitment, signature scheme and zero-knowledge proofs techniques. DAA protocol uses Camenisch-Lysyanskaya signature scheme [10] to provide a blind signature on

---

[7]www.trustedcomputinggroup.org

the committed value and allow the user to prove its possession, through a zero-knowledge proof of knowledge of the committed value. According to DAA protocol described in [8], communication between user and Audit Agency can be done using *Join Protocol* and communication between user and webserver can be done using *Sign/Verify Protocol*. Figure 3 shows the message flow for both protocols. Step *a* corresponds to steps 3 and 4 in the generic scheme. The rest of steps refer to the same order.

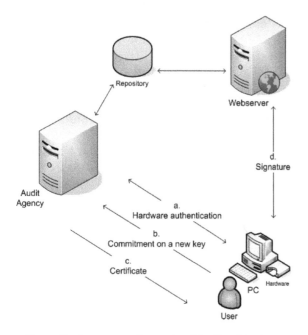

Figure 3. Anonymous Attestation Via User Hardware

The user gets authenticated to Audit Agency using EK (steps 3 and 4 in the generic scheme) and then receives a certificate as follows. In step 5, during Join Protocol, the user generates a secret key f and computes $U = z^f x^{v1} mod\ n$ where $v1$ is used to blind f and (n, x, y, z) is public key of Audit Agency. (z can be set-up as $x^{r2} mod\ n$ where $r2$ is random number so that the Audit Agency chosen random number will be multiplied by the secret f and added to the blind $v1$). Also, the user computes $N = Z^f mod\ p$ where Z is derived from Audit Agency identifier and p is a large prime. Then, the user sends (U, N) to the Audit Agency and convinces the Audit Agency that they are correctly formed using a proof knowledge of a discrete logarithm. We assume that the challenges and messages are securely chosen and constructed as specified in [8]. Then, in step 6 in the generic scheme, the Audit Agency computes $S = (y/(Ux^{v2}))^{1/e} mod\ n$ where $v2$ is random number and e is a random prime. Then, the Audit Agency sends $(S, e, v2)$ to the user to have (S, e, v) as a TPM certificate where $v = v1 + v2$. More than one secret can be generated here to guarantee unlinkability in case the Audit Agency is offline. The join phase is the heavy work phase of

the scheme and can be periodically done for different requirements.

In step 7 in the generic scheme, during Sign/Verify Protocol, the user signs messages using the secret key f and Audit Agency certificate (S, e, v) received in Join Protocol. The user also computes $N2 = Z_2^f mod\ p$ where Z_2 is a group generator that can be configured for a required anonymity level. Z_2 can be fixed for a limited period of time in synchronisation with Audit Agency certificate issuance to determine unique number of users. For example, to determine unique users for a period of one hour, the Audit Agency has to keep a record of hardware authentications so the user cannot generate another key f, and Z_2 has to be fixed, for that period of time. Also, Z_2 can be chosen by the webserver, reflecting its identity e.g. a code for its URL. The b bit can be specified in DAA protocol to indicate that the signed message was chosen by the user. Such feature makes DAA more flexible to different desirable web metering applications than ecash [11]. Furthermore, a signed hash chain as in [25] can be used to efficiently know the length of the visited session and set it to a desired length. Consequently, the user can use the hash chain result as a generator. If the user is still online after the session ends, a new signed message (of a new hash chain) is created. This new message cannot be linked to the previous one as the webserver is just convinced that these messages were signed by user secret keys generated during the Join Protocol without the need to know them (proof of knowledge). The different generators capturing webserver URL and time with different certificates can guarantee (and tune) such required unlinkability and session length.

The user can provide a proof that she has a certificate for the secret values (f and v) by providing a zero-knowledge proof of the secret values, such that the following equation holds: $S^e z^f x^v \equiv y\ mod\ n$. Then, the user sends the signature to the webserver and convinces the webserver that she knows f, S, e and v. The webserver checks the signature and if valid, the webserver stores $N2$ along the result of the zero-knowledge proof as web metering evidence, proving interactively the communicated user's TPM was genuine.

4. Security And Privacy Analysis Of Proposed Scheme

In this section, we analyse and compare web metering schemes, starting with the proposed one in this paper.

4.1. Security Analysis Of Proposed Scheme

Proposition 1. An adversary capturing all communicated messages, but not owning the device, cannot:

1. create fake web metering evidence (i.e. N2, see section 3.2);

2. cannot impersonate an existing user.

Therefore, the proposed scheme achieves integrity and security goals.

Proof. We assume that the user owns a secure hardware device and number of corrupt users is small (as in section 2.2). Thus, hardware authentication (as in Definition 2) can only succeed by interactively proving the ownership of the physical hardware device containing the built-in secret key. Without a successful hardware authentication, valid evidence cannot be created in the absence of the subsequent committed signature key in step 5 (i.e. f). Consequently, the adversary has to own a hardware device in order to create valid web metering evidence. Moreover, we assume that the challenges and messages in steps 5, 6, and 7 are securely chosen and constructed as specified in section 3. Therefore, evidential integrity goal is achieved.

Depending on the Audit Agency setup, x certificates can be issued to the user after the successful hardware authentication, and valid for a limited period and cannot be reused. We assume that user's secret keys are used to encrypt nonces or time stamps, as specified in section 3, to ensure freshness as a countermeasure against impersonation and replay attacks for an observed user. While producing the zero-knowledge proof in proposed DAA-based scheme, the user has to interactively convince the other entity (Audit Agency or webserver) of the knowledge of secret values, using freshly provided challenges. Any captured messages that are resent again during Join Protocol will be rejected by Audit Agency as they will not fit in the current window of acceptable responses. Similarly, captured and resent messages during Sign/Verify Protocol will not enable webserver to construct new valid evidence N2 as they will not fit in the required window. Therefore, security goal is achieved.

Using zero-knowledge proof of a discrete logarithm [39], the adversary will not be able to learn the built-in secret key to pass the required authentication in Join Protocol nor be able to learn the corresponding secret signature key in Sign/Verify Protocol. Therefore, observing messages sent by a user will not enable the adversary to get the secret values to impersonate a valid user or hijack the session. Therefore, communication integrity goal is achieved.

4.2. Privacy Analysis Of Proposed Scheme

Proposition 2. The proposed DAA-based web metering scheme protects any identifying information captured from the authentic certificate of the user's hardware secret key.

Proof. By Definition 2, after hardware authentication, the Audit Agency is assured that the communicating user can securely access the secret key inside the hardware device and consequently can confirm the user's identity. Then, the zero-knowledge protocol [39] is used to convince the Audit Agency that constructed commitment messages were formed correctly without disclosing the secret value f. Once the Audit Agency is convinced, the Audit Agency produces a certificate S for the user's committed secret value f which is later anonymously used during Sign/Verify Protocol. Therefore, the proposed scheme protects any captured user's identifying information.

We assume during Sign/Verify phase, the user keeps the Audit Agency certificate (S, e, v) secret and only uses it to convince the webserver of the knowledge of the chosen secret key f. Otherwise, the Audit Agency can match the hardware certificate identifier to user's visits as follows. The Audit Agency records all issued certificates for the received hardware certificates during the "blind" signature production. Then, once the webserver has the exact Audit Agency certificates along users' signatures, the webserver forwards them to Audit Agency. The Audit Agency can check and match the self-issued certificates to the recorded hardware certificates.

During the Join Protocol, the user computes and sends U which is "embedded" in the Audit Agency certificate S. The user then convinces the webserver the knowledge of S along other secret values without disclosing S. The user also computes and sends N as a provision for recording and possibly revoking the approved commitments, as proposed in [8]. If such *linkability* feature is not required and lifetime of all approved fs is designed to be limited, user's computation and submission of N can be skipped. Similarly, there has to be a non-predictable difference in time or no pattern between user committing to a new signature key and using it. This is initially achieved by the two roles of Audit Agency and webserver when their involvement is separated by time. For example, when the user machine boots up, new keys are generated, approved and stored. For the next immediate visit, the user can either use previously approved signature keys or have to wait a random time before using the new keys. Then, the user is always set to contact the Audit Agency for new signature keys once the number of user's available keys goes below a threshold, say two. The random delay should be minimal as not to affect the user browsing experience. Also, to limit the effect of an impersonation attack, we can assume the scheme needs to re-run daily or every hardware boot up to limit the number of fake evidences.

The proposed scheme does not stop an adversary from capturing other non-required private information about the user (e.g. IP address). A solution for such problem would be to use relevant security countermeasures (e.g. Network Addressing Translation [41] and Onion Network [19]) to prevent the capture of unrelated private data. In the rest of this section, we describe Network Address Translation and Onion Network.

Network Address Translation.

Using Network Address Translation (NAT), a unique address is changed to a different global one when communicated to another network. A NAT operation can be simplified as follows. A user inside a network makes a request to a webserver over the Internet. The user's request is first sent over the local network to the gateway (e.g. a router). The router changes the request machine IP address and source port into the global IP address and a new source port respectively. The router also records the request machine IP address and source port along the new assigned port. Then, the router does other required checks (e.g. integrity) before sending the user's request to the destination IP address and port specified in the request. When the router receives a reply, the router searches its record for the reply destination port address. If there is a match, the router extracts the corresponding user's IP address and port. Then, the router forwards the received reply to the extracted address and port.

Onion Network.

An *Onion Network* [19] is an alternative privacy-enhancing solution to proxies and home NATing devices especially in case the proxy is not trusted or not directly connected to the user (and consequently an adversary can observe the proxy's received requests). In such networks, the route from the user to the webserver is randomly set where each node en route only knows its predecessor and successor. The following are the protocol steps.

1. **User → Node 1** : $ENC_{pubkey} [x1 = g^{s1}]$

2. **Node 1 → User** : $y1 = g^{s2}$, $HASH [K1 = g^{s1*s2}]$

3. **User → Node 1 → Node 2**: $ENC_{K1} [x2 = g^{s3}]$

4. **Node 2 ⇒ Node 1 → User**: $ENC_{K1} [y2 = g^{s4}]$, $HASH [K2 = g^{s3*s4}]$

5. **User → Node 1 → ... → Node x**: $ENC_{K1,K2,.,Kx}$ [Connect To Webserver *identifier*]

6. **Node x → Webserver** : *TCP handshake*

7. **Node x → ... → Node 1 → User**: ENC_{Kx} [Successful Connection]

8. **User → Node 1 → ... → Node x**: $ENC_{K1,K2,.,Kx}$ [Get Webserver Object]

9. **Node x → Webserver** : Get Webserver Object

10. **Node x → ... → Node 1 → User**: ENC_{Kx} [Requested Webserver Object]

The user creates a secret key and negotiates Diffie Helmann parameters [17] with the first random node (Node 1). In step 1, the user sends $x1 = g^{s1}$ to Node 1 encrypted using Node 1 public key. In step 2, Node 1 sends back g^{s2} and a hash of the agreed Diffie Helmann key (g^{s1*s2}). Furthermore, from the agreed key, a key can be derived for each direction. In step 3, the user sends a request to Node 1 to extend the connection to Node 2. The request contains $x2$ for a new secret exponent, encrypted using the symmetric key $K1$. Once Node 1 receives the request, it decrypts it and forwards $x2$ to the next random node (Node 2). In step 4, Node 2 generates a secret ($s4$) and sends $y2$ and a hash of the agreed key ($K2$) to Node 1. Node 1 encrypts the received response using $K1$ and sends the encrypted message to the user. Then, the user calculates the new agreed key ($K2$) and checks the hash. Keys sharing and their forwarding in step 3 and 4 are repeated for further nodes e.g. Node 3. In step 5, the user sends a connect request to Node 1 encrypted using all agreed symmetric keys. In step 6, the end node (Node x) negotiates *TCP handshake* with the intended webserver, without encryption. In step 7, Node x sends a successful connection status message to the previous node in the path i.e. Node x-1, encrypted using the user and Node x agreed symmetric key. Similarly, the status message gets further encrypted down the path to user. Last, the user decrypts the received message using all agreed symmetric keys. In step 8, similar to step 6, the user sends webserver access request to Node 1 encrypted using all agreed symmetric keys. In step 9, Node x sends the received request to the webserver. In step 10, similar to step 7, Node x sends to the user the received webserver reply encrypted using the agreed key. Last, the user decrypts the received reply using using all agreed symmetric keys to reveal the requested webserver object.

The latest version of TOR (The Onion Router) browser should be used with its recommended settings. For example, Java[8] should be disabled so that Java circumventing attacks [34] are not successful. Otherwise an adversary's Java applet could surpass the onion network (or proxy) setup by making a direct connection to the webserver.

The following are two testing User Agents captured by a webserver for UK-based users. The locations of end nodes showed users' locations are instead Amsterdam (Netherlands) and Fremont (California) respectively. The operating system is the generic Windows NT 6.1 instead of MAC and Windows 7. Also, the browser was Mozilla instead of *Safari* and *Internet Explorer*.

% 31.5...141 is the user's IP address

---

[8]www.java.com

from a location in UK using Safari browser on iPad to access the webserver. The following is the user's request to get the webserver homepage.

31.5...141 – – [22/Apr/2014:12:40:33 –0400] "GET / HTTP/1.1" 200 1897 "–" "Mozilla/5.0 (iPad; CPU OS 7_0_4 like Mac OS X) AppleWebKit/537.51.1 (KHTML, like Gecko) Version/7.0 Mobile/11B554a Safari/9537.53"

% The following is the user's request to get the webserver homepage using a TOR browser. 77.2...162 is the IP address of the end node in Amsterdam. The end node's browser is instead Mozilla and operating system is Windows NT 6.1.

77.2...162 – – [22/Apr/2014:12:41:17 –0400] "GET / HTTP/1.1" 200 1897 "–" "Mozilla/5.0 (Windows NT 6.1; rv:24.0) Gecko/20100101 Firefox/24.0"

% 137.7...8 is the user's IP address from a location in UK using Internet Explorer browser (MSIE 9) on Windows 7 operating system. The following is the user's request to get the webserver homepage.

137.7...8 – – [24/Apr/2014:10:21:57 –0400] "GET / HTTP/1.0" 200 1897 "–" "Mozilla/5.0 (compatible; MSIE 9.0; Windows NT 6.1; Win64; x64; Trident/5.0)"

% The following is the user's request to get the webserver homepage using a TOR browser. 216.2...12 is the IP address of the end node in Fremont. The end node's browser is instead Mozilla and operating system is Windows NT 6.1.

216.2...12 – – [24/Apr/2014:10:39:11 –0400] "GET / HTTP/1.1" 200 1897 "–" "Mozilla/5.0 (Windows NT 6.1; rv:24.0) Gecko/20100101 Firefox/24.0"

4.3. Privacy Analysis And Comparison Of Schemes

The World Wide Web Consortium (W3C) Platform for Privacy Preferences Project (P3P) [15] provides a framework regarding privacy issues in accessing webservers by allowing them to express their privacy practices in a standard format. We have analysed representative web metering schemes according to relevant metrics described in P3P. Further details about the compared schemes are provided in section 1.2. The following are the relevant P3P metrics and detailed schemes analysis.

- **Identifiers** are issued to users by a third party, which can be the Government or generally a "trusted" third party, to identify the user e.g. a username. There can be various levels of identifiers. For example, it is reasonable not to consider the action of capturing an IP address as identifying as authenticating an audited hardware decoder. Each user has a private and public key pair that is used to produce and verify the signature in the category of signature-based schemes. Also, GA assigns an identifier to the user browser after capturing user's IP address. Also, unencrypted traffic that goes through HTTP proxy including identifiers can be captured. In Secret Sharing schemes, secrets, submitted by the user, cannot be used as identifiers, however, users may have to be authenticated to get or verify the secrets. User hardware decoders techniques require users to have decryption keys upon subscription. Also, users, in webserver voucher schemes, have to be authenticated to ensure that the webservers are not forwarding the coupons themselves. Users' identifiers are required in the proposed scheme during hardware authentication in the Join Protocol. However, during the Sign/Verify Protocol, users' identifiers are preserved as previously used identification information, during Join Protocol, cannot be related thanks to the used zero-knowledge protocol.

- **State Management Mechanisms** are used to maintain the state of the connection to the webserver e.g. cookies. If the state of the user is required or can be captured, unlinkability cannot be provided by the scheme. In signature-based schemes, the user continuously submits a signature (or a hash value) that can link his visits. In typical Secret Sharing schemes, each user continuously submits a share to the webserver for each visit or session which link them up until the threshold value is computed. In user hardware decoders techniques, the state of the user can be tracked while decrypting on-the-fly broadcast content. Similarly, in webserver voucher schemes, the state of the user can be tracked as the user frequently forwards the e-coupons. The state of the user is continuously tracked in processing-based schemes because of

Table 2. Privacy Comparison

Scheme	Identifiers	State	Interactive	Location	Computer	Navigation
Digital Signature [25]	✗	†	✓	✓	✓	✓
Secret Sharing [35]	✗	†	✓	✓	✓	✓
Webserver Voucher [29]	✗	†	✓	✓	✓	✓
Processing-Based [12]	✓	✗	✓	✓	†	†
Webserver Hardware [4]	✓	†	✗	✓	✓	✓
HTTP Proxy [2]	†	†	†	†	✓	†
Google Analytics [23]	†	✗	✓	†	†	†
This paper (DAA-Based [8])	✓	†	✓	✓	✓	†

the required participation from the user side. Also, user state can be figured out in HTTP Proxy schemes as all traffic goes through the proxy. Also, GA uses cookies to track the user state. The state of the user can be tracked in the proposed scheme depending on the key expiry date and the hash chain whenever used.

- **Interactive Data** includes data generated on the fly during user-webserver interaction e.g. a query to the webserver. User hardware decoders techniques can capture users' queries when encrypting particular responses (e.g. pay-per-view). Also, in webserver-centric hardware schemes, users' queries are occasionally captured and *MACed* before the result is forwarded to the Audit Agency. HTTP Proxy schemes can capture interactive data from the user as all traffic goes through the HTTP proxy.

- **Location Data** category covers information regarding the users location e.g. users' GPS location. Location of users is not directly captured in HTTP Proxy schemes, however, depending upon the location of the HTTP Proxy, location of users can be tracked. Also, GA captures IP address of the user which can have information about the user location. From the described user hardware decoders technique, we infer that users' location data cannot be captured.

- **Computer Information** is any information about the user computer e.g. IP address. Processing-based schemes and GA can capture computer information by analysing the time needed to return the solution (e.g. estimated CPU speed during the visit) and capturing the IP address respectively. Users' computer information can only be captured in the proposed scheme during the Join protocol by figuring out information regarding the platform from the hardware certificate.

- **Navigation and Click-stream Data** covers data about the user browsing behaviour e.g. user clickstream. Depending on the implementation of

the processing-based scheme, navigation data can be required (e.g. user presence is determined by mouse movements). Also, HTTP proxy returns to the user a tracking code that captures the user mouse movement. Also, GA captures navigation data about the user through the type of referral which is stored in the cookie. Depending on the proposed scheme implementation, navigation data can be captured. For example, if the signed webserver URL references various levels within the webserver content, navigation data can be captured during the lifetime of the used key f.

A summary of the P3P analysis is shown in Table 2. From two extremes, a particular private information can be either *required* by the scheme or *protected*. We use the symbol ✗ to denote the scheme cannot operate without the corresponding required private information in order to provide web metering result or evidence. On the other hand, we use the symbol ✓ to denote that the private information can be protected and not accessed by the adversary under secure user setup. Such setup can be achieved with countermeasures that can prevent the adversary from getting the private information. The countermeasures can be provided by the scheme itself or can be potentially provided by other techniques. An example of outside countermeasures that can prevent the adversary from getting protected information could be a user behind a firewall with anonymous browser settings. We use the symbol † to denote that the private information is not always or necessarily required by the web metering scheme; however, it is *available* and can still be captured by the adversary. Such available information can still be captured due to an efficient implementation (or a variation) of the scheme.

It can be observed from the analysis summary that the categories Identifiers and State are the least satisfied privacy categories. In particular, all schemes require or can capture the state of the user. Furthermore, once a user is identified or tracked, other private information e.g. user's preferences can be captured from the webserver content. If a scheme was able to capture the users' state but was not able to identify the user, the captured state alone is not considered a privacy concern. Identifier information is used to achieve security requirements however such authentication information inherently conflicts with privacy. Identifier information can be the determinant metric to provide a privacy-preserving scheme. Processing and webserver hardware-based schemes are the most satisfying privacy-preserving schemes as they do not require nor can capture users' identifiers.

5. Proof-Of-Concept Implementation Analysis

We tested[9] the proposed scheme from the user side. We did various optimised simulation tests on a user machine with 2727578 high resolution performance counter frequency, using standard *math*, *gmp* and *OpenSSL* libraries[10] and the bit length specified in [8]. The prime numbers were of 1024 bits and modulus and generators were of 2048 bits. The public key values can be precomputed and stored at the hardware device, or securely downloaded to the user.

The tests showed feasible results. It took around 1650 nanoseconds to execute the first part of the public commitment (U) and around 515 nanoseconds to execute the second part (N). Then, the certificate production equation (S) needs only to be executed at the Audit Agency side, possibly by utilising Montgomery reduction algorithm [33] since the equation requires floating number exponentiation and *div* operations. From the results, the operations throughout the scheme phases can be executed in a short time so that they are not noticeable to the user.

The kind of secure hardware device required by the scheme is commonly used today. For example, the BitLocker program[11] uses the TPM public key for disk encryption, allowing the decryption (by TPM private key) if baseline platform measurements are met again. Furthermore, all Windows systems are planned to be shipped with TPMs in order to pass hardware certification of the latest operating system [43]. The scope of the required hardware devices is not limited to TPMs as discussed in section 2.3.

6. Conclusion

Privacy-preserving web metering is a challenging problem, especially with current necessary trade-offs and in environments where the Audit Agency could collude with webservers to identify users and link their visits. In this paper, we proposed an alternative and new user-centric web metering scheme using hardware

to enhance users' privacy. We described a generic web metering scheme of a straightforward protocol and a special case using an existing protocol. We also used established privacy guidelines to analyse and compare representative categories of web metering schemes and showed the gained privacy benefits of the proposed scheme. The proposed scheme can provide different security countermeasures and users' privacy settings. However, denial of service attacks are still possible.

We built a proof of concept implementation on a traditional computer to evaluate the efficiency and transparency of the running web metering operations. Besides the operational cost from Audit Agency and webserver sides, the main barrier for a wide deployment is that users should accept the hardware device. However, in many contexts, the gain in privacy will offset the costs. We discussed how the user hardware assumption can be realistic in today's and future computing devices and showed different options.

Future work includes exploring techniques for discovering rogue hardware devices, and implementing the scheme with different settings to provide the evidential signature e.g. hash chaining. Various options for counting the number of unique users can be further explored for different advertising applications. Future work also includes analysing the performance of the proposed scheme using handheld devices. Formal validation of the proposed scheme is left for future work as well.

References

[1] F. Alarifi and M. Fernández. Towards privacy-preserving web metering via user-centric hardware. In *First Workshop on Secure Smart Systems (SSS 2014)*, September 2014.

[2] R. Atterer, M. Wnuk, and A. Schmidt. Knowing the user's every move: user activity tracking for website usability evaluation and implicit interaction. In *WWW '06: Proceedings of the 15th international conference on World Wide Web*, pages 203–212, New York, NY, USA, 2006. ACM.

[3] S. G. Barwick, W.-A. Jackson, and K. M. Martin. A general approach to robust web metering. *Des. Codes Cryptography*, 36(1):5–27, 2005.

[4] F. Bergadano and P. D. Mauro. Third party certification of http service access statistics (position paper). In *Security Protocols Workshop*, pages 95–99, 1998.

[5] C. Blundo, A. D. Bonis, and B. Masucci. Bounds and constructions for metering schemes. In *Communications in Information and Systems 2002*, pages 1–28, 2002.

[6] C. Blundo, S. Martìn, B. Masucci, and C. Padrò. A linear algebraic approach to metering schemes. Cryptology ePrint Archive, Report 2001/087, 2001.

[7] C. Boyd and D.-G. Park. Public key protocols for wireless communications. In *Proceedings of The 1st International Conference on Information Security and Cryptology, ICSCI '98, Seoul, Korea*, pages 47–57, December 1998.

---

[9]at National Center for Digital Certification (NCDC): Research & Development. www.ncdc.gov.sa

[10]www.openssl.org/docs/

[11]www.technet.microsoft.com/en-us/library/cc162804.aspx

[8] E. Brickell, J. Camenisch, and L. Chen. Direct anonymous attestation. In *Proceedings of the 11th ACM conference on Computer and communications security*, CCS '04, pages 132–145, New York, NY, USA, 2004. ACM.

[9] Calcutt, David. Committee on Privacy and Related Matters. *Report of the Committee on Privacy and Related Matters*. Cm (Series) (Great Britain. Parliament). H.M. Stationery Office, 1990.

[10] J. Camenisch and A. Lysyanskaya. Dynamic accumulators and application to efficient revocation of anonymous credentials. In *Proceedings of the 22nd Annual International Cryptology Conference on Advances in Cryptology*, CRYPTO '02, pages 61–76, London, UK, UK, 2002. Springer-Verlag.

[11] D. Chaum, A. Fiat, and M. Naor. Untraceable electronic cash. In *Proceedings on Advances in Cryptology*, CRYPTO '88, pages 319–327, New York, NY, USA, 1990. Springer-Verlag New York, Inc.

[12] L. Chen and W. Mao. An auditable metering scheme for web advertisement applications. In *ISC*, volume 2200 of *Lecture Notes in Computer Science*, pages 475–485. Springer, 2001.

[13] Y. Chen, M. Moinuddin, and Y. Yacobi. Mobile wallet and digital payment, Sept. 30 2011. US Patent App. 13/249,381.

[14] B. Chor, A. Fiat, M. Naor, and B. Pinkas. Tracing traitors. *IEEE Transactions on Information Theory*, 46(3):893–910, 2000.

[15] L. Cranor, B. Dobbs, S. Egelman, G. Hogben, J. Humphrey, M. Langheinrich, M. Marchiori, M. Presler-Marshall, J. Reagle, M. Schunter, D. A. Stampley, and R. Wenning. The platform for privacy preferences. W3C Recommendation, November 2006.

[16] A. W. Dent and C. J. Mitchell. *User's Guide To Cryptography And Standards (Artech House Computer Security)*. Artech House, Inc., Norwood, MA, USA, 2004.

[17] W. Diffie and M. E. Hellman. New directions in cryptography. *IEEE Transactions on Information Theory*, 22(6):644–654, 1976.

[18] X. Ding. A hybrid method to detect deflation fraud in cost-per-action online advertising. In *Applied Cryptography and Network Security, 8th International Conference, ACNS 2010, Beijing, China, June 22-25, 2010. Proceedings*, pages 545–562, 2010.

[19] R. Dingledine, N. Mathewson, and P. Syverson. Tor: The second-generation onion router. In *Proceedings of the 13th Conference on USENIX Security Symposium - Volume 13*, SSYM'04, pages 21–21, Berkeley, CA, USA, 2004. USENIX Association.

[20] D. Dolev and A. C. Yao. On the security of public key protocols. Technical report, Stanford, CA, USA, 1981.

[21] P. Dourish and K. Anderson. Collective information practice: Exploring privacy and security as social and cultural phenomena. *Human Computer Interaction*, 21(3):319–342, 2006.

[22] M. K. Franklin and D. Malkhi. Auditable metering with lightweight security. *Journal of Computer Security*, 6(4):237–256, 1998.

[23] Google. Google analytics blog. Official weblog offering news, tips and resources related to google's web traffic analytics service. analytics.blogspot.com.

[24] J. Grudin. Desituating action: Digital representation of context. *Human Computer Interaction*, 16(2-4):269–286, 2001.

[25] H. Harn, L. Lin. A non-repudiation metering scheme. *Communications Letters, IEEE*, 37(5):486–487, 2001.

[26] M. Hutter and R. Toegl. A trusted platform module for near field communication. In *Proceedings of the 2010 Fifth International Conference on Systems and Networks Communications*, ICSNC '10, pages 136–141, Washington, DC, USA, 2010. IEEE Computer Society.

[27] International Organization for Standardization. ISO 11889-1:2009. *Information technology – Trusted Platform Module – Part 1: Overview*, May 2009.

[28] International Organization for Standardization. ISO 7498-2:1989. *Information processing systems – Open Systems Interconnection – Basic Reference Model – Part 2: Security Architecture*, 1989.

[29] M. Jakobsson, P. D. MacKenzie, and J. P. Stern. Secure and lightweight advertising on the web. *Computer Networks*, 31(11-16):1101–1109, 1999.

[30] R. Johnson and J. Staddon. Deflation-secure web metering. *International Journal of Information and Computer Security*, 1(1/2):39–63, 2007.

[31] R. Kumar. *Human Computer Interaction*. Laxmi Publications, 2005.

[32] C.-S. Laih, C.-J. Fu, and W.-C. Kuo. Design a secure and practical metering scheme. In *International Conference on Internet Computing*, pages 443–447, 2006.

[33] P. L. Montgomery. Modular multiplication without trial division. *Mathematics of Computation*, 44(170):519–521, 1985.

[34] J. A. Muir and P. C. V. Oorschot. Internet geolocation: Evasion and counterevasion. *ACM Comput. Surv.*, 42(1):4:1–4:23, Dec. 2009.

[35] M. Naor and B. Pinkas. Secure and efficient metering. In *EUROCRYPT*, pages 576–590, 1998.

[36] T. P. Pedersen. Non-interactive and information-theoretic secure verifiable secret sharing. In *Proceedings of the 11th Annual International Cryptology Conference on Advances in Cryptology*, CRYPTO '91, pages 129–140, London, UK, 1992. Springer-Verlag.

[37] A. Pfitzmann and M. Köhntopp. Anonymity, unobservability, and pseudonymity - a proposal for terminology. In *Workshop on Design Issues in Anonymity and Unobservability*, pages 1–9, 2000.

[38] F. Roesner, T. Kohno, and D. Wetherall. Detecting and defending against third-party tracking on the web. In *Presented as part of the 9th USENIX Symposium on Networked Systems Design and Implementation (NSDI 12)*, pages 155–168, San Jose, CA, 2012. USENIX.

[39] C.-P. Schnorr. Efficient identification and signatures for smart cards. In *CRYPTO*, pages 239–252, 1989.

[40] A. Shamir. How to share a secret. *Commun. ACM*, 22(11):612–613, 1979.

[41] P. Srisuresh and K. Egevang. IP Network Address Translator (NAT) Terminology and Considerations, August 1999.

[42] R.-C. Wang, W.-S. Juang, and C.-L. Lei. A web metering scheme for fair advertisement transactions. *International Journal of Security and Its Applications*, 2(4):453–456, October 2008.

[43] Windows Certification Program. *Hardware Certification Taxonomy & Requirements*, April 2014.

12

Privacy Preserving Large-Scale Rating Data Publishing

Xiaoxun Sun[1], Lili Sun[2,*]

[1] Australian Council for Educational Research, Australia
[2] Department of Mathematics & Computing, University of Southern Queensland, Australia

Abstract

Large scale rating data usually contains both ratings of sensitive and non-sensitive issues, and the ratings of sensitive issues belong to personal privacy. Even when survey participants do not reveal any of their ratings, their survey records are potentially identifiable by using information from other public sources.

In order to protect the privacy in the large-scale rating data, it is important to propose new privacy principles which consider the properties of the rating data. Moreover, given the privacy principle, how to efficiently determine whether the rating data satisfied the required privacy principle is crucial as well. Furthermore, if the privacy principle is not satisfied, an efficient method is needed to securely publish the large-scale rating data. In this paper, all these problem will be addressed.

Keywords: Privacy preserving, anonymity

1. Introduction

The problem of privacy-preserving data publishing has received a lot of attention in recent years. Privacy preservation on relational data has been studied extensively. A major category of privacy attacks on relational data is to re-identify individuals by joining a published table containing sensitive information with some external tables. Most of existing work can be formulated in the following context: several organizations, such as hospitals, publish detailed data (called microdata) about individuals (e.g. medical records) for research or statistical purposes [22, 23, 28, 32].

Privacy risks of publishing microdata are well-known. Famous attacks include de-anonymisation of the Massachusetts hospital discharge database by joining it with a public voter database [32] and privacy breaches caused by AOL search data [16]. Even if identifiers such as names and social security numbers have been removed, the adversary can use linking [32], homogeneity and background attacks [23] to re-identify individual data records or sensitive information of individuals. To overcome the re-identification attacks, k-anonymity was proposed [25–27, 32]. Specifically, a data set is said to be k-anonymous ($k \geq 1$) if, on the quasi-identifier (QID) attributes (that is, the maximal set of join attributes to re-identify individual records), each record is identical with at least $k - 1$ other records. The larger the value of k, the better the privacy is protected. Several algorithms are proposed to enforce this principle [1, 7, 12, 18–21]. Machanavajjhala et al. [23] showed that a k-anonymous table may lack of diversity in the sensitive attributes.

To overcome this weakness, they propose the l-diversity [23]. However, even l-diversity is insufficient to prevent attribute disclosure due to the skewness and the similarity attack. To amend this problem, t-closeness [22] was proposed to solve the attribute disclosure vulnerabilities inherent to previous models.

Recently, a new privacy concern has emerged in privacy preservation research: how to protect individuals' privacy in large survey rating data. Though several models and many algorithms have been proposed to preserve privacy in relational data (e.g., k-anonymity [32], l-diversity [23], t-closeness [22], etc.),

*Corresponding author. Email: xiaoxun.sun@gmail.com

| ID | non-sensitive | | | sensitive |
	issue 1	issue 2	issue 3	issue 4
t_1	6	1	*null*	6
t_2	1	6	*null*	1
t_3	2	5	*null*	1
t_4	1	*null*	5	1
t_5	2	*null*	6	5

(a)

| name | non-sensitive issues | | |
	issue 1	issue 2	issue 3
Alice	excellent	so bad	-
Bob	awful	top	-
Jack	bad	-	good

(b)

Table 1. (a) A published survey rating data set containing ratings of survey participants on both sensitive and non-sensitive issues. (b) Public comments on some non-sensitive issues of some participants of the survey. By matching the ratings on non-sensitive issues with public available preferences, t_1 is linked to Alice, and her sensitive rating is revealed.

most of the existing studies are incapable of handling rating data, since the survey rating data normally does not have a fixed set of personal identifiable attributes as relational data, and it is characterized by high dimensionality and sparseness. The survey rating data shares the similar format with transactional data. The privacy preserving research of transactional data has recently been acknowledged as an important problem in the data mining literature [14, 37].

2. Motivation

On October 2, 2006, Netflix, the world's largest online DVD rental service, announced the $1-million Netflix Prize to improve their movie recommendation service [15]. To aid contestants, Netflix publicly released a data set containing 100,480,507 movie ratings, created by 480,189 Netflix subscribers between December 1999 and December 2005. Narayanan and Shmatikov shown in their recent work [24] that an attacker only needs a little information to identify the anonymized movie rating transaction of the individual. They re-identified Netflix movie ratings using the Internet Movie Database (IMDb) as a source of auxiliary information and successfully identified the Netflix records of known users, uncovering their political preferences and other potentially sensitive information.

We consider the privacy risk in publishing anonymous survey rating data. For example, in a life style survey, ratings to some issues are non-sensitive, such as the likeness of book "Harry Potter", movie "Star Wars" and food "Sushi". Ratings to some issues are sensitive, such as the income level and sexuality frequency. Assume that each survey participant is cautious about his/her privacy and does not reveal his/her ratings. However, it is easy to find his/her preferences on non-sensitive issues from publicly available information sources, such as personal weblog or social networks. An attacker can use these preferences to re-identify an individual in the anonymous published survey rating data and consequently find sensitive ratings of a victim.

Based on the public preferences, person's ratings on sensitive issues may be revealed in a supposedly anonymized survey rating data set. An example is given in the Table 1. In a social network, people make comments on various issues, which are not considered sensitive. Some comments can be summarized as in Table 1(b). People rate many issues in a survey. Some issues are non-sensitive while some are sensitive. We assume that people are aware of their privacy and do not reveal their ratings, either non-sensitive or sensitive ones. However, individuals in the anonymoized survey rating data are potentially identifiable based on their public comments from other sources. For example, Alice is at risk of being identified, since the attacker knows Alice's preference on issue 1 is 'excellent', by cross-checking Table 1(a) and (b), s/he will deduce that t_1 in Table 1(a) is linked to Alice, the sensitive rating on issue 4 of Alice will be disclosed. This example motivates us the following research questions:

(Satisfaction Problem): Given a large scale rating data set T with the privacy requirements, how to efficiently determine whether T satisfies the given privacy requirements?

Although the satisfaction problem is easy and straightforward to be determined in the relational databases, it is nontrivial in the large scale rating data set. The research of the privacy protection initiated in the relational databases, in which several state-of-art privacy paradigms [22, 23, 32] are proposed and many greedy or heuristic algorithms [12, 19, 20, 28] are developed to enforce the privacy principles. In the relational database, taking k-anonymity as an example [26, 32], it requires each record be identical with at least $k-1$ others with respect to a set of quasi-identifier attributes. Given an integer k and a relational data set T, it is easy to determine if T satisfies k-anonymity requirement since the equality has the transitive property, whenever a transaction a is identical with b, and b is in turn indistinguishable with c, then a is the same as c. With this property, each transaction in T only needs to be check once and the time complexity is at most $O(n^2d)$, where n is the number of transactions in T and d is the size of the quasi-identifier attributes. So the satisfaction problem is trivial in relational data sets. While, the situation is different for the large rating data. First of all, the survey rating data normally does not have a fixed set of personal identifiable attributes as relational data. In addition, the survey rating data is characterized by high dimensionality and sparseness. The lack of a clear set of personal identifiable attributes together with its high dimensionality and sparseness make the determination of satisfaction problem challenging. Second, the defined dissimilarity distance between two transactions (ϵ-proximate) does not possess the transitive property. When a transaction a is ϵ-proximate with b, and b is ϵ-proximate with c, then usually a is not ϵ-proximate with c. Each transaction in T has to be checked for as many as n times in the extreme case, which makes it highly inefficient to determine the satisfaction problem. It calls for smarter technique to efficiently determine the satisfaction problem before anonymizaing the survey rating data. To our best knowledge, this research is the first touch of the satisfaction of privacy requirements in the survey rating data.

How to preserve individual's privacy in the large scale rating data set?

Though several models and algorithms have been proposed to preserve privacy in relational data, most of the existing studies can deal with relational data only [22, 23, 31?]. Divide-and-conquer methods are applied to anonymize relational data sets due to the fact that tuples in a relational data set are separable during anonymisation. In other words, anonymizing a group of tuples does not affect other tuples in the data set. However, anonymizing a survey rating data set is much more difficult since changing one record may cause a domino effect on the neighborhoods of other records, as well as affecting the properties of the whole data set. Hence, previous methods can not be applied to deal with survey rating data and it is much more challenging to devise anonymisation methods for large scale rating data than for relational data.

3. Related work

Privacy preserving data publishing has received considerable attention in recent years. especially in the context of relational data [1, 7, 12, 18–20, 23, 25, 36]. All these works assume a given set of attributes QID on which an individual is identified, and anonymize data records on the QID. Their main difference consist in the selected privacy model and in various approaches employed to anonymize the data. The author of [1] presents a study on the relationship between the dimensionality of QID and information loss, and concludes that, as the dimensionality of QID increases, information loss increases quickly. Transactional databases present exactly the worst case scenario for existing anonymisation approaches because of high dimension of QID. To our best knowledge, all existing solutions in the context of k-anonymity [26, 27], l-diversity [23] and t-closeness [22] assume a relational table, which typically has a low dimensional QID.

There are few previous work considering the privacy of large rating data. In collaboration with MovieLens recommendation service, [11] correlated public mentions of movies in the MovieLens discussion forum with the users' movie rating histories in the internal MovieLens data set. Recent study reveals a new type of attack on anonymized data for transactional data [24]. Movie rating data supposedly to be anonymized is re-identified by linking non-anonymized data from other source. No solution exists for high dimensional large scale rating databases.

Privacy-preservation of transactional data has been acknowledged as an important problem in the data mining literature. There us a family of literature [5, 6] addressing the privacy threats caused by publishing data mining results such as frequent item sets and association rules. Existing works on topic [4, 34] focus on publishing patterns, The patterns are mined from the original data, and the resulting set of rules is sanitized to present privacy breaches. In contrast, our work addresses the privacy threats caused by publishing data for data mining. As discussed above, we do not assume that the data publisher can perform data mining tasks, and we assume that the data must be made available to the recipient. The two scenarios have different assumptions on the capability of the data publisher and the information requirement of the data recipient. The recent work on topic [14, 37] focus on high dimensional transaction data, while our focus is preventing linking individuals to their ratings.

This paper is also related to the work on anonymizing social networks [8], and the large scale rating data can be considered as a special case of the complex social network. A social network is a graph in which a node represents a social entity (e.g., a person) and an edge represents a relationship between the social entities. Although the data is very different from transaction data, the model of attacks is similar to ours: An attacker constructs a small subgraph connected to a target individual and then matches the subgraph to the whole social network, attempting to re-identify the target individual's node, and therefore, other unknown connection to the node. [8] demonstrates the severity of privacy threats in nowadays social networks, but does not provide a solution to prevent such attacks.

4. Privacy models

The auxiliary information of an attacker includes: (i) the knowledge that a victim is in the survey rating data; (ii) preferences of the victims on some non-sensitive issues. The attacker wants to find ratings on sensitive issues of the victim.

In practice, knowledge of Types (i) and (ii) can be gleaned from an external database [24]. For example, in the context of Table 1(b), an external database may be the IMDb. By examining the anonymous data set in Table 1(a), the adversary can identify a small number of candidate groups that contain the record of the victim. It will be the unfortunate scenario where there is only one record in the candidate group. For example, since t_1 is unique in Table 1(a), Alice is at risk of being identified. If the candidate group contains not only the victim but other records, an adversary may use this group to infer the sensitive value of the victim individual. For example, although it is difficult to identify whether t_2 or t_3 in Table 1(a) belongs to Bob, since both records have the same sensitive value, Bob's private information is identified.

Intuitively,in order to avoid such attack, a two-step protection model is needed. The first step is to protect individual's identity, which is to make sure that in the released data set, every transaction should be "similar" to at least to $(k-1)$ other records based on the non-sensitive ratings so that no survey participants are identifiable. For example, t_1 in Table 1(a) is unique, and based on the preference of Alice in Table 1(b), her sensitive issues can be re-identified in the supposed anonymized data set. Jack's sensitive issues, on the other hand, is much safer. Since t_4 and t_5 in Table 1(a) form a similar group based on their non-sensitive rating.

The second step is to prevent the sensitive rating from being inferred in an anonymized data set. The idea is to require that the sensitive ratings in a similar group should be diverse. For example, although t_2 and t_3 in Table 1(a) form a similar group based on their non-sensitive rating, their sensitive ratings are identical. Therefore, an attacker can immediately infer Bob's preference on the sensitive issue without identifying which transaction belongs to Bob. In contrast, Jack's preference on the sensitive issue is much safer than both Alice and Bob.

In our previous work, two privacy models have been proposed. The first one is (k, ϵ)-anonymity model, which targets at protecting individual's identity and the second model is (k, ϵ, l)-anonymity model, which not only protects individual's identity, but also the personal sensitive information. In the next, section, these two models will be discussed.

4.1. (k, ϵ)-anonymity

Let $T_A = \{o_{A_1}, o_{A_2}, \cdots, o_{A_p}, s_{A_1}, s_{A_2}, \cdots, s_{A_q}\}$ be the ratings for a survey participant A and $T_B = \{o_{B_1}, o_{B_2}, \cdots, o_{B_p}, s_{B_1}, s_{B_2}, \cdots, s_{B_q}\}$ be the ratings for a participant B. We define the dissimilarity between two non-sensitive ratings as follows.

$$Dis(o_{A_i}, o_{B_i}) = \begin{cases} |o_{A_i} - o_{B_i}| & \text{if } o_{A_i}, o_{B_i} \in \{1 : r\} \\ 0 & \text{if } o_{A_i} = o_{B_i} = null \\ r & \text{otherwise} \end{cases} \quad (1)$$

Definition 1 (ϵ-proximate). Given a survey rating data set T with a small positive number ϵ, two transactions T_A, $T_B \in T$, where $T_A = \{o_{A_1}, o_{A_2}, \cdots, o_{A_p}, s_{A_1}, s_{A_2}, \cdots, s_{A_q}\}$ and $T_B = \{o_{B_1}, o_{B_2}, \cdots, o_{B_p}, s_{B_1}, s_{B_2}, \cdots, s_{B_q}\}$. We say T_A and T_B are ϵ-proximate, if $\forall 1 \le i \le p$, $Dis(o_{A_i}, o_{B_i}) \le \epsilon$. We say T is ϵ-proximate, if every two transactions in T are ϵ-proximate.

If two transactions are ϵ-proximate, the dissimilarity between their non-sensitive ratings is bounded by ϵ. In our running example, suppose $\epsilon = 1$, ratings 5 and 6 may have no difference in interpretation, so t_4 and t_5 in Table 1(a) are 1-proximate based on their non-sensitive rating. If a group of transactions are in ϵ-proximate, then the dissimilarity between each pair of their non-sensitive ratings is bounded by ϵ. For example, if $T = \{t_1, t_2, t_3\}$, then it is easy to verify that T is 5-proximate.

Definition 2 ((k, ϵ)-anonymity). A survey rating data set T is said to be (k, ϵ)-anonymous if every transaction is ϵ-proximate with at least $(k-1)$ other transactions. The transaction $t \in T$ with all the other transactions that ϵ-proximate with t form a (k, ϵ)-anonymous group.

For instance, there are two (2,5)-anonymous groups in Table 1(a). The first one is formed by $\{t_1, t_2, t_3\}$ and the second one is formed by $\{t_4, t_5\}$. The idea behind this privacy principle is to make each transaction contains non-sensitive attributes are similar with other transactions in order to avoid linking to personal identity. (k, ϵ)-anonymity well preserves identity privacy. It guarantees that no individual is identifiable with the probability greater than the probability of $1/k$. Both parameters k and ϵ are intuitive and operable in real-world applications. The parameter ϵ captures the protection range of each identity, whereas the parameter k is to lower an adversary's chance of beating that protection. The larger the k and ϵ are, the better protection it will provide.

Although (k, ϵ)-anonymity privacy principle can protect people's identity, it fails to protect individuals' private information. Let us consider one (k, ϵ)-anonymous group. If the transactions of the group have the same rating on a number of sensitive issues, an attacker can know the preference on the sensitive issues of each individual without knowing which transaction belongs to whom. For example, in Table 1(a), t_2 and t_3 are in a $(2,1)$-anonymous group, but they have the same rating on the sensitive issue, and thus Bob's private information is breaching.

4.2. (k, ϵ, l)–anonymity

This example illustrates the limitation of the (k, ϵ)-anonymity model. To mitigate the limitation, we require more diversity of sensitive ratings in the anonymous groups. In the following, we define the distance between two sensitive ratings, which leads to the metric for measuring the diversity of sensitive ratings in the anonymous groups.

First, we define dissimilarity between two sensitive rating scores as follows.

$$Dis(s_{A_i}, s_{B_i}) = \begin{cases} |s_{A_i} - s_{B_i}| & \text{if } s_{A_i}, s_{B_i} \in \{1:r\} \\ r & \text{if } s_{A_i} = s_{B_i} = null \\ r & \text{otherwise} \end{cases} \quad (2)$$

Note that there is only one difference between dissimilarities of sensitive ratings $Dis(s_{A_i}, s_{B_i})$ and dissimilarities of non-sensitive ratings $Dis(o_{A_i}, o_{B_j})$, that is, in the definition of $Dis(o_{o_i}, o_{o_j})$, $null - null = 0$, and for the definition of $Dis(s_{A_i}, s_{B_i})$, $null - null = r$. This is because for sensitive issues, two $null$ ratings mean that an attacker will not get information from two survey participants, and hence are good for the diversity of the group.

Next, we introduce the metric to measure the diversity of sensitive ratings. For a sensitive issue s, let the vector of ratings of the group be $[s_1, s_2, \cdots, s_g]$, where $s_i \in \{1:r, null\}$. The means of the ratings is defined as follows:

$$\bar{s} = \frac{1}{Q} \sum_{i=1}^{g} s_i$$

where Q is the number of non-$null$ values, and $s_i \pm null = s_i$. The standard deviation of the rating is then defined as:

$$SD(s) = \sqrt{\frac{1}{g} \sum_{i=1}^{g} (s_i - \bar{s})^2} \quad (3)$$

For instance in Table 1(a), for the sensitive issue 4, the means of the ratings is $(6+1+1+1+5)/5 = 2.8$ and the standard deviation of the rating is 2.23 according to Equation (3).

Definition 3 $((k, \epsilon, l)$–anonymity$)$. A survey rating data set is said to be (k, ϵ, l)-anonymous if and only if the standard deviation of ratings for each sensitive issue is at least l in each (k, ϵ)-anonymous group.

Still consider Table 1(a) as an example. t_4 and t_5 is 1-proximate with the standard deviation of

2. If we set $k = 2, l = 2$, then this group satisfies $(2,1,2)$-anonymity requirement. The (k, ϵ, l)-anonymity requirement allows sufficient diversity of sensitive issues in T, therefore it could prevent the inference from the (k, ϵ)-anonymous groups to a sensitive issue with a high probability. The following theorem gives the upper bound of the parameter l in the (k, ϵ, l)-anonymity model. The proof of the following theorem can be found in [30].

Theorem 1. Let S be the set of ratings of the sensitive issue of T. Suppose S_min and S_max be the minimum and maximum ratings in S, then the maximum standard deviation of S is $\frac{(S\_max - S\_min)}{2}$.

5. Validating privacy requirements

In this section, we formulate the satisfaction problem and develop a slicing technique to determine the following *Satisfaction Problem*.

Problem 1 (Satisfaction Problem). Given a survey rating data set T and privacy requirements k, ϵ, l, the satisfaction problem of (k, ϵ, l)-anonymity is to decide whether T satisfies the k, ϵ, l privacy requirements.

The satisfaction problem is to determine whether the user's given privacy requirement is satisfied by the given data set. It is a very important step before anonymizing the survey rating data. If the data set has already met the requirements, it is not necessary to make any modifications before publishing. As follows, we propose a novel slice technique to solve the satisfaction problem.

5.1. Search by slicing

The slicing technique is proposed to efficiently search for the neighbor within distance ϵ in high dimension. As we shall see, the complexity of the proposed algorithm grows very slowly with dimension for small ϵ. We illustrate the proposed slicing technique using a simple example in 3-D space, as shown in Figure 1. Given $t = (t_1, t_2, t_3) \in T$, our goal is to slice out a set of transactions T $(t \in T)$ that are ϵ-proximate. Our approach is first to find the ϵ-proximate of t, which is the set of transactions that lie inside a cube C_t of side 2ϵ centered at t. Since ϵ is typically small, the number of points inside the cube is also small. The ϵ-proximate of C_t' can then be found by an exhaustive comparison within the ϵ-proximate of t. If there are no transactions inside the cube C_t, we know that the ϵ-proximate of t is empty, so as the ϵ-proximate of the set C_t'.

The transactions within the cube can be found as follows. First we find the transactions that are sandwiched between a pair of parallel planes X_1, X_2 (See Figure 1) and add them to a *candidate set*. The planes are perpendicular to the first axis of coordinate

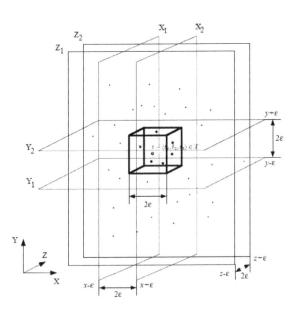

Figure 1. The slicing technique finds a set of transactions C_t inside a cube of size 2ϵ within the ϵ-proximate of t. The ϵ-proximate of the set C_t can then be found by an exhaustive search in the cube.

frame and are located on either side of the transaction t at a distance of ϵ. Next, we trim the candidate set by disregarding transactions that are not also sandwiched between the parallel pair of Y_1 and Y_2, that are perpendicular to X_1 and X_2, again located on either side of t at a distance of ϵ. This procedure is repeated for Z_1 and Z_2 at the end of which, the candidate set contains only transactions within the cube of size 2ϵ centered at t.

Since the number of transactions in the final ϵ-proximate is typically small, the cost of the exhaustive comparison is negligible. The major computational cost in the slicing process occurs therefore in constructing and trimming the candidate set.

6. Anonymous survey rating data

In this section, we describe our modification strategies through the graphical representation of the (k, ϵ)-anonymity model. Given a survey rating data set $T = \{t_1, t_2, \cdots, t_n\}$, its graphical representation is the graph $G = (V, E)$, where V is a set of nodes, and each node in V corresponds to a record t_i ($i = 1, 2, \cdots, n$) in T, and E is the set of edges, where two nodes are connected by an edge if and only if the distance between two records is bounded by ϵ with respect to the non-sensitive ratings.

Two nodes t_i and t_j are called connected if G contains a path from t_i to t_j ($1 \leq i, j \leq n$). The graph G is called connected if every pair of distinct nodes in the graph can be connected through some paths. A connected component is a maximal connected subgraph of G. Each node belongs to exactly one connected component, as

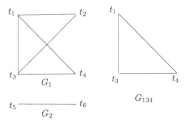

Figure 2. Graphical representation example

does each edge. The degree of the node t_i is the number of edges incident to t_i ($1 \leq i \leq n$).

We say G is a clique if every pair of distinct nodes is connected by an edge. The k-clique is a clique with at least k nodes. The maximal k-clique is the a k-clique that is not a subset of any other k-clique. We say the connected component $G = (V, E)$ is k-decomposable if G can be decomposed into several k-cliques $G_i = (V_i, E_i)$ ($i = 1, 2, \cdots, m$), and satisfies $V_i \cap V_j = \emptyset$ for ($i \neq j$), $\bigcup_{i=1}^{m} V_i = V$, and $\bigcup_{i=1}^{m} E_i \subseteq E$. The graph is k-decomposable if all its connected components are k-decomposable.

Theorem 2. Given the survey rating data set T with its graphical representation G, if G is k-decomposable, then T is (k, ϵ)-anonymous.

The proof of Theorem 2 can be found in [29]. For instance, the survey rating data shown in Table 2 is $(2, 2)$-anonymous since its graphical representation (Figure 2(a)) is 2-decomposable. With Theorem 2, to make the rating data satisfy privacy requirement, it only needs to make its graphical representation k-decomposable.

6.1. Distortion Metrics

In this section, we define a measure to capture the information loss.

Definition 4 (Tuple distortion). Let $t = (t_1, t_2, \cdots, t_m)$ be a tuple and $t' = (t'_1, t'_2, \cdots, t'_m)$ be an anonymized tuple of t. Then, the distortion of this anonymisation is defined as:

$$\text{Distortion}(t, t') = \sum_{i=1}^{m} |t_i - t'_i|$$

For example, if the tuple $t = (5, 6, 0)$ is generalized to $t' = (5, 5, 0)$, then the distortion of this anonymisation is $|5 - 5| + |6 - 5| + |0 - 0| = 1$.

Definition 5 (Total distortion). Let $T' = (t'_1, t'_2, \cdots, t'_n)$ be the anonymized data set from $T = (t_1, t_2, \cdots, t_n)$. Then, the total distortion of this anonymisation is defined as:

$$\text{Distortion}(T, T') = \sum_{i=1}^{n} \text{Distortion}(t_i, t'_i)$$

	non-sensitive			sensitive
ID	issue 1	issue 2	issue 3	issue 4
t_1	3	6	*null*	6
t_2	2	5	*null*	1
t_3	4	7	*null*	4
t_4	5	6	*null*	1
t_5	1	*null*	5	1
t_6	2	*null*	6	5

Table 2. Sample survey rating data (I)

	non-sensitive			sensitive
ID	issue 1	issue 2	issue 3	issue 4
t_1	3	6	*null*	6
t_2	2	5	*null*	1
t_3	4	7	*null*	4
t_4	5	6	*null*	1
t_5	1	*null*	5	1
t_6	2	*null*	6	5
t_7	6	*null*	6	3
t_8	5	*null*	5	2

Table 3. Sample survey rating data (II)

For example, let $T = (t_1, t_2, t_3, t_4)$, where $t_1 = (5, 6, 0)$, $t_2 = (2, 5, 0)$, $t_3 = (4, 7, 0)$ and $t_4 = (5, 6, 0)$. Let $T' = (t'_1, t'_2, t'_3, t'_4)$ be anonymization of T, where $t'_1 = (5, 5, 0)$, $t'_2 = (3, 5, 0)$, $t'_3 = (3, 7, 0)$ and $t'_4 = (5, 7, 0)$. Then, the distortion between the two data sets is $1 + 1 + 1 + 1 = 4$.

For ease, we first illustrate our approach in the scale of single attribute, and then we extend it to multiple attributes.

Let $t = (t_1, t_2, \cdots, t_n)$ be the ratings of some issue from n survey participants with the privacy requirement ϵ. We assume that some ratings in t are not bounded by ϵ, and our aim is to modify t to make every pair of ratings is bounded by ϵ while minimizing the distortion. The idea of the approach is as follows. Order all ratings for the issue t, and find the minimum rating Min and maximum rating Max. Find all intervals of the size ϵ between Min and Max. Change the ratings that does not fit in this interval such that the distortion is minimized. In the case of some tuples with the same minimum distortion, randomly pick up one of them as the anonymization. The process is described in **Algorithm 1**.

ALGORITHM 1: $single\_anonymizer(t, \epsilon)$

```
 1   Input: an ascended tuple t = (t_1, · · · , t_n), and ε
 2   Output: t' = (t'_1, · · · , t'_n) with minimum distortion
 3   /* Computing distortions for all intervals */
 4   for i ← 1 to (t_n−t_1)/ε
 5     do for j ← 1 to n
 6       do if t_j ∈ (t_i, t_i + ε)
 7         then t'_j ← t_j
 8         else if t_j < t_i
 9            t'_j ← t_i
10         else t'_j ← t_i + ε
11       D(i) ← Distortion(t', t);
12   /* Finding minimum distortion */
13   k ← 1; D_min ← D(k);
14   for i ← 2 to (t_n−t_1)/ε
15     do if D(i) < D_min
16       then D_min ← D(i);
17          k ← i;
18   /* Retrieving t' with minimum distortion */
19   for i ← 1 to n
20     do if t_i ∈ (t_k, t_k + ε)
21       then t'_i ← t_i
22       else if t_i < t_k
23          t'_i ← t_k
24       else t'_i ← t_k + ε
25   return t'
```

For example, if $t = (3, 4, 5, 6, 7, 7, 8, 8)$ and $\epsilon = 2$. The Min is 3 and Max is 8. Build all the intervals with the size of 2, which are $(3,5)$, $(4,6)$, $(5,7)$ and $(6,8)$. Following Algorithm 1, the anonymization of t is shown in Table 4, in which the vector in bold is the anonymisation we choose.

Intervals	Anonymization	Distortion
(3, 5)	(3, 4, 5, 5, 5, 5, 5, 5)	11
(4, 6)	(4, 4, 5, 6, 6, 6, 6, 6)	7
(5, 7)	**(5, 5, 5, 6, 7, 7, 7, 7)**	**5**
(6, 8)	(6, 6, 6, 6, 7, 7, 8, 8)	6

Table 4. Example of the anonymization algorithm

Let us take Table 1(a) as an example with $k = 2, \epsilon = 1$. There are two groups $HG_1 = \{t_1, t_2, t_3, t_4\}$ and $HG_2 = \{t_5, t_6\}$. HG_2 has already satisfied the privacy requirement, but HG_1 does not. The anonymization of HG_1 is shown in Table 5, in which the vector in bold is the anonymisation we choose.

	Intervals	Anonymization	Distortion
	(2,3)	(3,3,3,2)	4
Issue 1	(3,4)	**(4,3,4,3)**	**3**
	(4,5)	(5,4,4,4)	4
	(5,6)	(6,5,5,5)	6

	Intervals	Anonymization	Distortion
	(1,2)	(1,2,2,2)	10
	(2,3)	(2,3,3,3)	8
Issue 2	(3,4)	(3,4,4,4)	6
	(4,5)	**(4,5,5,5)**	**4**
	(5,6)	(5,6,5,5)	4

Table 5. Anonymizing HG_1 of Table 1(a)

6.2. Complexity analysis

Recall that our objective is to anonymize data consisting of a set of transactions $T = \{t_1, t_2, \cdots, t_n\}$, $|T| = n$. Each transaction $t_i \in T$ contains m issues. The computation cost consists of three parts, which are sorting, finding intervals and computing distortion. The complexity of the sorting is $O(mn\log n)$. During the next phrase of the algorithm, for each attribute, we find the Min and Max and all the possible intervals with size ϵ, which incur the amount of $O(2(n-1))$ overhead, and the cost for comparisons to search the one with least distortion is $O(n)$. So, the total complexity of all attributes in this phrase is $O(mn)$. The last phrase to compare original and anonymous data sets to estimate the distortion has the cost of $O(mn)$. The computational complexity of this alternative approach is $O(mn\log n + mn)$.

7. Experimental study

In this section, we experimentally evaluate the efficiency of the proposed slicing algorithm and the proposed anonymization algorithm. Our objectives are two-fold. First, we verify that our slice algorithm is fast and scalable for the satisfaction problem. Second, we show that the slicing technique is not only time efficient, but also space efficient compared with the heuristic pairwise algorithm.

7.1. Data sets

Our experimentation deploys two real-world databases. MovieLens[1] and Netflix data sets[2]. MovieLens data set was made available by the GroupLens Research Project at the University of Minnesota. The data set contains 100,000 ratings (5-star scale), 943 users and 1682 movies. Each user has rated at lease 20 movies. Netflix data set was released by Netflix for competition. The movie rating files contain over 100,480,507 ratings from 480,189 randomly-chosen, anonymous Netflix customers over 17 thousand movie titles. The data were collected between October, 1998 and December, 2005 and reflect the distribution of all ratings received during this period. The ratings are on a scale from 1 to 5 (integral) stars. In both data sets, a user is considered as an object while a movie is regarded as an attribute and many entries are empty since a user only rated a small number of movies. Except for rating movies, users' ratings some simple demographic information (e.g., age range) are also included. In our experiments, we treat the users' ratings on movies as non-sensitive issues and ratings on others as sensitive ones.

7.2. Efficiency

Data used for Figure 3(a) is generated by re-sampling the Movielens and Netflix data sets while varying the

percentage of data from 10% to 100%. For both data sets, we evaluate the running time for the (k, ϵ, l)-anonymity model with default setting $k = 20, \epsilon = 1, l = 2$. For both testing data sets, the execution time for (k, ϵ, l)-anonymity is increasing with the increased data percentage. This is because as the percentage of data increases, the computation cost increases too. The result is expected since the overhead is increased with the more dimensions.

Next, we evaluate how the parameters affect the cost of computing. Data set used for this sets of experiments are the whole sets of MovieLens and Netflix data and we evaluate by varying the value of ϵ, k and l. With $k = 20, l = 2$, Figure 3(b) shows the computational cost as a function of ϵ, in determining (k, ϵ, l)-anonymity requirement of both data sets. Interestingly, in both data sets, as ϵ increases, the cost initially becomes lower but then increases monotonically. This phenomenon is due to a pair of contradicting factors that push up and down the running time, respectively. At the initial stage, when ϵ is small, more computation efforts are put into finding ϵ-proximate of the transaction, but less used in exhaustive search for proper ϵ-proximate neighborhood, and this explains the initial decent of overall cost. On the other hand, as ϵ grows, there are fewer possible ϵ-proximate neighborhoods, thus reducing the searching time for this part, but the number of transactions in the ϵ-proximate neighborhood is increased, which results in huge exhaustive search for proper ϵ-proximate neighborhood and this causes the eventual cost increase. Setting $\epsilon = 2$, Figure 4(a) displays the results of running time by varying k from 10 to 60 for both data sets. The cost drops as k grows. This is expected, because fewer search efforts for proper ϵ-proximate neighborhoods needed for a greater k, allowing our algorithm to terminate earlier. We also run the experiment by varying the parameter l and the results are shown in Figure 4(b). Since the rating of both data sets are between 1 and 5, then according to Theorem 1, 2 is already the largest possible l. When $l = 0$, there is no diversity requirement among the sensitive issues, and the (k, ϵ, l)-anonymity model is reduced to (k, ϵ)-anonymity model. As we can see, the running time increases with l, because more computation is needed in order to enforce stronger privacy control.

In addition to show the scalability and efficiency of the slicing algorithm itself, we also experimented the comparison between the slicing algorithm (Slicing) and the heuristic pairwise algorithm (Pairwise), which works by computing all the pairwise distance to construct the dissimilarity matrix and identify the violation of the privacy requirements. We implemented both algorithms and studied the impact of the execution time on the data percentage, the value of ϵ, the value of K and the value of L.

---

[1] http://www.grouplens.org/taxonomy/term/14.
[2] http://www.netflixprize.com/.

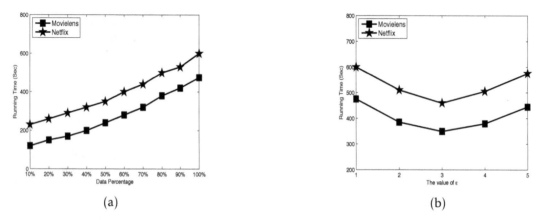

(a)   (b)

Figure 3. Running time comparison on Movielens and Netflix data sets vs. (a) Data percentage varies (b) ϵ varies

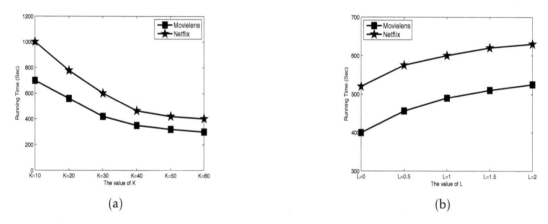

(a)   (b)

Figure 4. Running time comparison on Movielens and Netflix data sets vs. (c) k varies (d) L varies

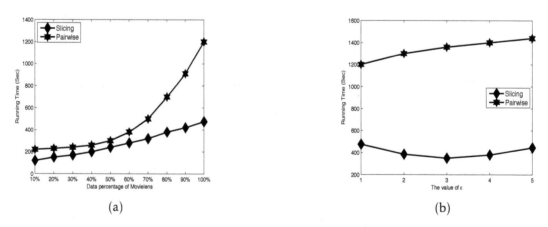

(a)   (b)

Figure 5. Running time comparison of Slicing and Pairwise methods on Movielens data set vs. (a) Data percentage varies (b) ϵ varies

Figure 5 plots the running time of both slicing and pairwise algorithms on the Movielens data set. Figure 5(a) describe the trend of the algorithms by varying the percentage of the data set. From the graph we can see, the slicing algorithm is far more efficient than the heuristic pairwise algorithm especially when the volume of the data becomes larger. This is

because, when the dimension of the data increases, the disadvantage of the heuristic pairwise algorithm, which is to compute all the dissimilarity distance, dominates the most of the execution time. On the other hand, the smarter grouping technique used in the slicing process makes less computation cost for the slicing algorithm. The similar trend is shown in Figure 5(b) by varying the

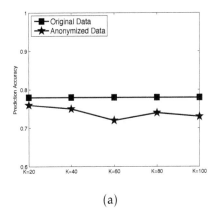

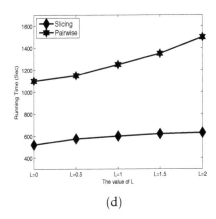

(c)   (d)

Figure 6. Running time comparison of Slicing and Pairwise methods on Netflix data set vs. (c) k varies (d) L varies

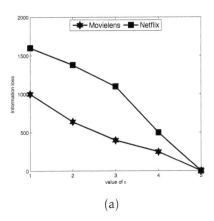

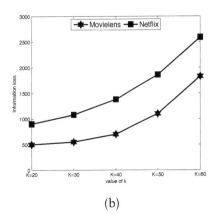

(a)   (b)

Figure 7. Information loss comparison on Movielens and Netflix databases vs. (a) k varies; (b) ϵ varies

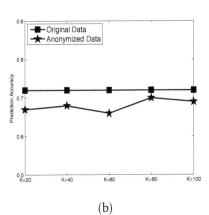

(a)   (b)

Figure 8. Prediction Accuracy: (a) Movielens; (b) Netflix

value of ϵ, in which the slicing algorithm is almost 3 times faster than the the heuristic pairwise algorithm. The running time comparisons of both algorithms in Netflix data set by varying the value of K and L are shown in Figure 6(a) and (b). Even on a larger data set, the slicing algorithm outperformed the pairwise

algorithm, and the running time of Slicing is quick enough to be used in practical.

7.3. Data Utility

Having verifying the efficiency of the slicing technique, we proceed to test its effectiveness. We measure the utility by the distortion metric defined in previous sections. Generally speaking, the more the distortion is, the less useful the anonymized data would be.

We first study the influence of ϵ (i.e., the length of a proximate neighborhood) on data utility. Towards this, we set k to 40. Concerning $(40, \epsilon)$-anonymity, Figure 7(a) plots the information loss on both data sets as a function of ϵ. The anonymization algorithm incurs less distortion as ϵ increases. This is expected, since a smaller ϵ demands stricter privacy preservation, which reduces data utility. When $\epsilon = 5$, there will be no anonymization required, and therefore the information loss reaches 0. Next, we examine the utility of $(k, 2)$-anonymous solution with different k. Figure 7(b) presents the information loss as a function of k. The error grows with k because a larger k demands tighter anonymity control requiring much more data modification.

Figures 8(a) and (b) evaluate the classification and prediction accuracy of the greedy anonymization algorithm. Our evaluation methodology is that we first divide data into training and testing sets, and we apply the anonymization algorithm to the training and testing sets to obtain the anonymized training and testing sets, and finally the classification or regression model is trained by the anonymized training set and tested by anonymized testing set. The Weka implementation [35] of simple Naive Bayes classifier was used for the classification and prediction. Using the Movielens data, Figure 8(a) compares the predictive accuracy of classifier trained on Movielens data produced by the greedy anonymization algorithm. In these experiments, we generated 50 independent training and testing sets, each containing 2000 records, and we fixed $\epsilon = 2$. The results are averaged across these 50 trials. For comparison, we also include the accuracies of classifier trained on the (not anonymized) original data. From the graph, we can see that the average prediction accuracy is around 75%, very close to the original accuracy, which preserves better utility for data mining purposes. Similar results are obtained by using the Netflix rating data in Figure 8(b).

8. Conclusion and future work

We have studied the problems of protecting sensitive ratings of individuals in a large scale rating data. Such privacy risk has emerged in a recent study on the de-identification of published movie rating data. We proposed a novel (k, ϵ, l)-anonymity privacy principle for protecting privacy in such survey rating data. We theoretically investigated the properties of this model, and studied the satisfaction problem, which is to decide whether a survey rating data set satisfies the privacy requirements given by the user. A greedy anonymization algorithm has been proposed to anonymize large scale rating data. Extensive

experiments confirm that our technique produces anonymized data sets that are useful.

This work also initiates the future investigations of approaches on anonymizing the survey rating data. Traditional approaches on anonymizing no matter relational data sets or transactional data set are by generalization or suppression, and the published data set has the same number of data but with some fields being modified to meet the privacy requirements. As shown in the literatures, this kind of anonymization problem is normally NP-hard, and several algorithms are devised along this framework to minimize the certain pre-defined cost metrics. Inspired by the research in this paper, the satisfaction problem can be further used to develop a different method to anonymizing the data set. The idea is straightforward with the result of the satisfaction problem. If the rating data set has already satisfies the privacy requirement, it is not necessary to do any anonymization to publish it. Otherwise, we anonymize the data set by deleting some of the records to make it meet the privacy requirement. The criteria during the deletion can be various (for example, to minimize the number of deleted records) to make it as much as useful in the data mining or other research purposes. We believe that this new anonymization method is flexible in the choice of privacy parameters and efficient in the execution with the practical usage.

References

[1] C. Aggarwal. On k-Anonymity and the curse of dimensionality. VLDB 2005.

[2] R. Agrawal and R. Srikant. Privacy-Preserving Data Mining. SIGMOD 2000.

[3] D. Agrawal and C. C. Aggarwal. On The Design and Qualification of Privacy Preserving Data Mining Algorithm. Proc. Symosium on Principles of Database Systems (PODS), pp247-255, 2001.

[4] M. Atzori, F. Bonchi, F. Giannotti and D. Pedreschi. Anonymity preserving pattern discovery. VLDB J. 17(4): 703-727 (2008)

[5] M. Atzori, F. Bonchi, F. Giannotti, and D. Pedreschi. Blocking anonymity threats raised by frequent itemset mining. ICDM 2005.

[6] M. Atzori, F. Bonchi, F. Giannotti, and D. Pedreschi. k-anonymous patterns. PKDD 2005.

[7] R. J. Bayardo and R. Agrawal. Data privacy through optimal k-anonymisation. ICDE 2005.

[8] L. Backstrom, C. Dwork and J. Kleinberg. Wherefore Art Thou R3579x?: Anonymized Social Networks, Hidden Patterns, and Structural Steganography. WWW 2007.

[9] R. Evfimievski, R. Srikant, R. Agrawal, and J. Gehrke. Privacy preserving mining of association rules. SIGKDD 2002.

[10] J. K. Friedman, J. L. Bentley, R. A. Finkel. An algorithm for finding best matches in logarithmic expected time, ACM Trans. on Math. Software, 3(1977), pp. 209–226.

[11] D. Frankowski, D. Cosley, S. Sen, L. G. Terveen and J. Riedl. You are what you say: privacy risks of public mentions. SIGIR 2006: 565-572

[12] B. C. Fung, K. Wang, and P. S. Yu. Top-down specialization for information and privacy preservation. ICDE 2005.

[13] M. R. Garey and D. S. Johnson. Computers and Intractability: A Guide to the Theory of $\mathcal{NP}$-Completeness. San Francisco. Freeman, 1979

[14] G. Ghinita, Y. Tao and P. Kalnis. On the Anonymisation of Sparse High-Dimensional Data, In Proceedings of International Conference on Data Engineering (ICDE) April 2008.

[15] K. Hafner. And if you liked the movie, a Netflix contest may reward you handsomely. New York Times, Oct 2 2006.

[16] S. Hansell. AOL removes search data on vast group of web users. New York Times, Aug 8 2006.

[17] R. W. Hamming. Coding and Information Theory, Englewood Cliffs, NJ, Prentice Hall (1980)

[18] V. Iyengar. Transforming data to satisfy privacy constraints. SIGKDD 2002.

[19] K. LeFevre, D. DeWitt, and R. Ramakrishnan. Incognito: efficient full-domain k-anonymity. SIGMOD 2005.

[20] K. LeFevre, D. DeWitt, and R. Ramakrishnan. Mondrian multidimensional k-anonymity. ICDE 2006.

[21] J. Li, Y. Tao and X. Xiao. Preservation of Proximity Privacy in Publishing Numerical Sensitive Data. ACM Conference on Management of Data (SIGMOD), 2008

[22] N. Li, T. Li and S. Venkatasubramanian. t-Closeness: Privacy Beyond k-anonymity and l-diversity. *ICDE 2007*: 106-115

[23] A. Machanavajjhala, J. Gehrke, D. Kifer, and M. Venkitasubramaniam. l-Diversity: Privacy beyond k-anonymity. ICDE 2006.

[24] A. Narayanan and V. Shmatikov. Robust De-anonymisation of Large Sparse Datasets. to appear in IEEE Security & Privacy 2008.

[25] P. Samarati and L. Sweeney. Generalizing data to provide anonymity when disclosing Information. PODS 1998.

[26] P. Samarati and L. Sweeney. Protecting privacy when disclosing information: k-anonymity and its enforcement through generalization and suppression. *Technical Report SRI-CSL-98-04*, SRI Computer Science Laboratory, 1998.

[27] P. Samarati. Protecting respondents' identities in microdata release. *IEEE Transactions on Knowledge and Data Engineering*, 13(6): pp: 1010-1027. 2001.

[28] X. Sun, H. Wang and J. Li. Injecting purposes and trust into data anonymization. in CIKM 2009.

[29] X. Sun, H. Wang, J. Li and J. Pei. Publishing anonymous survey rating data. Data Min. Knowl. Discov. 23(3): 379-406 (2011)

[30] X. Sun, H. Wang, J. Li and Y. Zhang. Satisfying Privacy Requirements Before Data Anonymization. The Computer Journal. doi : 10.1093/comjnl/bxr028. 2011.

[31] L. Sweeney. Weaving technology and policy together to maintain confidentiality. J. of Law, Medicine and Ethics, 25(2ï£¡C3), 1997.

[32] L. Sweeney. k-Anonymity: A Model for Protecting Privacy. *International Journal on Uncertainty Fuzziness Knowledge-based Systems*, 10(5), pp 557-570, 2002

[33] T. M. Traian and V. Bindu. Privacy Protection: p-sensitive k-anonymity Property. International Workshop of Privacy Data Management (PDM2006), In Conjunction with 22th International Conference of Data Engineering (ICDE), Atlanta, 2006

[34] V. S. Verykios, A. K. Elmagarmid, E. Bertino, E. Dasseni and Y. Saygin. Association Rule Hiding. IEEE Transactions on Knowledge and Data Engineering, vol. 16, no. 4, pp. 434-447, April 2004.

[35] I. Witten and E. Frank. Data Mining: Practical machine learning tools and techniques. Morgan Kaufmann, San Francisco, 2nd edition, 2005.

[36] X. Xiao and Y. Tao. Anatomy: simple and effective privacy preservation. VLDB 2006.

[37] Y. Xu, K. Wang, Ada Wai-Chee Fu and Philip S. Yu. Anonymizing Transaction Databases for Publication. KDD 2008.

13

Concurrent Operations of O_2-Tree on Shared Memory Multicore Architectures

Daniel Ohene-Kwofie[1,†], E. J. Otoo[1,*], Gideon Nimako[2]

[1]School Of Electrical and Information Engineering, University Of the Witwatersrand, South Africa
[2]School of Public Health, University Of the Witwatersrand, South Africa

Abstract

Modern computer architectures provide high performance computing capability by having multiple CPU cores. Such systems are also typically associated with very large main-memory capacities, thereby allowing them to be used for fast processing of in-memory database applications. However, most of the concurrency control mechanism associated with the index structures of these memory resident databases do not scale well, under high transaction rates. This paper presents the O2-Tree, a fast main memory resident index, which is also highly scalable and tolerant of high transaction rates in a concurrent environment using the relaxed balancing tree algorithm. The O2-Tree is a modified Red-Black tree in which the leaf nodes are formed into blocks that hold key-value pairs, while each internal node stores a single key that results from splitting leaf nodes. Multi-threaded concurrent manipulation of the O2-Tree outperforms popular NoSQL based key-value stores considered in this paper.

Keywords: Pessimistic Concurrency, Indexing, In-Memory Databases,Performance, Algorithms

1. Introduction

Indexes in database managements system (DBMS) facilitate fast query processing. Tree structured indexes in particular, are critical to database processing systems since they allow for both random and range query processing. Today's data processing tasks in transaction processing, scientific data management, financial analysis, network monitoring, data analytics, etc., handle large volumes of data which require fast accesses and very high throughput.

Recent advances in memory architectures, with 64-bit addressing, now allow for memory sizes of the order of hundreds of gigabytes and beyond at a reasonable cost. It is, therefore, feasible to have sufficiently large shared memory such that the entire index of either, a memory resident or disk-resident database, can be maintained in main memory. For instance, the latest Oracle Exadata X3-8 system ships with 4TB of main memory [25]. This has, therefore, motivated much research to exploit memory as well as the many-cores available on such architectures to provide fast application processing for main-memory databases.

Recently, there has been a flood of developments and implementations of in-memory data stores with associated index schemes. These are characterised in general as *NoSQL* databases. They are also referred to as *key-value* pair index structures [21]. Notably in this pack are index schemes such as BerkeleyDB [26], LevelDB [12], Kyoto Cabinet [10], RedisIO [30] and MongodB [1]. Such in-memory indexes, optimized for in-memory databases and running on multi-core processors, can support very high query processing rates. The challenge with such systems is how to efficiently ensure that the concurrently executing processes are isolated from each other in such an

---
*Corresponding author. Email: papaotu@gmail.com
†Corresponding author. Email: danielkwofie@gmail.com

environment. Current DBMS typically rely on locking but in a traditional implementation with a separate lock manager, the lock manager becomes a bottleneck and results in much overhead cost, especially at high transaction rates [17].

In this paper we present an in-memory index structure, referred to as O_2-Tree with emphasis on its implementation in a shared memory multi-core architecture. Such achitectures have a common shared memory that can be accessed by multiple programs running concurrently on several cores, as a result of the multiprocessing design capability, on the same node. We address primarily its concurrency control and fault recovery mechanism. The O_2-Tree is essentially a Red-Black Binary Search Tree in which the leaf nodes are data blocks that store multiple records of "key-value" pairs. The internal nodes contain copies of the keys that result from splitting the blocks of the leaf nodes in a manner similar to the B$^+$-Tree. However, the O_2-Tree is structurally different from the minimal order of a B$^+$-Tree also called the 2−3-Tree. A 2−3-Tree is a tree in which each internal or non-leaf node has either 2 or 3 children, all leaf nodes are at the same level, and every node may contain 1 or 2 keys. On the other hand, all leaf nodes of the O_2-Tree may not necessary be at the same level but the leaves contain $m \geqslant 2$ keys. One could question whether the AA-Tree [2] could not be used in place of the Red-Black-Tree (or RB-Tree) that the O_2-Tree uses for the internal nodes. The answer is yes and this leads to a generalization of the O_2-Tree where the internal nodes could be organized as an AVL-Tree [?], a 2−3-Tree [?], an AA-Tree [?] or a SkipList [?].

In the O_2-Tree, internal nodes are simply binary placeholders or routers to facilitate and guide the tree traversal. The tree index is fault tolerant in the sense that it is easily reconstructed by reading only the lowest key values of each leaf node that is always made persistent. It is inherently persistent and scales well in highly concurrent environment.

We use a pessimistic concurrency control, but allow multiple *readers* to proceed without blocking internal nodes except for leaf nodes where an updater needs to hold a lock. This allow us to reduce the lock overhead due to blocking of concurrent interleaved query operations. We achieve further performance gains by using the following mechanisms; search operations are interleaved using the hand-over-hand (also referred to as lock-coupling) locking technique; and mutations perform rebalancing separately which encompasses smaller fixed sized atomic regions.

We use the relaxed balance algorithm for Red-Black Tree presented by Hanke et al.[13], to maintain the invariants of the O_2-Tree. We have explored and evaluated the O_2-Tree, and done extensive experimental evaluations and comparisons with some of the well known key-value storage schemes, in multi-core environment under high contentions and index workloads. The experiments confirmed that the concurrent O_2-Tree has a superior performance compared to popular NoSQL key-value stores (*Tuple Store category*) which are often used as in-memory database indexes. These include the BerkeleyDB key-value store (BerkeleyDB), the TreeDB of Kyoto Cabinet and Google's LevelDB.

The major contribution being reported in this paper is the development, implementation and comparative experimental tests of the O_2-Tree main memory index structure. This is usable as a *NoSQL* key-value store for database systems that require a high performance concurrent access in shared memory multi-core architectures. We present results which show that the O_2-Tree in-memory index has high scalability in highly shared concurrent environment, and performs comparatively better than most popular *NoSQL* key-value storage schemes.

The remainder of this paper is organised as follows. Section 2 presents the background of our study. In Section 3, we describe the O_2-Tree in-memory index and present our basic algorithms for concurrency control. A mechanism for persistent storage and recovery is presented in Section 4. In Section 5, we describe our experimental setup and report the performance results of the experimental comparative study of the O_2-Tree with representative NoSQL key-value stores. We conclude in Section 6 and give some directions for future work.

2. Related Work

Tree structured index operations are fundamental in database management systems (DBMS). These provide for fast transaction processing in the DBMS. They allow for both efficient random as well as sequential processing of keys and are therefore widely used in DBMS. Recent advances in main memory technology and the availability of configured systems with memory sizes of the order of hundreds of gigabytes and tens of terabytes have motivated several research in developing main memory index schemes [4, 15, 18, 20]. The usage is such that the index of a main memory resident database or a disk-resident database is kept entirely in memory for high transaction throughput. Some of the widely used tree-based index structures include the B$^+$-Tree, and the T-Tree. However, recently a number of such index-driven databases have emerged under the banner of *NoSQL* databases. *NoSQL* stores consist basically of a key-value pair and and as such these databases are able to scale easily.

The B$^+$-Tree [3, 8] is one of the well studied and well understood index structure for database systems. It is generally characterised as a multi-way search tree of order m in which each node holds at least $\lceil m/2 \rceil - 1$ and at most $m - 1$ data item. B$^+$-Tree was specially designed to speed-up index searches on disk-based DBMS. In such DBMS the number of disk accesses to retrieve a record, is proportional to the height h of the tree, where $h \leq \log_{\lceil m/2 \rceil} N$ for a tree of order m or *fanout* of m. B$^+$-Tree therefore has a significantly low height for a high *fanout*.

An alternative to the B$^+$-Tree, designed specifically for main-memory indexing, is the T-Tree [18]. It was proposed as the preferred index structure for main-memory databases. Though the T-Trees has less storage overhead than the B$^+$-Tree, research in [27, 28] has shown that the B$^+$-Tree is able to efficiently utilise the cache line in modern processors to provide a better performance. Another index structure which has been widely studied is the Red-Black binary Tree (or RB-Tree) [9]. It is noteworthy that in the use of an RB-Tree as main-memory index, each internal node stores a key-value pair while external nodes are represented as NULL values. The RB-Tree provides an efficient scheme for main memory indexing. However, the performance deteriorates as the datasets become very large. This is due to the fact that, the height of the tree increases greatly and hence traversals and restructuring after updates become expensive especially in concurrent environment with high contention. Further, the CPU cache-line is poorly utilised since each node, either internal or leaf, is visited once for a single key-value access.

Restructuring of the RB-Tree after insertions and deletions can be done during the *top-down* traversal before the operation or *bottom-up* after the operation. One would expect that the concurrency control in RB-Tree would be efficiently implemented with *top-down* insertions and deletions algorithms. Unfortunately standard top-down restructuring algorithm, does not scale well with the RB-Tree and other index structures in general. The process of restoring the tree's invariant becomes a bottleneck for concurrent tree implementations. The mutating operations must acquire not only locks to guarantee the atomicity of their operations, but also locks to guarantee that no other mutation affects the balance condition of any nodes or the sub-tree that will be involved in the restoration process. The strict standard top-down algorithm limits the amount of concurrency of the index since every update will proceed with several top-down balancing steps before exiting. This difficulty led to the idea of relaxed balance trees [13, 16, 22].

The relaxed balance techniques, effectively uncouple the mutating operations from the restructuring operations by allowing the invariants to be violated but restored by separate rebalancing operations [5, 6, 13, 14, 16, 22, 23]. These separate rebalancing operations involve only local changes. Larsen [16], showed that for a relaxed RB-Tree the number of restructuring changes after update is bounded by $O(1)$ and the number of color changes by $O(\log n)$, where n is the size of the tree. The process of restoring the invariants in relaxed RB-Tree has an amortized constant of $O(1)$ [16].

Concurrent control algorithm for relaxed balance tree implementations based on fine-grain read-write locks provide good scalability for tree-indexes. Optimistic concurrency control (OCC) schemes using version numbers are also attractive for concurrency control especially for in-memory index. They naturally allow readers to proceed without locks, and thus avoid the coherence contention inherent in read-write locks. The readers simple read version numbers updated by writers to detect concurrent mutations since readers assume that no mutation will occur during access to a critical region. They retry if that assumption fails i.e if a mutation occurs. This could however, lead to spurious retries and wasted work. Software transactional memory (STM) provides a generic implementation of optimistic concurrency control. STM groups shared-memory operations into transactions that appear to succeed or fail atomically. The aim of STM is to deliver a simple parallel programming at an acceptable performance. However, performance gains and scalability are amongst the most important goal of a data structure library, and not just simplicity [7]. In practice STM systems also suffer a performance hit relative to fine-grain lock-based systems on small numbers of processors (1 to 4 depending on the application) [7].

In this paper, we present the concurrent operations of the O_2-Tree memory resident index structure that can be used also as a persistent key-value store. It utilizes an in-memory cache for the leaf nodes and a fine-grain relaxed balance concurrent algorithm in a manner similar to the approach in [16]. This effectively allows for greater degree of concurrency in the O_2-Tree. We discuss this in detail in Section 3. The distinctive differences in the T-Tree, B$^+$-Tree, RB-Tree and the O_2-Tree are clearly illustrated in Figure 1. The approach we advocate here where the RB-Tree is used as the memory resident index can easily be generalised to replace the internal RB-Tree with any of the following: a 2-3-Tree, an AA-Tree and a SkipList.

3. The O_2-Tree In-memory Index

3.1. Structure of the O_2-Tree

The O_2-Tree is basically a binary search tree, managed as a Red-Black Binary-Search Tree, whose leaf nodes are organised as index blocks, data pages, or chunks that store records of "key-value" pairs of the form ⟨key, value⟩. The "value" may also represent a pointer to the location where the record is held in memory. In which case we could also denote it as "⟨key, recptr⟩", where "recptr" denotes the record pointer.

The internal nodes contain copies of only the keys of the middle "key-value" pairs that split the leaf nodes when they become full. These internal nodes are formed into a simple binary search tree that is balanced using the RB-Tree rotation algorithms. Let K_s be the search key and let K_p be key stored at a node p. During a traversal from the root node to a leaf node, a left branch of the node p is followed if $K_s < K_p$ and the right branch is followed if $K_s \geq K_p$. The process continues until a leaf node is reached.

We adopt the RB-Tree balancing algorithm for the O_2-Tree since it is less complex than that of the AVL-Tree which has a more strict balancing condition. The RB-Tree has been widely studied and known for its excellent performance. The O_2-Tree structure, however has a number of advantages over existing indices such as the T-Tree and some of the recent NoSQL key-value stores. The O_2-Tree can easily be reconstructed by reading only the lowest "keys" of each of the leaf nodes. Further, by maintaining only the leaf nodes persistent, the index tree is inherently persistent. The height of the internal RB-Tree is also significantly reduced compared to the situation where each node stores a single "key-value" pair and the entire tree is maintained as a simple RB-Tree. By grouping multiple 'key-value" pairs in the leaf nodes, we optimise the tree so that it also exhibit much better cache sensitivity especially during operations of the leaf nodes. The leaf nodes are therefore able to utilise the cache-line architectural features of the machine, and as such reduce the number of cache misses which would have otherwise resulted from making single node comparison of "key-value" pair. We also achieve significant performance gains by doing single data comparison internally per node during traversal, unlike other structures such as the B$^+$-Tree and the T-Tree that require at most m comparisons.

The *order* of the tree, denoted by m, is the maximum number of "key-value" pairs a leaf node can hold. Data is stored in the leaf nodes; whiles the internal nodes

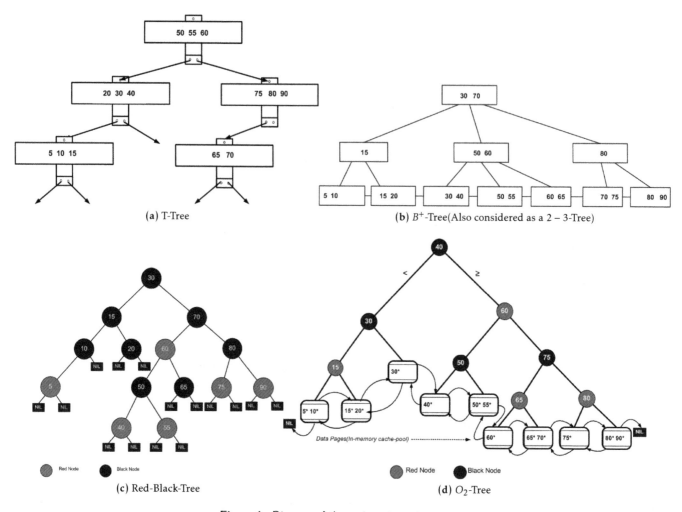

(a) T-Tree

(b) B^+-Tree(Also considered as a 2 − 3-Tree)

(c) Red-Black-Tree

(d) O_2-Tree

Figure 1. Diagram of the various tree structures

are simply binary place holders that facilitate or guide the tree traversal to reach a leaf node. All successful or unsuccessful searches always terminate at a leaf node. This is reminiscent of the search process in a B^+-Tree except that now internal nodes hold only single key values as opposed to m key values. Figure 1d illustrates the schematic layout of the O_2-Tree of order $m = 3$. We show only the keys in the leaf nodes. The corresponding equivalent Red-Black-Tree is shown side-by-side in Figure 1. Detailed explanation of the RB-Tree can be found in [9].

The properties of the O_2-Tree index include all of the RB-Tree [9] properties, plus the following:

1. Each internal node holds a single key value which is a copy of the minimum key value at the leaf node. These keys are equivalent to the middle keys after a leaf node splits.

2. Leaf-nodes are blocks that have between $\lceil m/2 \rceil$ and m "key-value" pairs.

3. If a tree has a single node, then it must be a leaf which is the root of the tree, and it can have between 1 to m key values.

4. Leaf nodes are doubly-linked in forward and backward directions. These links provide easy mechanism to traverse the tree in sorted order for key range searches.

We implemented the O_2-Tree index structure as a persistent key-value store by reading and writing the leaf-nodes using an in-memory cache pool in which the leaf nodes of blocks of key-values pairs are managed by the *BerkeleyDB Mpoolfile* subsystem. The BerkeleyDB Mpool subsystem is a general-purpose shared memory buffer pool which can be used for page-oriented, shared and cached file access. The BerkeleyDB Mpool library implementation uses the same on-disk format as its in-memory format as well. This provides a simple mechanism to flush cached pages since a page can be flushed from the cache without format conversion [21]. The internal nodes of the O_2-Tree provide simply binary place-holders for fast tree index traversal. New internal nodes are only added when leaf-nodes split as a result of overflows. The index tree may grow in height after a split of a leaf-block. The reverse occurs when there is an underflow resulting in the merging of leaf-nodes and the subsequent removal of the parent of the nodes that are merged. It should be noted here that this does not constitute an implicit use of BerkeleyDB Tree search. Only the cache functionality of the *Mpool* subsystem is used.

3.2. Some Analytical Results

We state some analytical properties of the O_2-Tree without formal proofs.

Proposition 3.1. In the O_2-Tree, the black leaf-nodes of blocks of "key-value" pairs remain as leaf nodes under all rotations of the internal nodes which are structured as a Red-Black tree.

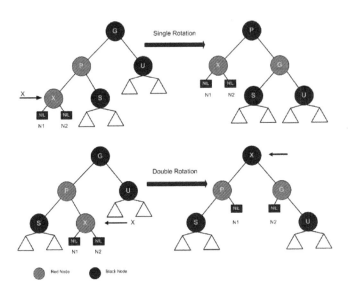

Figure 2. Single and double rotations in a Red-Black Tree

Proof. In a Red-Black Tree, the two rotations used to restore the tree's invariant after update operations are the single and double rotations. Rotations basically swap (using pointer manipulations) the roles of the parent and the child while maintaining the search order

of the binary tree. Single and double rotations are illustrated in Figure 2. Only leaf-nodes affected by the rotation are indicated. A single rotation between P and G restores the tree's balance after insertion of X caused a violation. It is evident from the illustration that, all leaf-nodes (NIL) remain leaves even after the single rotation. Similarly, leaf-nodes in a double rotation still remain leaf-nodes. Though node X, has become the new parent of the sub-tree, its leaves $N1$ and $N2$ still remain leaves but with different parents and the binary search order is still maintained. ☐

Proposition 3.2. An O_2-Tree with n black leaf-nodes will still maintain its n black leaf-nodes after single or double rotations.

Proposition 3.3. The O_2-Tree, supports the query operations of *Put()*, *Delete()*, and *Get()* in time $O(\log_2 N/\lceil m/2 \rceil)$, where N is the number of "key-value" pairs in the structure.

Proof. This follows from the fact that the number of leaf-node blocks is at most $n_b = N/\lceil m/2 \rceil$. The number of nodes of supporting internal RB-Tree is $n_b - 1$. The height h of the internal RB-Tree is given by $h \leq \log_2 n_b$ [9]. This implies that a search (given by *Get()*), and insertion (given by *Put()*) and a deletion (given by *Delete()*) is each computed in time $O(\log_2 N/\lceil m/2 \rceil)$. ☐

Proposition 3.4. Assuming the response set of key-value pairs retrieved in a range search is s, such a range search can be carried out in an O_2-Tree of order m and N key-value pairs in time $O(\log_2 N/\lceil m/2 \rceil + s)$.

Proof. Given a lower bound k_l and an upper bound k_u values of keys, a range search retrieves the set of key-value pairs whose keys lie in the interval $[k_l, k_u]$. Using the key k_l the search for the leaf node bucket B_l that should contain the key k_l is first retrieved. This takes time $O(\log_2 N/\lceil m/2 \rceil)$. Once the bucket B_l is retrieved, the forward pointer from this bucket, and all subsequent buckets, can followed to retrieve all the key-value pairs whose their keys are less than k_u. The process stops when the maximum key value k_u is retrieved. The sequential scan performed, retrieves only leaf-buckets that contain s key-value pairs satisfying the range search. The total time to conduct the range search is therefore $O((\log_2 N/\lceil m/2 \rceil) + s)$. □

3.3. Concurrency Control in the O_2-Tree

We present our concurrent control scheme based on the relaxed balance RB-Tree algorithm by Larsen [16], but we manage our index structure such that the number of restructuring steps after mutation operations is further reduced. To achieve maximum concurrency, we implement the thread-safe algorithm with *page-level* or *node-level* locking. In this case, each node can be locked and unlocked. This simple fine-grain lock-coupling technique ensures that multiple threads can proceed concurrently as long as they don't interfere with each other at the same node. We use three locks as in [23, 24] which we denote as *rlock, wlock, and xlock*. Several user processes can *rlock* a node at the same time, whereas, only one process can *wlock* a node at a time but can coexist with other processes with *rlock* on the same node. *xlock* on the other hand ensures exclusive access to a node and cannot coexist with any other process.

The entire process of handling contentions in the tree is also handled by a rebalancing process which we denote as the *rebalancer()* process and runs in the background. The *rebalancer()* process locates nodes in the tree with conflicts and resolves them appropriately. We adopt the *problem queue* approach to manage contentions instead of random traversal by the *rebalancer()* which could result in several interferences with other query processes and causes degradation in the performance of the index. Let a user operation intending to insert/delete a "key-value" pair be denoted as an *updater()* process. In the problem queue approach, when a lock conflict situation is created in the tree, a pointer to the parent of the node involved is placed in the *problem queue*. The *rebalancer()* continuously reads the queue and purposefully proceeds to the exact location to fix the imbalance. The tree is balanced if the *problem queue* is empty. We implemented a concurrent problem queue to allow for simultaneous *push()* and *pop()* operations such that neither the *rebalancer()* nor the *updater()* processes are blocked while traversing and manipulating the O_2-Tree. An *updater()* appends requests to the *tail* of the *problem*

queue, while the *rebalancer()* pops these request from the *head* of the queue. This prevents interference as much as possible and guarantees consistency between *updaters()* and the *rebalancer()* processes. The problem queue is temporarily locked and released both by an *updater()* and a *rebalancer()* during times that they access the problem queue only.

Before presenting the algorithm for the concurrent operations, we first define the following notations. Let T denote an O_2-Tree. The root node will be designated as $Root(T)$ whose parent is the *header* of the index. If z denotes an *Internalnode* in T, then $z.left$ and $z.right$ refer to the left and right child respectively of z. Let $z.parent$ denote the parent of z and let $z.sibling$ refer to the sibling of z such that z and $z.sibling$ have the same parent (i.e., if z is a left child of its parents then $z.sibling$ will be the right child of the parent and vice versa). Also $z.key$ is the value of the key in z, if z is an internal node (i.e., $nodeType \neq leaf$). In addition, $z.key[i]$ and $z.value[i]$ refers to the key and value respectively in the ith position of z given that z is a page block (i.e., $nodeType = leaf$).

Exact–Match Search Algorithm: $Get(x, T)$. The $Get(key\ x)$ function returns the exact-match key-value pair $\langle x, val_x \rangle$ associated with the key x from the data store T, if x exist, otherwise a null value is returned. The search traverses nodes from the root by lock-coupling with *rlocks* until the the leaf page with the given x is found. Once the leaf-page z, in which the search key x resides, is located, we utilise a binary search function $binarySearch(x, z)$ to locate the "key-value" pair $\langle x, val_x \rangle$ from z. Unlike the T-Tree and the B^+-Tree, the search proceeds with only one key comparison in the internal node. The T-Tree and the B^+-Tree do on the average $m/2$ comparisons before continuing with the search for a given key. The thread-safe search algorithm for the O_2-Tree is given in Algorithm 1.

Range Search Algorithm (*Get Next, Get Previous*). The range search traverses nodes from the root by lock-coupling with *rlocks* just as in the exact-match search query. The search begins by locating the minimum key, x_{min} in the given range e.g., $x_{min} \leq x \leq x_{max}$, where x_{max} is the maximum key in the range. The search returns the range of key-value pairs $\langle x, val_x \rangle$ within the specified

---

Algorithm 1: *Get(key x)*

Data: *key x, T*

Result: corresponding $\langle x, val_x \rangle$ pair if *found*,
otherwise *null*

1 **begin**
2   /* *node is the current node pointer for traversal*/
3   *node* ← *root(T)*
4   *node.rlock()*
5   **while** *node.nodeType ≠ leaf* **do**
6   **if** *x < node.key* **then**
7   *node.left.rlock()*
8   *node.unlock()*
9   *node ← node.left*
10   **else**
11   *node.right.rlock()*
12   *node.unlock()*
13   *node ← node.right*
14   *done ← binarySearch(x, node)*
15   *node.unlock()*
16   **return** *done*

Algorithm 2: Range Scan(*key x_{min}, key x_{max}*)

Data: *key x_{min}, key x_{max}*

Result: corresponding key-value for each existing
key in the range

1 **begin**
2   *x ← x_{min}*
3   *node ← root(T)*
4   *node.rlock()*
5   **while** *node.nodeType ≠ leaf* **do**
6   **if** *x < node.key* **then**
7   *node.left.rlock()*
8   *node.unlock()*
9   *node ← node.left*
10   **else**
11   *node.right.rlock()*
12   *node.unlock()*
13   *node ← node.right*
14   **return** *rangeScan(x_{min}, x_{max}, node)*
15   *∗Range search starting from the min key in the range from the current leaf*
16   *∗Continue scan in the next leaf if the maximum key in the range is not encountered*

---

range. Once the leaf page with key x_{min} is located, the algorithm proceeds with a sequential scan of the leaf node until the last key in that node is reached. If the maximum key x_{max} is still not located, the scan continues with an *rlocks* to the next leaf following the *forward-link* pointer between the leaf nodes. This continues until the last key x_{max} in the range is found. The algorithm is illustrated in Algorithm 2.

Insert and Update Algorithm: *Put(x, val_x, T)*. The *Put()* operation proceeds with a traversal similar to that of the *Get()*. However, a much more elegant approach is to use a *wlock*, which allows several *rlock* by other threads on the resource but not another *wlock* or *xlock*. This allows for interleaved *Get()* operations to overtake *updater()* operations if necessary and not be blocked. To insert the key-value pair $\langle x, val_x \rangle$, the leaf page (denoted as *node*) in which the key-value pair belongs is first located. Once the page is located, it is locked exclusively with an *xlock* and if there is room, the new key-value pair $\langle x, val_x \rangle$, is inserted in order

by the function *insertInOrder(x, val_x, node)* into the page, based on the value of the key *x*. If the page is already full, then a split is performed using the function *splitInsert(x, val_x, node)* (see Algorithm 4), where *node* is the leaf-node to be split. A split basically allocates a new page in the in-memory cache pool and assigns half of the key-value pair $\langle x, val_x \rangle$ from the overflow page to the new page. The *previous* and *next* page pointers are updated appropriately. After the split, a new internal node is inserted which becomes the parent of the two page blocks. The new internal node is coloured *Red*. The tree may grow in height only when a page (leaf-node) overflows. If the operation results in the violation of the invariant condition, the parent of the new parent node is pushed to the problem queue. The thread-safe *Put* algorithm is presented in Algorithm 3.

Delete Algorithm: *Delete(x, T)*. The delete algorithm follows a similar pattern as the insert algorithm. However, the delete may result in page underflow. In

Algorithm 3: Put(*key x, value val_x, T*)

Data: *key x, value val_x T*
Result: *true* for success *false* otherwise

```
1  begin
2      /*node is the current node pointer for traversal */
3      parent ⟵ header
4      node ⟵ root(T)
5      parent.wlock()
6      node.wlock()
7      while node.nodeType ≠ leaf do
8          if x < node.key then
9              node.left.wlock()
10             parent.unlock()
11             parent ⟵ node
12             node ⟵ node.left
13         else
14             node.right.wlock()
15             parent.unlock()
16             parent ⟵ node
17             node ⟵ node.right
18     /* Upgrade both node and parent locks to xlock */
19     parent.xlock()
20     node.xlock()
21     if !(node.isfull) then
22         done ⟵ insertInOrder(x, val_x, node)
23     else
24         done ⟵ splitInsert(x, val_x, node)
25         Update problem queue if invariant is violated
26     parent.unlock()
27     node.unlock()
28     return done
```

Algorithm 4: splitInsert(*Key x, value val_x, O2node node*)

Data: *key x, value val_x, O2node node*
Result: *true* for success *false* otherwise

```
1  begin
2      newPage ⟵ new leafPage()
3      newNode ⟵ new internalNode()
4      midpoint ⟵ m/2
5       where m is the order of the tree */
6      j ⟵ 0
7      for i ⟵ midpoint to m − 1 do
8          newPage[j ++] ⟵ node.remove(i)
9      /* insert "key, value" into the appropriate leaf page */
10     newNode.key ⟵ newPage.key[0]
11     newPage.parent ⟵ newNode
12     node.parent ⟵ newNode
13     /* reset forward and backward links of leaf nodes */
```

3.4. Correctness

The concurrent protocol presented guarantees linearisability as well as deadlock freedom. This ensures correctness of all transactions. The algorithm does define lock order for traversals such that all request are made in the same top-down approach. This ensures freedom from deadlock. For instance, a request by one thread for a lock on a child node can only be granted after a lock request on the parent node has been granted. Each critical region preserves the binary search tree property. The lock ordering ensures that there is no deadlock cycle loop where a thread, T_1 waits on a lock by another thread, T_2 whiles T_2 waits on a lock held by T_1. Since no such loop exists in the tree structure, and all parent-child relationships are protected by the required locks to make them consistent, the concurrent protocol algorithm is deadlock free.

In order for the algorithms to behave as expected in a concurrent environment, they require that their implementations be linearisable. This implies that operations for a particular key produce results consistent with sequential operations on the tree-index structure. Atomicity and ordering is trivially provided between *Put()* and *Delete()* operations by the *wlock* hand-over-hand tree traversal. This ensures that no

this case, either key-value pairs ⟨*x, val_x*⟩ are borrowed from adjacent pages (*previous or next pages*) or pages are merged with the leaf-node that underflowed and the other page is deallocated or released into the cache pool. A merger of pages also results in the the subsequent removal of the parent node. If this results in the violation of the invariant condition, the grandparent of the new parent node is pushed to the problem queue. The thread-safe delete algorithm is as given in Algorithm 5.

Algorithm 5: Delete(*key x, T*)

Data: *key x T*

Result: *true* for success *false* otherwise

1 **begin**
2   /* *minKeys ensures that node is at least half full* */
3      $minKeys \longleftarrow \frac{m}{2}$
4      $parent \longleftarrow header$
5      $node \longleftarrow root(T)$
6   *parent.wlock()*
7   *node.wlock()*
8   **while** *node.nodeType ≠ leaf* **do**
9   **if** *x < node.key* **then**
10   *node.left.wlock()*
11   *parent.unlock()*
12   *parent ← node*
13   *node ← node.left*
14   **else**
15   *node.right.wlock()*
16   *parent.unlock()*
17   *parent ← node*
18   *node ← node.right*
19   /* *Upgrade both node and parent locks to xlock* */
20   *parent.xlock()*
21   *node.xlock()*
22   *done ← removeKey(x, node)*
23   **if** *done* **and** *node.underflow()* **then**
24   *sibling.xlock()*
25   **if** *node.sibling.keys > minKeys* **then**
26   /* *Borrow from sibling to keep occupancy* */
27   *done ← node.appendKeyFrom(sibling)*
28   **else**
29   /* *Merge leaf and sibling into the left node; release page block and delete parent node* */
30   *done ← mergeLeaf(node, node.sibling)*
31   *sibling.unlock()*
32   *parent.unlock()*
33   *node.unlock()*
34   **return** *done*

---

two of such operations overtake or interfere with each other. It is not possible for two threads, T_1 and T_2, to lock the same node resource simultaneously. This ensures that the updates are serialised. More over,

each critical region during a mutation operation, only changes child and parent links after acquiring all of the required locks, hence guaranteeing the atomicity of the transaction.

3.5. Storage Utilisation

The expected storage utilisation the O_2-Tree, from the fact that it grows and shrinks from block splitting and merging respectively, is $O(\ln 2)$. It can easily be shown using a similar approach as in the approximate storage utilisation of B-Trees [19]. Let N be the total number of keys in the tree and let n denote the number of index blocks at the leaves of the tree. Let m be the *order* of the tree. Each leaf block has at least $\lceil m/2 \rceil$ and m keys. The storage utilisation denoted by μ is the total number of keys stored divided by the total storage capacity of all the nodes.

$$\mu = \frac{N}{m \times n}$$

The expected storage utilisation is

$$E(\mu) = \frac{N}{m} E\left(\frac{1}{n}\right)$$

To evaluate $E(1/n)$ we note that n lies in the interval $[N/m, 2N/m]$. By approximating the distribution as a continuous random rectangular distribution over the interval, we have

$$E(\mu) \approx \frac{N}{m} \int_{N/m}^{2N/m} \frac{dn}{n} = \ln 2.$$

4. Persistence and Recovery

A major concern with main-memory databases and and their memory resident indexes is the guarantee of the database persistence, recovery and fault-tolerance. Since main memory is volatile, it is essential that one adopts recovery techniques for the entire database as well as the index, such that the mechanism to restore the database to a consistent and operational state is not expensive and time consuming. An expensive and time consuming recovery index technique will obviously become a bottleneck in the overall performance of the database. Fast recovery mechanisms are essential to ensure that the database and its associated index can be quickly repaired and restored into a *usable state* from which normal processing can resume. The faster the index can be restored or recovered, the less impact it will have on the performance of the entire database recovery process. Generally, transactional logging, check-pointing and reloading techniques are employed. Logging maintains a log of transactions that occur during normal execution, whereas check-pointing takes a snapshot of the database periodically and copies it onto persistent storage for backup purposes. After a system failure, the persistent copy of the database is reloaded into main memory. The indexes are rebuilt and the database is then restored to a consistent state by applying information in the undo and redo logs to the reloaded copy.

Since disk (persistent storage) reads are expensive, reducing the disk overhead during recovery from persistent dumps is very crucial in designing the recovery techniques for in-memory databases. The O_2-Tree in-memory key-value store ensures persistence by accessing the leaf-pages through the in-memory cache pool. A separate thread periodically flushes dirty pages to the persistent store asynchronously.

The O_2-Tree persistent store provides an efficient and simply approach for index recovery. The reason being that rebuilding the index structure of the O_2-Tree from persistent store, unlike the $B^+ - Tree$ and the T-Tree structures, requires reading only the first key values in each of the leaf-page. This eliminates the performance bottleneck of traversing the entire "key-value" pairs of data in the leaf-pages. In systems where the index data is too large to fit into available memory, pages are paged-in and paged-out of the in-memory cache using a cache replacement policy such as the least recently-used protocol. In addition, bulk-loading the index from the persistent pages provides a much faster approach to restoration as the amount of restructuring is minimal.

Besides storing the leaf-pages by a background process, such that the entire RB-Tree structure can be rebuilt from the minimum key values of each leaf-page, the internal-nodes of the O_2-Tree that form the RB-Tree can be occasionally dumped onto disk during checkpoint or after each session of usage. Just before a session starts and as part of the initialisation phase, the RB-Tree can be restored from the persistent store.

5. Performance Evaluation

We evaluated the performance of the O_2-Tree index as a key-value persistent store, on the Intel Xeon E5630 CPU machine. We enabled hyper-threading for all performance evaluations. We conducted all the implementations and code compilation with the GNU GCC/G++ compiler on a 64-bit machine having a *72GB* of RAM and running the Scientific Linux release 5.4 Operating system. We generated 32-bit uniform distributed keys with which we formed key-value pairs where the values were also uniform random generated values. We also performed some experiments with live data read from the flight statistics datasets [11] as well as the records of the *Order* table generated from the TPC-H dbgen data generator [31].

5.1. Sequential Evaluations

For completeness, we present the comparative results of the performance of the O_2-Tree with the basic index structures such as the B^+-Tree, T-Tree, AVL-Tree, and the Top-Down Red-Black-Tree. Figure 3 shows the performance of the five data structures for a simple build of the index. We performed up to 50M unique key insertions. The *order* of the T-Tree, B^+-Tree, and the O_2-Tree used was $m = 512$. The graphs show the times for building the respective data structures in memory. As can be observed from the graphs, T-Tree performed worst among the index structures considered while the O_2-Tree had the best performance. The Top-down Red-Black Tree also performed better than the AVL-Tree. AVL's strict balancing requirement accounts for its worst performance. The O_2-Tree on the other hand, required fewer splits and rotations which accounts for its superior performance. The B^+-Tree performed better than the T-Tree due to its significant low depth and less complexity in restoring the tree's invariants.

The O_2-Tree, however, outperformed the B^+-Tree from the simple fact that the B^+-Tree makes multi-way-decision during its traversal down the tree while the O_2-Tree makes single data comparison to determine the search path during traversal. Splitting and redistribution of keys in the nodes of a B^+-Tree may continue all the way to the root of the tree. In the O_2-Tree only colour changes may propagate to the top. These operations are less expensive compared to the splitting and redistribution in B^+-Trees. The poor performance of the T-Tree and the B^+-Tree, compared to the O_2-Tree, is due to the fact that several data comparisons are required to locate a bounding node for child node to be visited during traversals.

Figure 4 illustrates the performance evaluation of these structures when subjected to interleaved mix of insertions, deletions and searches with different percentage of each operation for a total of 50 million (50M) query operations with a single thread. Figure 5 shows the operational throughput (*query operations per second*) for varying workloads with 50% updates. The query mix operations involved generating either an update (insertion or deletion) and conducting a lookup with varying probabilities. We refer to the probability of generating an update multiplied by 100 as the update ratio. Thus, a 0% update ratio indicates only data look-ups whiles a 100% update ratio indicates only update operations. Each update ratio consist of 30% deletions and 70% insertions. The preliminary results from the graphs indicates that the O_2-Tree clearly outperforms all the basic structures considered.

We however observed that for lower update ratios such as 0% and 10% , the T-Tree and the B^+-Tree provided better and comparable performance to the O_2-Tree. Figures 5 and 6 show the operational throughput (*query operations per second*) for varying workloads with 50% and 100% updates respectively. We observed a general decline in throughput as the workload and update ratios increase. However, the O_2-Tree index provided the best operational throughput in all cases.

We also present the results with the single threaded persistent implementation of the O_2-Tree index structure, where the key values and their associated data are kept persistent through an in-memory cache, with some NoSQL(key-value) data stores such as the BerkeleyDB [29], the Kyoto Cabinet TreeDB [10] and LevelDB [12]. These experiments were conducted primarily to compare the O_2-Tree index structure to other popular NoSQL(key-value) data stores that use tree structured access methods and have been reported in the literature as being extremely fast. The data operations are performed with the leaf nodes read and written through the memory resident cache. The leaf nodes are then written to disk using the Least Recently Used (LRU) cache replacement algorithm. At the end of a session, the cache is flushed so that all memory resident leaf-nodes are written to disk and the overall time to complete the operation is reported. We repeated this for varying data sizes.

We refer to the persistent O_2-Tree implementation as O_2-Tree-KV. This scheme was compared with other popular NoSQL key-value stores. Each key-value store was initialised with a page size of 4K, as well as a 2.5GB in-memory cache size. We applied the tuning mechanisms to the NoSQL databases as indicated and recommended in their respective documentations. Furthermore, we did not enable compression. The TPC-H generated dataset from the *Order* table was used for all experiments involving the persistent key-value stores.

Our results indicate that the O_2-Tree, using BerkeleyDB Mpool subsystem as a cache, outperformed the BerkeleyDB and the Kyoto Cabinet when using the TreeDB, by several orders of magnitude. However, O_2-Tree-KV performance is very comparable to that of the LevelDB. We actually did run the LevelDB in asynchronous mode in which a separate thread concurrently flushed the cache contents to disk. Even though in comparing the LevelDB and the BerkeleyDB with with O_2-Tree, the O_2-Tree had the disadvantage that it does not have asynchronous back-end persistent store. Our objective however, was to evaluate how the O_2-Tree-KV, in writing to disk through an in-memory cache, compared with these popular industry-standard NoSQL key-value stores without multi-threading.

The O_2-Tree-KV with a cache support however, performed over $5X$ faster than the Kyoto Cabinet when using the TreeDB access method of the Kyoto Cabinet and about $1.5X$ as fast as BerkeleyDB using the *B-Tree* access method. The results are as shown in Figure 7. The second set of experiment with the NoSQL databases, illustrated in Figure 8, show the performance of each key-value store under different mix of queries where the update ratio was varied from 0% to 100%. Again,

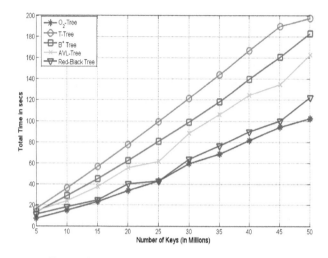

Figure 3. Index build with randomly generated keys

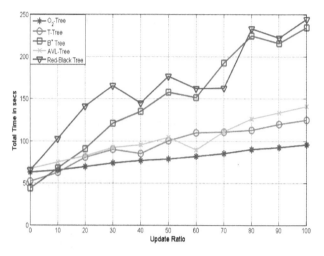

Figure 4. Mixed Operations of Searches, Inserts and Deletes

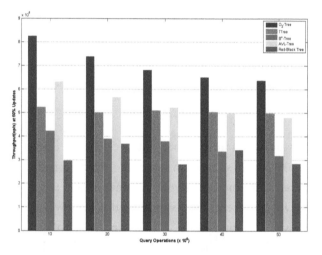

Figure 5. Operational Throughput with 50% Updates for Basic Indexes using TPC Dataset

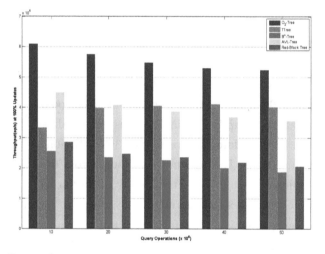

Figure 6. Operational Throughput with 100% Updates for Basic Indexes using TPC Dataset

the O_2-Tree, when reading and writing through a cache shows performance characteristics that are superior to the NoSQL databases considered. It outperformed the LevelDB and the Kyoto Cabinet that used the TreeDB access method, by several orders of magnitude.

Figure 9 and Figure 10 show the corresponding operational throughput results for mixed query operations on the NoSQL data stores with varying data sizes from 10M to 50M operations for 100% and 50% updates respectively. The O_2-Tree-KV demonstrated a superior operational throughput at high update ratio and large dataset. The efficient index mechanism of the O_2-Tree-KV accounts for its superior performance.

5.2. Multi-threaded Evaluations

We evaluated the average time for a multi-threaded insertion of "key-value" pairs of generated data into each of the following storage schemes: the O_2-Tree persistent store, which we refer to as O_2-Tree-KV, the BerkeleyDB and Kyoto-Cabinet TreeDB using the $B-Tree$ access method as well as the LevelDB NoSQL key-value store. These experiments were conducted primarily to compare the performance of O_2-Tree with these key-value stores where the data blocks are written and read through an in-memory cache to a disk file using multiple threads. We evaluated the average time it takes to perform 20 million (20M) insertions of "key-value" pairs concurrently with the number of threads

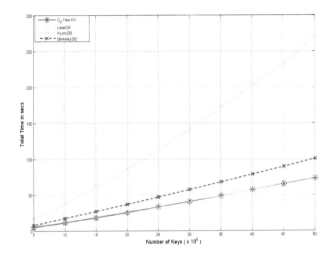

Figure 7. Persistent store: Insertions With In-Memory Cache using TPC dataset

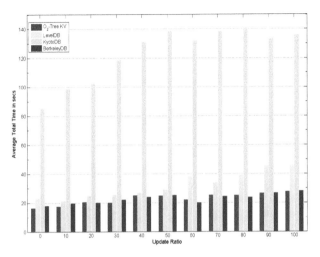

Figure 8. Persistent store: With In-Memory Cache and 50M Mixed Queries

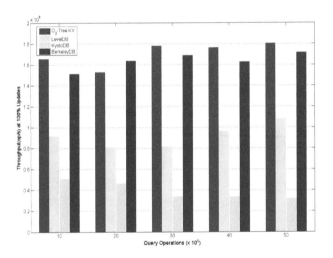

Figure 9. Operational Throughput for Persistent Storage with 100% Updates

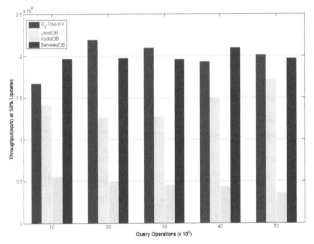

Figure 10. Operational Throughput for Persistent Storage with 50% Updates

varying from 2 to 16. The page size as well as the in-memory cache size for each key-value store was set to 4k and 2.5GB respectively for all experiments in this group. We ensured that the operations were performed with the index tree in memory while the data pages were maintained in the in-memory cache pool. The data pages were read and written to disk according to the *Least Recently Used* (LRU) cache replacement policy of the *Mpool*. The results are shown in Figure 11. The general observation was that the average time to perform insertions decreased with increasing number of threads. This was due to the fact that the degree of access contentions increased as the number of threads increased. Threads must block to ensure correctness of query operations. Therefore, more threads result

in lock contention as several threads are blocked. This eventually affects the overall performance of the structures. However, the O_2-Tree-KV performed better than the other "key-value" storage schemes discussed in the paper. The O_2-Tree-KV employs a simple index mechanism which accounts for its better performance. The O_2-Tree-KV, performed about $2 \sim 3X$ faster than the KyotoDB and BerkeleyDB both of which use the B-Tree access method. The results are shown in Figure 11.

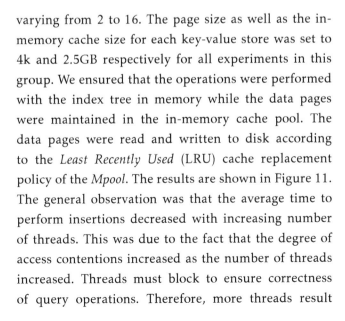

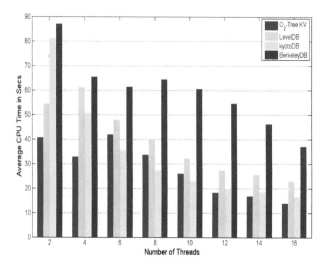

Figure 11. Index Construction with Varying Number of Threads

Figure 12 shows the operational throughputs of each of the key-value stores under different workloads. Each workload consisted of a mix of look-ups, insertions and deletions referred to as update ratio from the previous discussion. For each update ratio, we interleaved all operations such that a thread performed either an update or a lookup. All operations were performed by a maximum 16 threads we had on the machine. We observed a general decrease in throughput as the update ratio increased. This was due to the fact that, updates require restructuring of the index which affects the overall performance. The O_2-Tree-KV did record the highest throughput which was about *1.9M* operations per second (*op/s*). This rate later dropped to *1.3M op/s* at 100% updates. A similar trend was observed for all the other key-value stores considered.

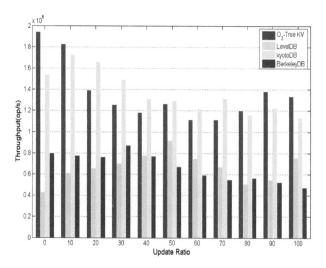

Figure 12. Operational Throughput for Different Mix of Workloads

We also compared the average time to conduct a search or lookup for all key-value stores. One objective of *NoSQL* key-value store is to provide effective lookup without the bottlenecks of traditional Relational database systems (RDBMS). We conducted the experiments with 20M $32 - bit$ keys. We gradually increased the number of threads to ascertain the effect of shared memory multi-threaded concurrent access of these different data storage systems. The results show that, as the number of threads increased, the lookups proceeded faster since there was relatively little work per thread. During lookups, threads do not block and thus, can proceed immediately with expected linearisable results. Though the O_2-Tree-KV outperformed all the key-value stores considered, it rather exhibited a poor performance gain as the number of worker threads increased. This could be due to the cache coherence problem associated with single node traversals. We anticipate a much better performance with a lock-free protocol such as Software transactional memory *STM*.

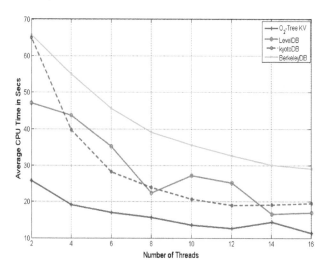

Figure 13. Concurrent Look-ups for 20M "key-value" using Varying Workloads

Additionally, we performed multi-threaded scalability evaluations on the O_2-Tree-KV as well as the BerkeleyDB, Kyoto-Cabinet TreeDB and the LevelDB NoSQL key-value stores. We adopted the *strong scalability* test approach in which we doubled the dataset as well as the number of threads for each run of the experiment. The dataset was varied from 5M with 2 threads and doubled for each run to 40M with 16 threads for the last run.

The first set of scalability test shown in Figure 14 illustrated the results with only insertions (*Puts*). Figure 15 however, indicates similar experiment but this time for a mix of query operations in which 50% were look-ups and 50% updates (of which 70% were insertions and 30% deletions). We observed a comparable and even better performance for the O_2-Tree-KV which exhibited a high level of scalability. Generally, a gradual increase in CPU times for all the key-value stores considered was observed as the number of threads and datasets were doubled.

We also evaluated the total size of the *problem queue* which is used by the relax balance algorithm of the O_2-Tree. We varied the data size as well as the number of threads in each run of the experiment. We observed that the total problem queue size was a function of the size of the dataset used to build the index. Large datasets resulted in larger problem queue size. Figure 16 shows the graph for the total problem queue sizes for the tree index. We observe a rapid increase in the problem queue size given a small increase in the dataset. For instance, the queue increases rapidly from about $6X10^4$ when the dataset is $3M$ to an average of about $12X10^4$ (double the previous size) when the dataset is increased by $1M$.

However, a series of experiments conducted indicated that the average problem queue size was comparatively small at any instance using a single *rebalancer* thread. Since, the *rebalancer* thread does not traverse the index from the root but goes directly to the offending node, it is able to process problem queue items faster than the update threads. This accounts for the minimal average problem queue size observed in the experiment. Further, as the number of the *rebalancer* threads increases, the problem queue size reduces significantly as more threads are able to concurrently process the queue with minimal interference to ensure that the tree is balanced.

Finally, we evaluated the performance of each key-value store using real life flight statistics data [11] that consisted of 32bit keys and their corresponding data values. The physical size of the file was about 600MB. We loaded 10M keys and their corresponding values into each key-value store using varying concurrent threads up to to 16 threads. The operational throughput

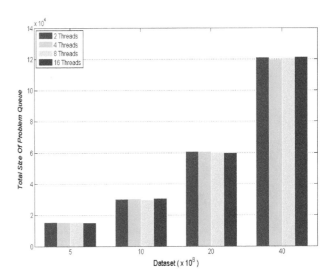

Figure 16. Total *Problem Queue* size for Varying Data sizes and Threads

to load the data from the persistent dump was then reported. The primary objective of this experiment was to measure the performance with real life data besides the synthetic data used in the previous experiments. We observed a comparable performance between all the key-value stores considered. The O_2-Tree-KV exhibited a much better throughput even though the others were comparable. Figure 17 illustrates the results.

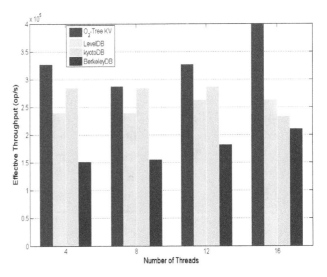

Figure 17. Concurrent Loading of Real-life Persistent data

6. Conclusion and Future Work

In this paper we have presented the O_2-Tree as an in-memory resident index for a persistence key-value store. It delivers high performance and exhibits good

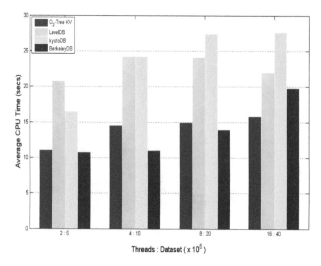

Figure 14. Scalability Test with 100% Insertions

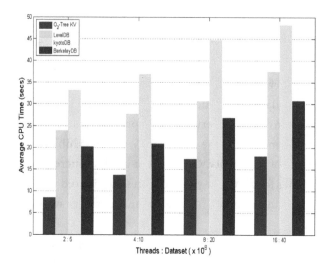

Figure 15. Scalability Test with 50% Update Ratio

scalability while being tolerant of contention. We have also presented a concurrent access protocol based on the relax balance tree technique which allows the scheme to attain high performance as well.

We compared our index persistent O_2-Tree implemented through an in-memory cache against popular high performance and widely used *NoSQL* key-value stores such as the BerkeleyDB, Google's LevelDB and Kyoto-Cabinet TreeDB. Our experiments show that O_2-Tree key-value store outperforms both BerkeleyDB, and Kyoto-Cabinet TreeDB by $2 - 3X$. It also performs comparatively well against Google's LevelDB for many access patterns. More importantly, the experimental results show that O_2-Tree index structure exhibit a good scalability and tolerates contention. It also exhibit superior performance especially during high updates. It's therefore the index structure of choice in applications with frequent updates such as Online Transaction Processing (OTP).

Future work anticipated involves using optimising techniques to make the structure much more cache aware using blocking techniques to improve CPU cache usage as well as bulk loading techniques for greater throughput. We are also exploring the use of GPU traversals for O_2-Tree for even higher throughput. Other future work include replacing the RB-Tree internal structure of the O_2-Tree with other

tree structures such as the *AA-Tree*, *2−3-Tree* and the *SkipList*, and compare their performance measures.

6.1. Acknowledgements

This work was supported with funding from the Centre For High Performance Computing (CHPC), and the Department of Science and Technology (DST), South Africa.

References

[1] 10GEN, INC (2011) Mongodb: High-performance, open source NoSQL database., http://www.mongodb.org//.

[2] ANDERSSON, A. (1993) Balanced search trees made simple. In *In Proc. 3rd Workshop on Algorithms and Data Structures* (Springer): 60–71.

[3] BAYER, R. and MCCREIGHT, E.M. (1972) Organization and Maintenance of Large Ordered Indices. *Acta Inf.* **1**: 173–189.

[4] BOHANNON, P. ET AL. (1997) The Architecture of the DALA Main-Memory Storage Manager . *Multimedia Tools Appl.* **4**: 115–151.

[5] BOYAR, J., FAGERBERG, R. and LARSEN, K.S. (1995) Amortization results for chromatic search trees, with an application to priority queues. In *Proceedings of the 4th International Workshop on Algorithms and Data Structures*, WADS '95 (London, UK, UK: Springer-Verlag): 270–281.

[6] BOYAR, J. and LARSEN, K.S. (1993) Efficient rebalancing of chromatic search trees. *Journal of Computer and System Sciences* **49**: 667–682.

[7] BRONSON, N.G., CASPER, J., CHAFI, H. and OLUKOTUN, K. (2010) A Practical Concurrent Binary Search

tree. In *Proc. of the 15th ACM SIGPLAN symposium on Principles and practice of parallel programming*, PPoPP '10 (New York, NY, USA: ACM): 257–268. doi:http://doi.acm.org/10.1145/1693453.1693488.

[8] Comer, D. (1979) Ubiquitous B-Tree. *ACM Comput. Surv.* **11**: 121–137. doi:http://doi.acm.org/10.1145/356770.356776.

[9] Cormen, T., Leiserson, C., Rivest, L. and Stein, C. (2009) *Introduction to Algorithm*, 1 (Cambridge, Massachusetts London, England: MIT Press), 3rd ed.

[10] FAL Labs (2011) Kyoto cabinet: a straightforward implementation of dbm, `http://fallabs.com/kyotocabinet/`.

[11] FlightStats: (2005) FlightStats Database, http://dl.flightstats.us/, `http://dl.flightstats.us/`.

[12] Google.com (2011) Leveldb: A fast key-value storage library written at google, `http://code.google.com/p/leveldb/`.

[13] Hanke, S., Ottmann, T. and Soisalon-Soininen, E. (1997) Relaxed balanced red-black trees. In *Proc. of the Third Italian Conference on Algorithms and Complexity*, CIAC '97 (London, UK: Springer-Verlag): 193–204.

[14] Hanke, S. (1998) *The Performance of Concurrent Red-Black Tree Algorithms*. Tech. rep.

[15] Kong-Rim, C. and Kyung-Chang, K. (1996) T *-Tree: A Main Memory Database Index Structure for Real Time Applications. In *Proc. IEEE Real-Time Computing Systems and Applications* (South Korea): 81 – 84.

[16] Larsen, K.S. (1998) Amortized constant relaxed rebalancing using standard rotations. *Acta Informatica* **35**: 35–10.

[17] Larson, P.A., Blanas, S., Diaconu, C., Freedman, C., Patel, J.M. and Zwilling, M. (2011) High-performance concurrency control mechanisms for main-memory databases. *Proc. VLDB Endow.* **5**(4): 298–309.

[18] Lehman, T.J. and Carey, M.J. (1986) A Study of Index Structures for Main Memory Database Management Systems. In *Proc. of the 12th International Conference on Very Large Data Bases*, VLDB '86 (San Francisco, CA, USA: Morgan Kaufmann Publishers Inc.): 294–303.

[19] Leung, C.H. (1984) Approximate storage utilisation of B-trees: a simple derivation and generalisations. *Inf. Process. Lett.* **19**: 199–201.

[20] Lu, H., Ng, Y.Y. and Tian, Z. (2000) T-Tree or B-Tree: main memory database index structure revisited. In *Proc. of the Australasian Database Conference*, ADC 2000 (Washington, DC, USA: IEEE Computer Society): 65– 73.

[21] Marcus, A. (2012) The architecture of open source applications: Chapter 13. the NoSQL ecosystem, `http://www.aosabook.org/en/NoSQL.html`.

[22] Nurmi, O., Soisalon-Soininen, E. and Wood, D. (1987) Concurrency control in database structures with relaxed balance. In *Proc. of the sixth ACM SIGACT-SIGMOD-SIGART symposium on Principles of database systems*, PODS '87 (New York, NY, USA: ACM): 170–176. doi:http://doi.acm.org/10.1145/28659.28677.

[23] Nurmi, O. and Soisalon-Soininen, E. (1991) Uncoupling updating and rebalancing in chromatic binary search trees. In *Proceedings of the tenth ACM SIGACT-SIGMOD-SIGART symposium on Principles of database systems*, PODS '91 (New York, NY, USA: ACM): 192–198.

[24] Nurmi, O. and Soisalon-Soininen, E. (1996) Chromatic binary search trees: A structure for concurrent rebalancing. *Acta Informatica* **33**: 547–557. 10.1007/BF03036462.

[25] Oracle (2013) Oracle exadata database machine x3-8 datasheet, `http://www.oracle.com/us/products/database/exadata/database-machine-x3-8/overview/index.html`.

[26] Oracle.com (2011) Oracle berkeleydb 11g, `http://www.oracle.com/technetwork/products/berkeleydb/overview/index-085366.html`.

[27] Rao, J. and Ross, K.A. (1999) Cache Conscious Indexing for Decision-Support in Main Memory. In *Proc. of the 25th International Conference on Very Large Data Bases*, VLDB '99 (San Francisco, CA, USA: Morgan Kaufmann Publishers Inc.): 78–89.

[28] Rao, J. and Ross, K.A. (2000) Making B^+-trees cache conscious in main memory. In *Proc. of the 2000 ACM SIGMOD international conference on Management of data*, SIGMOD '00 (New York, NY, USA: ACM): 475–486. doi:http://doi.acm.org/10.1145/342009.335449.

[29] Seltzer, M. and Bostic, K. (2012) The Architecture of Open Source Applications: Chapter 4. BerkeleyDB, `http://www.aosabook.org/en/bdb.htm`.

[30] Sponsored by VMWARE (2011) Redis: An open source, advanced key-value store., `http://redis.io/`.

[31] TPC-H (2001) Transaction Processing Council, `http://www.tpc.org/tpch/`.

14

End-to-End Key Exchange through Disjoint Paths in P2P Networks

Daouda Ahmat[1], Damien Magoni[1,*], Tegawendé F. Bissyandé [2]

[1]LaBRI, University of Bordeaux, France
[2]SnT, University of Luxembourg, Luxembourg

Abstract

Due to their inherent features, P2P networks have proven to be effective in the exchange of data between autonomous peers. Unfortunately, these networks are subject to various security threats that cannot be addressed readily since traditional security infrastructures, which are centralized, cannot be applied to them. Furthermore, communication reliability across the Internet is threatened by various attacks, including usurpation of identity, eavesdropping or traffic modification. Thus, in order to overcome these security issues and allow peers to securely exchange data, we propose a new key management scheme over P2P networks. Our approach introduces a new method that enables a secret key exchange through disjoint paths in the absence of a trusted central coordination point which would be required in traditional centralized security systems.

Keywords: P2P networks, key management, Diffie-Hellman algorithm, MITM attacks, multipath routing, backtracking.

1. Introduction

Due to their inherent characteristics, including self-organization, availability, reliability, scalability, and their potential for managing load balancing and dynamic topology changes, P2P networks remain one of the prime choices to provide affordable means for sharing data, publishing information and streaming media.

Distributed systems, and especially P2P networks, can effectively operate in dynamic environments because they are free from relying on a single point of failure that a central infrastructure would represent. However, although these systems provide interesting properties, they are not suitable for integrating approved traditional centralized security infrastructures. Therefore, in the context of P2P system setups, security remains challenging to implement and maintain.

Without guarantee of provision of a central infrastructure, P2P networks are bound to face security and reliability issues that existing common policies and techniques fail to take into account in their implementations. For example, standard security measures for communications involve the use of key-based encryption which is often implemented with public-key cryptography. However, a central problem with the use of such type of cryptography lies in confidence that a given public key is authentic, this means that it is correct and belongs to the entity claimed, that it has not been tampered with or that is has not been replaced by malicious third party. Usually, this confidence is guaranteed by a central Public-Key Infrastructure (PKI), in which one or several certification authorities certify ownership of the public keys. There is therefore a requirement for a strong centralization to manage cryptographic keys, a luxury that P2P networks cannot practically afford [26].

In order to overcome the limitations imposed by the incompatibility between traditional security policies and decentralized P2P networks, we propose in this paper a new approach for key management in the absence of a centralized infrastructure. Our approach takes into account the specific features of P2P networks and the security and reliability requirements for exchanging information in this area. In opposition to the traditional PKI, the key management in our scheme is *decentralized*. Indeed, this management is distributed among both nodes of a communicating pair of nodes, meaning that the key negotiation model is based upon an End-to-End (E2E) exchange scenario. In practice, we rely on a scheme that uses multiple paths (hereafter referred to as *multipath*) for performing a Diffie-Hellman (DH) key exchange where each source node can select several disjoint paths inside the P2P network in order to prevent Man-In-The-Middle (MITM) attacks. Accordingly, key parts (hereafter referred to as *subkeys*) can then be forwarded safely through these disjoint paths to ensure communication security.

In our related work [2], we have proposed an approach to address the security issues of communication sessions when users are mobile across networks. As this approach, which is targeting user-level applications, is based on a P2P network, it can leverage the solution proposed in this paper. Indeed, this solution solves the security issues arising from the intermediate P2P network nodes and provides a PKI-less solution to securely create a secret key whose distribution was undefined in [2]. This paper is an extended version of our previous work [1] which was presented at the 5th EAI International Conference on e-Infrastructure and e-Services for Developing Countries in 2013. We have increased the description of our solution and detailed its implementation and we have added simulation results showing that our solution can secure a communication with high probability depending on the network conditions.

The main contributions of this paper are:

- We discuss the challenges that arise in P2P networks, focusing on the safety and reliability of communications. We then discuss the opportunities of multipath key exchange for P2P networks: how they can be harnessed to deliver secure communication in a truly beneficial way. We expose the challenges for implementing the authentication of source nodes in a P2P network. We emphasize on why traditional PKI, which are currently successful on the Internet, are inadequate for P2P systems (Section 2).

- We describe the design of our solution in detail, including the discovery of multiple paths and the use of backtracking (Section 3).

- We present a security analysis of our solution, which is based on the probability of having a given set of paths intercepted by malicious nodes (Section 6).

- We evaluate the efficiency of our solution by carrying out experiments by simulations and we provide detailed results for assessing its usefulness (Section 5).

The remainder of this paper is organized as follows. Section 2 presents background information including the challenges that must be overcome to secure communications in P2P networks as well as some insights from our previous work on P2P overlay networks including our implementation of mobile secure sessions. We present the design of our approach in Section 3. Section 6 provides an analysis about the weaknesses and strengths of our scheme. Experiments carried out by simulation are detailed in Section 4 and the results are outlined in Section 5. Finally, we discuss the related work in Section 7.

2. Background

Securing communications in P2P networks is a challenging endeavor, especially with regards to the standard practice of encrypting information with the assurance that receivers have knowledge of the public key for deciphering the data payload. In this section, we precisely detail some obvious and non-obvious challenges to highlight the constraints of finding a solution to key management for P2P systems. We then introduce the concept of Hyperbolic coordinates which are leveraged in our approach.

2.1. Challenges

- *Key distribution.* The first challenge that we encounter is the mode of distributing cryptographic keys in P2P networks. Indeed, in traditional networks, a management-friendly central infrastructure, called PKI, is relied upon for this task. In the absence of such infrastructure for P2P networks, it is important to devise a fully distributed approach to spread keys reliably. This approach should take into account the volatility of P2P networks but could leverage their self-organization property.

- *Assurance of alternate infrastructure.* Relying on the P2P networks to assure the forwarding of cryptographic keys also comes with problems that a traditional PKI was able to easily handled. Indeed, there is a new need to ensure in a distributed system that the construction of the overlay network where each peer is properly identified will guarantee the robustness of the exchange scenario with little possibility of corruption by any intermediate peer.

- *Exchange of keys.* When initiating a communication in a P2P network, there will be a need for the peers to agree on the generation of a session key that the two peers will shared. A challenge in this requirement lies in the negotiation which should also be secured. Although usage of cryptography asymmetric algorithm to secure information is expensive, it can be leveraged in the initial step that is the key negotiation phase.

- *Detection of attacks.* Last but not least, we note the issue of providing a mechanism for detecting corrupted keys. Indeed, when a corrupted cryptographic key is introduced, it should be detected and then revoked by peers who are aware of the corruption. This information should be included in a notification message that must be sent by broadcast through the network. Thus, the system will assure that all corrupted keys are flushed out of the memory of connected peers. Similarly, a challenging endeavor will be to prevent MITM attacks during session key negotiation phase

2.2. Hyperbolic Coordinates

In order to address the security challenges of P2P networks, we propose a solution that will be used on top of the CLOAK [1] P2P architecture defined in our previous work [25]. The CLOAK architecture is appealing because it provides an addressing scheme based on virtual hyperbolic coordinates and a greedy routing based on the hyperbolic distance between nodes computed by using those coordinates. Thus CLOAK does not use routing tables and does not impose any specific type of topology upon the nodes of the network.

We have demonstrated in [5] that the hyperbolic geometry can be efficiently used for addressing and routing in P2P overlay networks. Indeed, the hyperbolic plane provides a means for distributing unique coordinates taken from an infinite q-regular tree (with q being an integer chosen at will). These coordinates are used as addresses by the network nodes. When the P2P network expands, starting from a root node, any joining node will ask for an address from one of the existing nodes. The joining of new nodes thus create an *addressing* spanning tree. Upon obtaining an address, a newly included node will then be able to independently compute the addresses that it will be able to provide to its potential children and so on. The degree q of the regular tree determines how many addresses each node in the network will be able to give to future joining nodes.

3. System Design

In this section, we describe our key exchange model and its corollaries. We present the design of the variety of features that are in play for realizing the goals of a robust key exchange. These include the multipath routing approach, a backtracking algorithm, an authentication mechanism, the use of proxies in case of negotiation failure, and the addition of extra properties to the generated keys.

3.1. Approach Overview

In recent years, a number of research efforts have emphasized that P2P networks are subject to various security threats due to their inherent features [6, 9, 18, 21, 22]. To overcome the identified security issues in P2P networks, we propose to design a new key exchange scheme that provides end-to-end (E2E) authentication and confidentiality capabilities. The proposed key exchange protocol is an extension of the well known DH

cryptographic algorithm. Our approach mainly aims at preventing MITM attacks by using multipath key exchange technique.

In our approach, the main key which must be transferred between the source and destination nodes is split into several parts called subkeys. Each subkey is sent to the destination node through separate paths over the P2P network. Thus, a malicious node eavesdropping on a particular overlay link will only recover one subkey but would not be be able to recover all other subkeys. Therefore, the main key, which consists of all subkeys, will remain unknown to the malicious node. Similarly, in the case of a coordinated attack involving many nodes eavesdropping on different links, they must be able to intercept all subkeys for their attack to be successful as all subkeys are needed to reconstruct the main key. An incomplete subkey collection will not allow any node (intended recipient or attacker) to infer the main key.

In order to fulfill the goal of diversifying the links used to transfer the subkeys, we propose a routing algorithm based on the use of multipath to route each subkey separately. The marking algorithm is an essential subroutine in our routing algorithm to route split key parts via disjoint paths between the source and destination nodes. Thus, a multipath key exchange scenario could prevent MITM attacks and ensure confidential communication. We further complete our architecture with features such as the designation of *trusted peers* and backtracking capabilities.

3.2. Multipath Routing and Backtracking Algorithm

Goals. For reasons of simplicity, in the remainder of this paper, we represent a P2P network by means of an undirected graph $G = (V, E)$, where V is a non-empty set of vertices and E the set of edges between pairs of vertices. Given two distinct nodes $\{s, t\} \in V$, we want to find k disjoint paths $P_0, ..., P_{k-1}$ over G from s to t. We also assume that each node knows only its immediate neighbors and has no knowledge of the global topology of the graph.

Each P_i is a collection of n selected consecutive hops $h_{i_0}, ..., h_{i_{n-1}}$. Each hop h_{i_j} in path P_i represents an optimal hop at step j that meets routing protocol criteria. A path P_i is formally defined as follow:

$$P_i = \{\langle h_{i_0}, ..., h_{i_{n-1}} \rangle_{0 \leq i \leq k-1} \mid for\ 0 \leq j \neq l \leq k-1,$$
$$and\ for\ 0 \leq p, q \leq n-1,\ h_{j_p} \neq h_{l_q}\} \quad (1)$$

The main key is composed of a set K of subkeys, of size ω. Formally, let $K' \subset K$ such that $|K'| < \omega$, then the combination of all subkeys $\in K'$ cannot determine the main key. Thus, at the end of the transmissions, to recompute the main key, all ω subkeys are needed.

---

[1] Covering Layers Of Abstract Knowledge

This property is needed in our security infrastructure to introduce a robust key exchange policy.

Finding k Disjoint Paths. In our context, the *disjoint paths* term means vertex/node-disjoint paths. Two paths are vertex-disjoint if they have no vertex in common except for the first (source) and last (destination) vertices. In order to find k disjoint paths over graph G, we proceed by relying on a graph marking algorithm that enables to mark all hops which make up a separate path P that fulfills properties presented by equation (1). Concretely, information about each visited node is stored into a *visited list* (V) contained within the subkey transfer packet. Thus, step by step, all visited node will be recorded into V. On receipt of a subkey packet, the destination node replies to the source node through an acknowledgement (ACK) message that contains among other information an updated list V and the sequence number of the received packet. This process is iterated k times for the requested k separate paths. In the ideal case described by equation (2), all used paths between the source and destination nodes are fully pairwise disjoint.

$$P_i \cap P_{j\,0 \leq i \neq j \leq n-1} = \emptyset \qquad (2)$$

Furthermore, although some paths may intersect at some nodes, if these paths verify equation (3), MITM attacks could be prevented. Unfortunately, in this scenario, key exchange would be more vulnerable to coordinated MITM attacks launched by malicious nodes placed on the intersection points.

$$\bigcap_{i=0}^{k} P_i = \emptyset \qquad (3)$$

A node N is qualified to become a hop if, and only if, $N \notin V$ and N is an *optimal choice* at current step to reach the destination node.

In order to avoid redundancy, we assume that all paths P_i do not contain source and destination peers because these nodes are already recorded in the packet to indicate the origin and the final recipient of the subkey being transmitted.

Options for Routing Policies. Several multipath routing algorithms can be used to exchange security keys over P2P networks. However, these routing algorithms have different robustness and complexity characteristics. We now describe some common routing methods to highlight various choice-impacting criteria.

1. *combining routing and marking.* A first routing algorithm enables to perform both marking and routing at the same time. Key components are successively sent from source to destination through nodes that are marked as soon as they are used as hops (not *just visited* nodes) to reach the destination node. Thus, each subkey packet

is augmented with information of all hops that it goes through. This process is repeated for all the k subkeys. In the worst case, the running time of this algorithm is $\tau_1 = \mathcal{O}(k(|E| + |V| log |V|))$, where E represents the set of edges, V is the set of vertices and k represents the number of subkeys. The space complexity of this algorithm amounts to $\sigma_1 = \mathcal{O}(k(|V|))$. Unfortunately, this approach is unable to determine proportionality between the number of subkeys and the existence of enough disjoint paths. Indeed, the key is split before the process for finding disjoint paths is launched.

2. *pre-routing then routing.* This method requires two successive steps. In the first step, the source node tries to find disjoint paths between the source and destination nodes over the network by marking visited nodes. In the second step, refers to the obtained list of visited nodes to split the key and send them separately. The running time for this approach is $\tau_2 = |V| * \tau_1 = \mathcal{O}(k|V|(|E| + |V| log |V|))$. IN the worst case scenario, its space complexity amounts to $\sigma_2 = k|V| * \sigma_1 = \mathcal{O}(k^2(|V|^2))$.

Despite such high complexity costs, this method has a major benefit: it enables to safely split the key in a number of subkeys that is in adequacy with the number of disjoint paths that exist in the network. Unfortunately, as the topology may change in the meantime, the considered paths may become invalid when the subkeys are finally forwarded.

3. *discovery and routing.* This method consists in a combination of the algorithms described above for creating a scheme where after each path discovery a subkey packet is sent. A packet discovery is indeed sent first to discover a path from source to destination. Then, a subkey is sent. This process is repeated in k times. The algorithm thus has the same complexity costs as the second method.

4. *reduction of the disjoint paths problem to a max-flow problem in undirected graphs.* The idea behind this approach is based on Menger's theorem[2] (1927) [4].

The problem of finding disjoint paths can be reduced to a *Max-flow* finding problem. Double reduction is needed to achieve this goal: (1) reduction from vertex-disjoint paths problem to edge-disjoint paths problem and (2) reduction from edge-disjoint paths problem to *Max-flow* finding problem. This approach assigns unity

---

[2]"Let a finite undirected graph $\mathcal{G}$ and two distinct nodes $\mathcal{S}$ and $\mathcal{T}$, the maximum number of edge-disjoint $\mathcal{S} - \mathcal{T}$ paths is equal to the minimum number of edges whose removal disconnects $\mathcal{S}$ from $\mathcal{T}$."

capacity to every edge and then determines *Max-flow* between source(S) and destination(D) over the network. The *Max-flow* value computed is then precisely equal to the number of disjoint paths between S and D. Although theoretically interesting, the approach is currently impractical, as it requires a global knowledge on network topology. This requirement is not fulfilled in P2P fully decentralized systems.

Recently, Ken-ichi *et al.* have proposed an algorithm that improves the time complexity of the disjoint paths problem by solving it in quadratic time, i.e., in $\mathcal{O}(n^2)$ [13].

Thanks to its robustness, the second algorithm emerges as the best choice, even though it is more costly than other solutions. Thus, our key exchange scheme leverages the *pre-routing then routing* method described above.

Routing Policy Illustration. Figure 1 illustrates in simplicity the case of two disjoint paths between two nodes, namely *1* and *9*. First, the source node sends a first packet (with *visited list* $V= \emptyset$) that determines P_1 consisting of hops 4 and 7. These hops are added into V. Then, on receipt of the acknowledgement message which contains the updated list V, the source node sends a second message (with $V= \{4,7\}$) that selects optimal hops without contacting any node already visited. Thus, P_2 will consist e.g. of nodes 3, 6 and 8 (with $V= \{4, 7, 3, 6, 8\}$), such that $P_1 \cap P_2 = \emptyset$: two disjoint paths between source node *1* and destination node *2* have been found, and the two subkeys can be transferred.

Subsequently to receiving all subkeys, the destination node *9* can apply the symmetric operation of re-assembling the key in order to complete the key exchange process. At this time, and unless a coordinated attack have recovered all subkeys from all disjoint paths, both nodes *1* and *9* can start a secure communication using the exchanged key. The transmitted packets can even be routed through any nodes even those that were not included in V.

Description of our Routing Policy. When a source node S is attempting to exchange a key with a destination node D, it splits this key into several subkey components that must be dispatched through separate channels[3]. We propose a straightforward technique to overcome the challenge of selecting the different paths that are necessary to route separately each subkey towards the destination node. As illustrated in the previous simple example, in the graph representing the nodes and connections in a P2P network, the degree of a node is the cardinal number of the set of neighbors which are

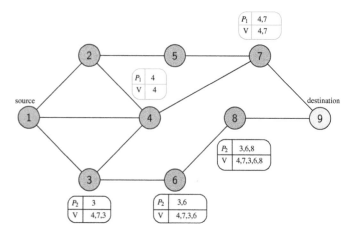

Figure 1. Routing across disjoint paths: the case of 2 disjoint paths.

known from start. Each connected node has a degree that may be less or equal to the network degree, which is the maximum number of neighbors.

Thus, given a network degree n, each node wishing to forward a key must split it into at most n components that will be routed in potential separate paths. The selection of paths is iterative to route each subkey through different neighbor. Thus, each node numbers the different nodes it is connected to and considers each of these nodes as the beginning of a different path. When a node has a degree k, it sends the first component to its first connected node, and the second component to the second connected node, and so on until all connected nodes are used if the number of disjoint paths is equal to the number of key components. We formally describe this path selection in Algorithm 1.

Algorithm 1: Neighbor selection and subkey routing.

Input: keyToTransmit, nodeNeighbors
usedNodesList ← ∅;
nodeNeighbors ← sortConnectedNodes(nodeNeighbors);
keyComponents ← splitKey(keyToTransmit);
node ← firstOf(nodeNeighbors);
while keyComponents ≠ ∅ **do**
  if *node* ∉ usedNodesList **then**
    component ← firstOf(keyComponents);
    transmitViaSeparatePath(component, *node*);
    keyComponents ← keyComponents \ {component};
    usedNodesList ← usedNodesList ∪ {*node*};
  node ← nextOf(nodeNeighbors, *node*);

Our solution can be used by any P2P network, whether unstructured or structured (such as DHT-based P2P networks). Indeed, the algorithm 1 is generic to any P2P network. However, the algorithm 2 is specific to our CLOAK P2P network.

---

[3]Algorithm 2 illustrates how the dispatching of key components is implemented over our infrastructure.

Algorithm 2: Routing a subkey in a CLOAK network.

Input: CurrentPeer, SubKeyPacket, VisitedList
Output: Success
$w \leftarrow SubKeyPacket.DestinationPeer.Coords$;
$m \leftarrow CurrentPeer.Coords$;
$d_{min} \leftarrow \mathrm{argcosh}\left(1 + 2\frac{|m-w|^2}{(1-|m|^2)(1-|w|^2)}\right)$;
$p_{min} \leftarrow CurrentPeer$;
foreach $Neighbor \in p_{min}.Neighbors$ **do**
  IsAlreadyVisited $\leftarrow VisitedList.contains(Neighbor)$;
  if IsAlreadyVisited $= false$ **then**
    $n \leftarrow Neighbor.Coords$;
    $d \leftarrow \mathrm{argcosh}\left(1 + 2\frac{|n-w|^2}{(1-|n|^2)(1-|w|^2)}\right)$;
    if $d < d_{min}$ **then**
      $d_{min} \leftarrow d$;
      $p_{min} \leftarrow Neighbor$;

if $p_{min} \neq CurrentPeer$ **then**
  VisitedList$.add(p_{min})$;
  routeSubKey $(p_{min},$ SubKeyPacket$)$;
  if $p_{min} = SubKeyPacket.DestinationPeer$ **then**
    return $success$;
  else
    algorithm 2 $(p_{min},$ SubKeyPacket, VisitedList$)$;

else
  backtracking$()$;

Hop Number Threshold. In order to address the overheads in packets sizes, the number of hops is limited as illustrated in figure 2. Indeed, each packet used to transfer a subkey contains information on all nodes that are already visited in the current session, so as to allow the selection of paths that are will be disjoints. Thus, when the number of hops is significant then the portion of the packet that records the visited nodes grows, and can lead to an overflow in packet size. Consequently, it is necessary to limit the number of hops by a *threshold*.

Backtracking Algorithm. Backtracking algorithm is illustrated by a scenario depicted in Figure 3. First, path $P_1 = \{1, 5, 8\}$ between S and D is determined. Then, an attempt to determine a second path P_2 after performing hops on nodes 2, 3 and 7 fails at node 7, since continuing will not produce a disjoint path: nodes 5 and 8 is already used by P_1 and there are no other ways from node 7. As a result, a backtracking routine must be switched on. Discovery path packet will be sent back to node 3 and finally to node 2, where it will attempt an alternate hop. Finally, the algorithm finds a second disjoint path P_2 such that $P_2 = 246$ ($P_1 \cap P_2 = \emptyset$). Note that all path

Formally, the backtracking algorithm enables to go back when the destination node is unreachable from current node or when the only path to reach it will

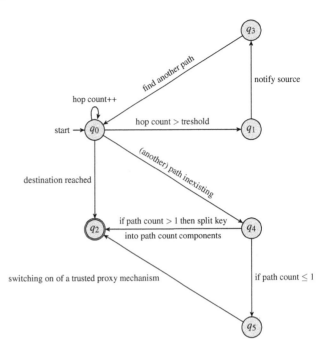

Figure 2. A threshold in the number of hops should prevent packet overhead; switching to a trusted proxy (see subsection 3.3) in unsuccessful cases.

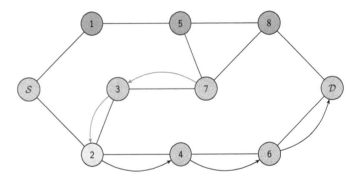

Figure 3. Backtracking illustration: there is no way out from node 7 to destination D, because nodes 5 and 8 are marked.

intersect with previously selected paths. In such cases, the packet must sent backwards until either source node S is reached or until another relevant node for alternate paths is reached.

3.3. E2E Key Exchange and Re-Authentication

E2E Key Exchange Scheme. In this section, we propose an extension that improves both the state-of-the-art models designed by Takano *et al.* [23] and Diffie & Hellman [8]. Scalable, decentralized and self-organized networks such as P2P systems enable many users to join them. Thus, each node can be connected to many other nodes; there can then be several paths between two endpoints of the network. For the above reasons, network topology can be transformed into a graph: each

Alice				sends each k_{a_i}/b_i via a disjoint path	Bob			
Data		Actions			Actions		Data	
secret	public	chooses randomly	computes		computes	chooses randomly	public	secret
s_a	p,g			$p,g \rightarrow$				s_b
...	p,g,A		$A = g^{s_a}(mod\ p)$				p,g	...
...	...	$k_{a_0},...,k_{a_{q-1}}$ such that	$\sum_{i=0}^{q-1} k_{a_i}(mod\ p) = A$	$k_{a_i}\ (0 \le i < q) \rightarrow$			p,g,A	...
...	...				$B = g^{s_b}(mod\ p)$		p,g,A,B	...
...	p,g,A,B			$\leftarrow k_{b_i}\ (0 \le i < q)$	$\sum_{i=0}^{q-1} k_{b_i}(mod\ p) = B$	$k_{b_0},...,k_{b_{q-1}}$ such that	...	...
s_a,s	...		$s = B^{s_a} = g^{s_a s_b}(mod\ p)$	$\leftarrow$ Untrusted channels $\rightarrow$	$s = A^{s_b} = g^{s_a s_b}(mod\ p)$		...	s_b,s

Figure 4. Data owned and actions performed by the source and destination nodes to negotiate the keys. Illustration of Algorithm 3.

node represents an edge of the graph and each link indicates a connection between two nodes.

Although the DH cryptographic algorithm has been widely used to share secrets on insecure communication channels, it is vulnerable to MITM attacks. In order to overcome this limitation of DH-based scenarios, we propose to use a multipath key exchange scheme. Building upon a scalable P2P overlay [5, 24] which uses virtual coordinates taken from the hyperbolic plane that is indifferent to underlying P2P network topology. Figure 4 summarizes the data and actions performed by a pair of nodes during a key negotiation phase. The multipath key negotiation is described more formally in Algorithm 3 and in figure 4. Takano et al. [23] have previously proposed a similar key negotiation mechanism. However, unlike our approach, their method is restricted only to Symphony and Chord P2P networks with a ring topology.

Trusted Peers Used as Proxies. Using *trusted peers* provides an alternative security solution when several disjoint paths between a source peer and a destination peer cannot be found. It is a key concept for addressing issues caused by the potential lack of disjoint paths in some network topologies. Thus, when a node fails to find disjoint paths through network, it activates the mechanism that enables to select one *trusted peer* from a *trusted peers* list and uses it as a proxy in order to securely reach the destination peer.

A node can be chosen from a list of nodes as *trusted peer* if it meets two conditions: (1) it is near to destination node than source peer except there is not another *trusted peer* and (2) it already exists a shared key between proxy node and source node. Thus, source peer will then delegate key negotiation process to its proxy node (see *request* message in figure 5).

Figure 5 presents a topology that shows that there is no disjoint paths between both nodes *1* and *2*. Therefore, using of a *trusted peer* (node *4* in this case) is then needed in this context.

trusted peer applies algorithm that enables to find disjoint paths (paths P_1 and P_2 use respectively hop 6 and hop 7 in figure 5) in order to exchange key with destination. Subsequently, Node *7* forwards key to source node by *reply* message sent over secure

---

Algorithm 3: Multipath key exchange scheme.

public parameters:
  p : a prime number
  g : a generator
secret parameters:
  Alice : secret key s_a
  Bob : secret key s_b

1. **Alice** selects a random number s_a and she then computes $Key_a = g^{s_a}(mod\ p)$;

2. **Alice** selects q random numbers $k_{a_0},...,k_{a_{q-1}}$, such that
   $$Key_a = \sum_{k=0}^{q-1} k_{a_k}(mod\ p) ;$$

3. **Alice** routes all k_{a_k} to **Bob** through q potential separate channels ;

4. **Bob** receives q Key_a's components $k_{a_0},...,k_{a_{q-1}}$ sent by **Alice** and computes $\quad Key_a = \sum_{k=0}^{q-1} k_{a_k}(mod\ p)$;

5. **Bob** selects a random number s_b and he then computes $Key_b = g^{s_b}(mod\ p)$;

6. **Bob** chooses q random numbers $k_{b_0},...,k_{b_{q-1}}$, such that
   $$Key_b = \sum_{k=0}^{q-1} k_{b_k}(mod\ p) ;$$

7. **Bob** sends all k_{b_k} to **Alice** via q potential disjoint paths through network ;

8. **Alice** receives q Key_b's components $k_{b_0},...,k_{b_{q-1}}$ from **Alice** and computes
   $$Key_b = \sum_{k=0}^{q-1} k_{b_k}(mod\ p) ;$$

9. **Alice** computes $\quad Key = Key_b{}^{s_a} = g^{s_a \cdot s_b}(mod\ p)$;

10. **Bob** computes $\quad Key = Key_a{}^{s_b} = g^{s_a \cdot s_b}(mod\ p)$;

11. **Alice** generates a cryptographic challenge and sends it to **Bob**;

12. **Bob** receives the challenge message, resolves it and replies to **Alice**;

---

tunnel. Finally, node *1* and node *8* could then securely communicate without a *trusted proxy*.

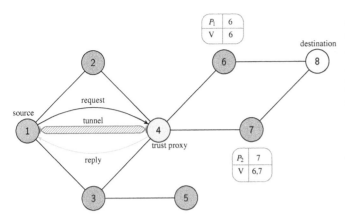

Figure 5. Key negotiation delegated to a trusted proxy (here represented by node 4).

Re-Authentication. Frequent key exchange causes both network overhead and latency. In order to overcome these issues, we propose to introduce re-authentication feature in our key exchange scheme. Indeed, in order to refresh tunnel and authenticate a correspondent node, a peer submit quite simply to its correspondent node a challenge to resolve. If the correspondent node successfully resolves the challenge and replies to the source node, then secure communication can be restarted over the old tunnel. Otherwise, key renegotiation will be needed.

Technically, challenge message is consisted of two random operands (O_1 and O_2) and one random operator op destined to be computed. Correspondent node then performs this arithmetical operation (O_1 op O_2) and sends the result, encapsulated within an encrypted packet, to source node.

Intrusion Detection Mechanism. *Trusted peers* can be used as proxies in order to join nodes that are unreachable via multipath (at least two disjoint paths). However, it could happen that a proxy node is compromised and can thus launch MITM attacks. In this scenario, the malicious node should be excluded from the list of *Trusted peers* and this event should be broadcast over the network.

Intrusion is detected when a peer fails to decipher data sent by its *legitimate* correspondent. In this case, *trusted peers* used as proxies for the negotiation phase will be repudiated from the list of *trusted peers* and this information will also be broadcast over the network.

4. Simulation Settings

To assess our approach, we use the *nem* simulator[4]. We use it for traveling along the paths between any pair of nodes in the studied network. In our experiments,

[4]http://www.labri.fr/perso/magoni/nem/

different parameters are variable. These include the topology of the network, the size of the network, the proportion of network nodes that are considered to be compromised and the source and destination nodes. Each experiment follows a specific set of steps and is designed based on realistic network settings.

4.1. Experimental Steps

We succinctly present the steps that are carried out for the experiments:

1. *Definition of the network*: a P2P network is first created. In this step, we set the type of the topology (real map, synthesized topology such as Erdõs-Rényi, Internet-like, etc.) and the size of this network.

2. *Selection of the set of compromised nodes*: in the second step, we select a subset of $X\%$ nodes which will act as attackers. These nodes are supposed to coordinate their actions.

3. *Identification of source and destination nodes*: we then select, among the non-compromised nodes, a pair of source and destination nodes for the data exchange.

4. *Data packet transfer*: we launch the transmission by transferring a data packet through the shortest path towards the destination node. All intermediate nodes will be marked and may not be used for another packet between the same pair of source/destination nodes.

5. *Check for attacks*: at the end of the packet transfer, we check whether the packet was intercepted by an attacker.

6. *Use of alternative paths*: at this time, we start over from step 4, using the same source node but a different path to reach the same destination.

   (a) after we have selected a new path and started the routing process, if we arrive at a node where we can no longer move forward, we backtrack to the preceding node.

   (b) if we backtrack through all nodes until we reach the source node, we conclude that no other disjoint path exists in the network between the selected pair of source and destination nodes.

   (c) we also stop the search for alternate paths when we have tried all neighboring nodes of the source node. We also limit the search to a maximum of 10 disjoint paths between a given pair of nodes.

7. *Confirmation of the validity of the generated key*: we check whether the packet was potentially intercepted on all disjoint paths. If this is the case, then this attempt to generate a key is a failure. It is a success otherwise.

8. *Change of source/destination nodes*: we repeat the experiments starting from step 3 with a new pair of source and destination nodes.

9. *Change of compromised nodes set*: we repeat the experiments starting from step 2 with a new subset of compromised nodes. Basically, we change the percentage of attackers.

10. *Change of network settings*: we start over the experiments from step 1 with a new network topology and/or a new size value for the network.

4.2. Network Topologies

In our experiments, we have used two maps obtained by real Internet measurements. We have used an IPv4 map created in 2004 by traceroutes from multiple vantage points. It contains about 12.9k nodes and 60k links. We have also used a BGP-4 map created in 2010 by using the *route-views* BGP observer, which contains about 34k nodes and 70k links. We also use the Erdõs-Rényi [12] (ER) and the Magoni-Pansiot [16] (MP) models to build synthesized graphs of 4 different sizes (2.5k, 5k, 10k and 20k) for each model. The ER model creates random graphs while the MP model creates Internet-like graphs (i.e., following power laws and small world properties). In each map, we remove the tree part and keep the mesh only. We then proceed to execute 100 runs (i.e., each time with different attacker placements) for each scenario. Each run evaluates 1000 random pairs of source and destination nodes. Attackers are supposed to be coordinated (i.e., they know each other).

5. Simulation Results

We now describe the output of our experiments and the insights that they provide for assessing the performance of our approach. As suggested by our experimental steps, we explore the impact that each of the different parameters involved in the P2P network setup may have on the success of a key exchange which is defined as follows:

A key exchange is defined as successful if at least one subkey (i.e., key part) has not been intercepted by an attacker. This means that at least one disjoint path does not contain any attacking node and this guarantees that the key cannot be properly reconstructed by the attackers.

We then define the success rate as follows:

The success rate is equal to the number of successful key exchanges divided by the total number of key

exchanges, the total consisting in the sum of the successful ones plus the unsuccessful (i.e., intercepted) ones.

5.1. Influence of Network Topology

The topology of a P2P network is a characteristic which influences various performance metrics. Because our approach of key distribution is at the overlay layer using coordinates from a hyperbolic plane, we are compatible with any topology for the underlying P2P network. However, the nature of this topology may affect the success of the key distribution scenario. Unless otherwise indicated, the settings of our experiments are: 10 disjoint paths and a maximum of 64 hops per packet.

Figure 6 depicts the success rates of key exchanges for various topologies, namely Ring, MP and ER topologies. We have plotted here the results of the experiments carried out with networks of 2.5k peers, and with a varying number of attackers defined as a % of the nodes.

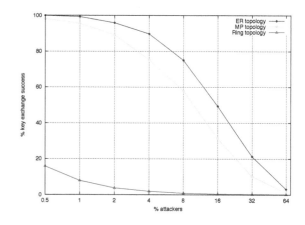

Figure 6. Success rate of the key exchanges for different network topologies.

The graph shows that, for all topologies, the success rate steadily drops with the number of attackers. Beyond 60% for the number of attackers, it becomes virtually impossible to distribute keys in a safe way within the network. The Ring topology appears to perform poorly in a significant way. Indeed, while other topologies can still achieve close to a 100% key exchange success rate in front of 0.5% attackers, i.e., 100 infected nodes out of 20,000, the Ring topology achieves less than 5% exchange rate.

5.2. Influence of Network Size

We then proceed to investigate the performance in each network when varying the size of the network, on the one hand, and the degree to which the network is

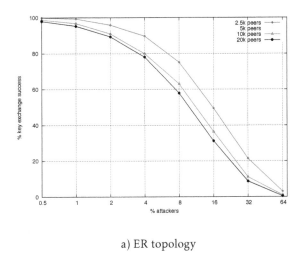

a) ER topology

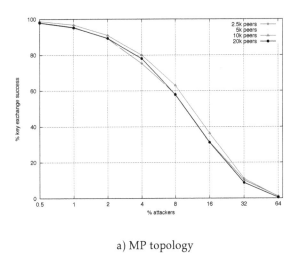

a) MP topology

Figure 7. Success rates for different network sizes.

infected, i.e., the proportion of attackers. Figure 7(a) illustrates the case of the er topology. The experiments were conducted with networks of alternative 2.5K, 5K, 10K and 20K peers. The ordering of the different curves suggest a strong correlation between the network size and the success rate of key exchanges, for all proportions of attackers.

In Figure 7(b) however, the graph shows that for the ER topology, the size of the network does not strongly influence the key exchange rate. Indeed, the evolution curves for the different network sizes are close and often intersect. Further, they are not ordered, with the top curve showing the evolution of a network of 10K peers, which is neither the smallest nor the biggest size considered.

Experimental results for the ring topology are depicted in Figure 8. As the network size grows, the success rate of key exchange drops. For large networks (e.g., 20K peers), even a small proportion of attackers (e.g., 2%) leads to virtually null performance in key exchange. We thus experimentally confirm with our implementation a well-known limitation of the Ring topology: multipath key exchange based on this topology is vulnerable to coordinated MITM attacks [23].

We can see that the network size has a small influence on the success rate of the key exchange and this depends on the network topology. In the ring topology this influence is the most important as increasing the number of nodes automatically increases the probability of having both paths intercepted.

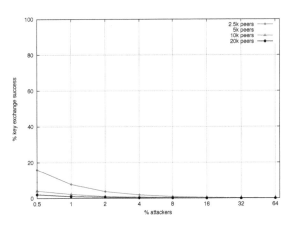

Figure 8. Success rates for different ring topology sizes.

5.3. Influence of the Chosen Number of Disjoint Paths

To reliably transfer a key between a source and a destination nodes, one must choose the number of disjoint paths to use for the different parts of the key (as the key split is done beforehand). We have performed experiments by varying the chosen number of disjoint paths from 1 to 10 for a network of 20k peers based on the ER model. The results are shown in Figure 9.

We observe that increasing the number of disjoint paths leads to an increase in the success rate of the key exchanges, which is expected, especially in the medium area where attackers are between 4% to 32%. We also notice a diminishing return when increasing the number of paths above 5 paths. There is a trade-off between increasing the success rate and decreasing the performances (as more paths means longer overall

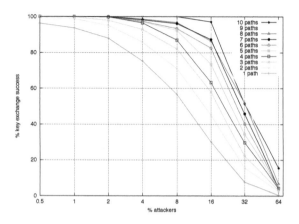

Figure 9. Success rates comparison by disjoint paths number (20k-node ER-model graphs with rerouting capped at 64 hops.

exchange time). Using from 2 to 5 paths already gives significant improvements on the success rate. The average number of available disjoint paths that can be found in a network between any given pair of nodes is thus important to exchange keys reliably.

5.4. Average Number of Available Disjoint Paths

Figure 10 shows the average number of disjoint paths found in 20k-node ER-model networks. We see that any source-destination pair has on average 7 paths. Many of them are intercepted on average but more than 16% of coordinated attackers are needed to be highly successful in intercepting all the subkeys, which makes the scheme efficient over this topology model.

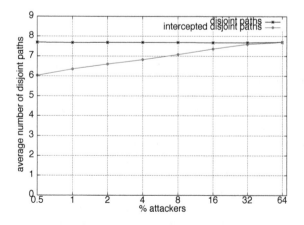

Figure 10. Average number of disjoint paths (total and intercepted) per source-destination pair in ER 20k-node networks.

Figure 11 shows the average number of disjoint paths in the 12.9k-node IPv4 map. We see that any source-destination pair has on average 3 paths, about half as what is found in the ER topologies. This is due to the fact that in the IP topology, the degree distribution of the nodes obeys a power law, which means that many nodes have only a few neighbors, while a few have a lot of neighbors. Despite this fact, much less paths are intercepted on average and more than 32% of coordinated attackers are needed to be highly successful in intercepting all the subkeys.

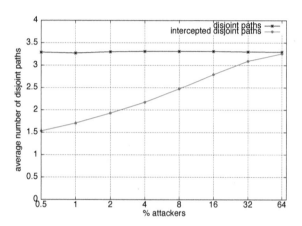

Figure 11. Average number of disjoint paths (total and intercepted) per source-destination pair in the IPv4 12.9k-node map.

The average number of available disjoint paths is thus highly dependent on the topology but also on the routing mechanism. Using a MP model topology with a strict greedy routing can lead to situations where those conditions render the key exchange inefficient or even impossible due to the lack of alternate paths.

5.5. Maximum Number of Hops

During a key exchange, the routing process requires that the identifiers of all nodes through which a packet transits must be added to this packet. Furthermore, each packet containing a subkey that leaves the source node, must contain the identifiers of all visited by previous packets carrying the first subkeys. Thus, it is important in practice to limit the number of hops between the source and the destination nodes, so as to prevent excessively huge packet sizes.

To measure the impact of the maximum number of hops, we have performed an experiment with two different values for the maximum. The experiments are based on a network of 20K peers and is built following an ER topology. Figure 12 illustrates the success rate of key exchange when the number of hops is capped at 64

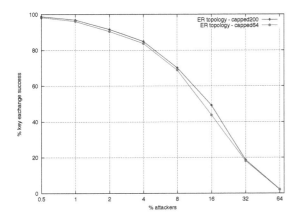

Figure 12. Influence of the maximum number of authorized hops.

and 200. We observe two curves that evolve similarly and are extremely close.

The number of hops has no significant influence on the success rate of the key exchange within the scope of the settings used in our experiments.

5.6. Influence of Routing Modes

We now discuss two modes of routing packets within the P2P network during a multipath key exchange process. In the *relax* mode, the process requires many disjoint paths that are relatively long. On the other hand, the *strict* mode corresponds to a reduced number of disjoint paths that are also short.

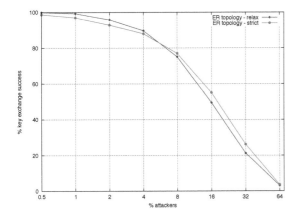

Figure 13. Comparison between the relax and strict routing modes.

Figure 13 depicts the results of our experiments in a network based on the ER topology. We observe that although the two curves are close, an interesting pattern emerges: for smaller proportions of attackers, the strict

mode provides lower success rates for key exchange; for higher proportions of attackers, the relax mode is the one that leads to lower performances. This could be explained by the fact that when the number of attackers are small, a large number of paths (*relax mode*) increases the chance of avoiding them. On the other hand, when the number of attackers gets large, shorter paths (*strict mode*) are more effective.

Large numbers of disjoint paths and short paths favorably influence the success rates of the multipath key exchange scheme. The former works well in presence of small scale coordinated attacks while the latter is recommended for largely infected networks.

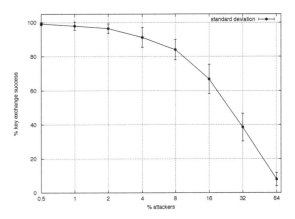

Figure 14. Standard deviation in the BGP-4 network map [17].

6. Security Analysis

In a multipath key exchange scheme, a malicious node that wishes to compromise a key being exchanged must be able to collect each of all key components routed over the network. Formally, when paths $\mathcal{P}_0, ..., \mathcal{P}_{k-1}$ are used to send several distinct subkeys from source $\mathcal{S}$ to destination $\mathcal{D}$, the only malicious nodes that could compromise the key should be located at the intersection of all paths. In other words, all the malicious node belong to set $M = \bigcap_{i=0}^{k} \mathcal{P}_i$ which represents the set of intersection points of all paths $\mathcal{P}_i$. $\mathcal{S}$ and $\mathcal{D}$ are obviously ignored in this set. Thus, when $\bigcap_{i=0}^{k} \mathcal{P}_i = \emptyset$ (*bigon criterion* is respected [11, Lemma 2.5]), then all paths are disjoint and any MITM attack attempt cannot succeed. In such a desirable case, there exists a k-connected subgraph between $\mathcal{S}$ and $\mathcal{D}$ in the network topology. When $|\bigcap_{i=0}^{k} \mathcal{P}_i| \geq 1$, there exists a real risk that MITM attacks will be committed on exchange transmitted between $\mathcal{S}$ and $\mathcal{D}$.

Consequently, the probability to have a MITM attack is estimated by $\sigma = \frac{|\bigcap_{i=0}^{k} \mathcal{P}_i|}{|\bigcup_{i=0}^{k} \mathcal{P}_i|}$ (where each path $\mathcal{P}_i$ is constituted of a set of consecutive hops from source $\mathcal{S}$ to destination $\mathcal{D}$). When all used paths are pairwise disjoint, the probability of *isolated* MITM attack (no *coordinated* MITM attack) is then: $\sigma = 0$ (i.e $|\bigcap_{i=0}^{k} \mathcal{P}_i| = 0$).

The number of distinct paths is also dependent on the source node degree. Thus, for a given *q-regular tree*, if q is a large number, then there is a probability to have several disjoint transmission channels. Nonetheless, despite the robustness of our multipath negotiation approach, cooperative (i.e., coordinated) MITM attacks, where several nodes maliciously cooperate to compromise a key, are possible. However, it is very hard, and excessively costly to execute such an attack in a real environment, especially in distributed systems where network topology changes dynamically.

In order to improve performances, re-authentication feature is introduced (see subsection 3.3). However, this challenge message used in this phase could be replayed. Furthermore, when a malicious node caches a challenge message, it can then create its copies and send them successively to target node. Thus, target node tries to resolves each challenge request because it does not know what packet is more fresh than other. Consequently, it will be rapidly saturated with requests from malicious node. Therefore, this causes a Denial-of-Service (DoS) attack.

In order to avoid such an attack from malicious nodes, a timestamp is assigned to each encrypted challenge message. Thus, the target node could distinguish between fresh packets and replayed packets.

Furthermore, during the key negotiation phase, all packets are exchanged in a clear text mode. Thus, traffic analysis attacks could reveal details about captured packets such as *sequence number* or *payload* which is nothing other than the transported subkey. Hence, multipath key exchange is needed to prevent the knowledge of all subkeys.

Figure 15 illustrates the slight decrease of key exchange success rate when the size of the network grows. In other words, the number of successful MITM attacks increases with the network size. As the number of attackers was set in our simulations to be proportional to the network size, this means that the size has indeed a small negative impact on the success rate. This is due to the consecutive average increase of the length of the paths themselves which increase their chance of being intercepted.

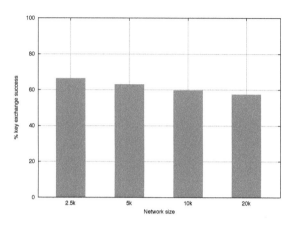

Figure 15. Variations of key exchange success rates for different network sizes (ER topology, capped by 64 hops max per path).

7. Related Work

Previous work have proposed several security infrastructures over fully decentralized or ad hoc networks [6, 7, 10, 15, 23, 27]. Although they are designed to be suitable in such environments, proposed approaches fulfill this gaol with more or less success and mostly with many caveats. In this section, we describe some models proposed in the literature to highlight the benefits of our approach.

Srivasta and Liu have relied on the Diffie-Hellman algorithm to deliver a solution that prevents threats in DHT networks [21]. Their scheme, however, remained sensitive to Man-in-the-middle attacks. Wang *et al.* have built a distributed PKI on top of the Chord structured overlay network [3]. They have used threshold cryptography to distribute the functionality of the PKI across the nodes of the DHT network. This Chord-PKI provides traditional PKI features such as certification, revocation, storage and retrieval.

Jiejun K. *et al.* propose to distribute certification authority functions through a threshold secret sharing mechanism [14]. In this system, private key is computed by k neighbor nodes and public key is derived from node identity.

Threshold cryptography is also used in identity-based key managements [7]. Nutshell, the key idea for identity-based cryptography is to define public keys derived from communicating nodes identities [20]. This method is really interesting however, the process of authentication could cause network overhead and the achieving of the value of threshold is not still guaranteed.

Takano *et al.* have designed a Multipath Key Exchange [23] similar to that proposed in our work. Their techniques however were designed to fit the Symphony and Chord P2P systems that are based upon a ring topology. However, this inflexibility causes some drawbacks. Indeed, the proposed approach is based on clockwise/anticlockwise routing and this makes it sensitive to coordinated MITM attacks.

Jaydip Sen proposes a multipath certification protocol for MANETs that proceeds by broadcasting in order to discover the route between both source and destination nodes [19]. The key exchange protocol is based on this routing approach to retrieve the public keys of the nodes. However, broadcasting technique has proven that is not relevant in scalable networks such as fully decentralized P2P systems. Therefore,

Shehadeh E. *et al.* investigate secret key generation from wireless multipath channels [10]. The proposed protocol is based mainly on both the physical characteristics of the wireless channel and key pre-distribution schemes. This solution is implemented within physical layer.

8. Conclusion

P2P networks are self-organizing systems that do not need to resort to any central coordination point. The flexibility of such networks thus allows them to operate effectively on the Internet. Unfortunately, the inherent features that make them desirable also make them vulnerable to various security threats such as eavesdropping, modification and usurpation attacks.

In order to address the security challenges of P2P networks, we propose an improvement built upon our CLOAK architecture defined in our previous work [25] on one hand and a new approach for key exchange that generalizes a model proposed by Takano *et al.* [23] on the other hand. The benefit of using CLOAK instead of another DHT is that it does not use routing tables and does not impose any topology structure upon its peers. Our solution presented in this paper allows two peers from a P2P network to generate a common secret key in order to secure their communication without the need of a trusted third party when multipath is possible. This key generation is based on the Diffie-Hellman method with several subkeys being exchanged through several disjoint paths. Simulation results show that the probability of securely generating a key is above 90% when the percentage of coordinated attackers is lower than 2%. Depending on the topology type, the network size and the amount of attackers, the success rate can remain around 80% when the percentage of coordinated attackers is around 4%. As the size of P2P networks is typically several orders of magnitude (from thousands to millions of nodes), we expect our solution to be efficient because a few percent of coordinated attackers will result in hundreds of those attackers being needed in the network and that may not be feasible in practice.

For future work, we will consider adding peer authentication which for the moment only relies on the CLOAK DHT not allowing a peer to usurp another peer's name. We will also evaluate the time which is necessary to exchange a key. Although this happens only at the beginning of the communication and needs not be done after that even when interruptions happen, we will use event driven temporal simulations for investigating those delays.

References

[1] Daouda Ahmat, Tegawende Bissyande, and Damien Magoni. Towards securing communications in infrastructure-poor areas. In *5th EAI International Conference on e-Infrastructure and e-Services for Developing Countries*, 2013.

[2] Daouda Ahmat and Damien Magoni. Muses: Mobile user secured session. In *5th IFIP Wireless Days Conference*, pages 1–6, 2012.

[3] Agapios Avramidis, Panayiotis Kotzanikolaou, Christos Douligeris, and Mike Burmester. Chord-pki: A distributed trust infrastructure based on p2p networks. *Comput. Netw.*, 56(1):378–398, January 2012.

[4] Thomas Bohme, Frank Goring, and Jochen Harant. Menger's theorem. *Journal of Graph Theory*, 37(1):35–36, 2001.

[5] Cyril Cassagnes, Telesphore Tiendrebeogo, David Bromberg, and Damien Magoni. Overlay addressing and routing system based on hyperbolic geometry. In *Proceedings of the 16th IEEE Symposium on Computers and Communications*, pages 294–301, 2011.

[6] Miguel Castro, Peter Druschel, Ayalvadi Ganesh, Antony Rowstron, and Dan S. Wallach. Secure routing for structured peer-to-peer overlay networks. *SIGOPS Oper. Syst. Rev.*, 36(SI):299–314, December 2002.

[7] Hongmei Deng, Anindo Mukherjee, and Dharma P. Agrawal. Threshold and identity-based key management and authentication for wireless ad hoc networks. In *Proceedings of the International Conference on Information Technology: Coding and Computing (ITCC'04) Volume 2 - Volume 2*, ITCC '04, pages 107–, Washington, DC, USA, 2004. IEEE Computer Society.

[8] W. Diffie and M.E. Hellman. New directions in cryptography. *IEEE Transactions on Information Theory*, 22(6):644–654, 1976.

[9] Antonio Gracia Dimitri Defigueiredo and Bill Kramer. Analysis of peer-to-peer network security using gnutella, 2002.

[10] Y. El Hajj Shehadeh, O. Alfandi, and D. Hogrefe. Towards robust key extraction from multipath wireless channels. *Communications and Networks, Journal of*, 14(4):385–395, 2012.

[11] D. B. A. Epstein. Curves on 2-manifolds and isotopies. In *Acta Math*, pages 15–16, 1966.

[12] Paul Erdős and Alfred Rényi. On the evolution of random graphs. *Publications of the Mathematical Institute of the Hungarian Academy of Sciences 5*, pages 17–61, 1960.

[13] Ken ichi Kawarabayashi, Yusuke Kobayashi, and Bruce A. Reed. The disjoint paths problem in quadratic time. *J. Comb. Theory, Ser. B*, 102(2):424–435, 2012.

[14] Jiejun Kong, Z. Petros, Haiyun Luo, Songwu Lu, and Lixia Zhang. Providing robust and ubiquitous security support for mobile ad-hoc networks. In *Network Protocols, 2001. Ninth International Conference on*, pages 251–260, 2001.

[15] Hyeokchan Kwon, Sunkee Koh, J. Nah, and Jongsoo Jang. The secure routing mechanism for dht-based overlay network. In *Advanced Communication Technology, 2008. ICACT 2008. 10th International Conference on*, volume 2, pages 1300–1303, 2008.

[16] Damien Magoni and Jean-Jacques Pansiot. Internet topology modeler based on map sampling. In *Proceedings of the 7th IEEE Symposium on Computers and Communications*, pages 1021–1027, 2002.

[17] Y. Rekhter. A border gateway protocol 4 (bgp-4). *RFC 4271*, http://www.ietf.org/rfc/rfc4271.txt ,january, 2006.

[18] J. Schafer and K. Malinka. Security in peer-to-peer networks: Empiric model of file diffusion in bittorrent. In *Internet Monitoring and Protection, 2009. ICIMP '09. Fourth International Conference on*, pages 39–44, 2009.

[19] J. Sen. A multi-path certification protocol for mobile ad hoc networks. In *Computers and Devices for Communication, 2009. CODEC 2009. 4th International Conference on*, pages 1–4, 2009.

[20] Adi Shamir. Identity-based cryptosystems and signature schemes. In *Proceedings of CRYPTO 84 on Advances in Cryptology*, pages 47–53, New York, NY, USA, 1985.

Springer-Verlag New York, Inc.

[21] M. Srivatsa and Ling Liu. Vulnerabilities and security threats in structured overlay networks: a quantitative analysis. In *Computer Security Applications Conference, 2004. 20th Annual*, pages 252–261, 2004.

[22] Jani Suomalainen, Anssi Pehrsson, and Jukka K. Nurminen. A security analysis of a p2p incentive mechanisms for mobile devices. In *3rd International Conference on Internet and Web Applications and Services*, pages 397–402, 2008.

[23] Y. Takano, N. Isozaki, and Y. Shinoda. Multipath key exchange on p2p networks. In *Availability, Reliability and Security, 2006. ARES 2006. The First International Conference on*, pages 8 pp.–, 2006.

[24] T. Tiendrebeogo, D. Ahmat, and D. Magoni. Reliable and scalable distributed hash tables harnessing hyperbolic coordinates. In *New Technologies, Mobility and Security (NTMS), 2012 5th International Conference on*, pages 1–6, 2012.

[25] Telesphore Tiendrebeogo, Daouda Ahmat, Damien Magoni, and Oumarou Sié. Virtual connections in p2p overlays with dht-based name to address resolution. *International Journal on Advances in Internet Technology*, 5(1):11–25, 2012.

[26] Guido Urdaneta, Guillaume Pierre, and Maarten Van Steen. A survey of dht security techniques. *ACM Comput. Surv.*, 43(2):8:1–8:49, February 2011.

[27] Peng Wang, Ivan Osipkov, and Yongdae Kim. Myrmic: Secure and robust dht routing. Technical report, University of Minnesota, 2007.

15

How did you know that about me?
Protecting users against unwanted inferences

Sara Motahari*, Julia Mayer, Quentin Jones

New Jersey Institute of Technology, Newark, New Jersey, NJ 07103-3513, USA

Abstract

The widespread adoption of social computing applications is transforming our world. It has changed the way we routinely communicate and navigate our environment and enabled political revolutions. However, despite these applications' ability to support social action, their use puts individual privacy at considerable risk. This is in large part due to the fact that the public sharing of personal information through social computing applications enables potentially unwanted inferences about users' identity, location, or other related personal information. This paper provides a systematic overview of the social inference problem. It highlights the public's and research community's general lack of awareness of the problem and associated risks to user privacy. A *social inference risk prediction framework* is presented and associated empirical studies that attest to its validity. This framework is then used to outline the major research and practical challenges that need to be addressed if we are to deploy effective social inference protection systems. Challenges examined include how to address the computational complexity of social inference risk modeling and designing user interfaces that inform users about social inference opportunities.

Keywords: inference problem, privacy, social computing, ubiquitous computing

1. Introduction

Social computing applications connect users to each other to support interpersonal communication (e.g. Twitter and Instant Messaging), social navigation (e.g. Facebook), and the sharing of user-generated content (e.g. YouTube and Flickr). While social computing applications gain incredible popularity, their use puts users' privacy at considerable risk. This is because social computing applications are often designed in a manner that impels their users to share personal data as publicly and as widely as possible [1]. Privacy in social computing is complex, multifaceted, and poorly understood, in part because both users' perceptions and the real privacy risks have to be considered when protecting users.

In this paper we discuss a rarely considered but very pervasive and serious privacy threat to users of social computing applications, namely the risk of people making unwanted social inferences. Social inferences occur when unauthorized information is deduced from authorized information about users' identity, location, or other related personal information. Such inferences are often enabled by the public sharing of personal information through social media. Unfortunately, people are often unaware of possible consequences of their personal information sharing, e.g. privacy invasions. This is due to the fact that users have a very limited understanding of the relationship between the information they share intentionally and the information this allows others to infer, e.g. the user's identity, location, or other related personal information (social inferences).

While researchers have mainly addressed the more evident privacy threats related to user access control, the problem of social inferences has not yet been addressed in detail by researchers, particularly in regard to:

(i) measuring social inference risk;

(ii) understanding how social inference risks vary by contexts of use and system types (e.g. mobile vs. online); and

(iii) designing effective social inference risk management systems.

*Corresponding author. Email: Sara.gatmir-motahari@sprint.com

The objective of this paper is to provide a systematic and comprehensive overview of these aspects. We aim at providing a broader understanding of the real-world privacy threats faced by users. This paper will start with a detailed definition and explanations of what social inferences are, when they happen, give examples of real-world incidences and impacts. In this way, we want to draw attention to people's and researchers' lack of understanding and awareness of the problem. We will also examine previous efforts into dealing with inferences as privacy threat and map out their shortcomings as well as their strengths on which we can build. After this, we will present an entropy-based framework to predict the risk followed by findings from studies exploring this method. Finally, we examine the key challenges to the development of effective social inference protection systems that the research community needs to address.

2. The social inference problem

Inferences are generally understood to result from 'the process of arriving at some conclusion that, though it is not logically derivable from the assumed premises possesses some degree of probability relative to the premises' [2]. In the privacy literature *inferences* are understood to result from the process of deducing unrevealed information from authorized information.

In the computing literature inferences have generally been thought of as a confidentiality threat to database security and privacy resulting from database querying or data mining [3, 4, 5, 6, 7]. A well-known example of this general inference problem relates to an organization's database of employees [8], where the relation <Name, Salary> is a secret, but user *u* requests the following two queries: 'List the *rank* and *salary* of all employees' and 'List the *name* and *rank* of all employees'. None of the queries contain the secured *<name; salary>* pair; however, an individual may utilize the known information <Rank, Salary> and <Rank, Name> to infer the private <Name, Salary> information through deductive reasoning. For example, the knowledge that Bob is a manager and all managers earn $x can help one deduce that Bob earns $x.

While it may be possible in some narrow circumstances to make social inferences using database queries or data mining, we will illustrate in this paper that the majority of cases do not result from such deductive reasoning processes. *Social inferences* are the subset of inferences that results from using social applications and can pose serious threats to the privacy of their users. They typically occur through linking authorized information from a social computing application with some background knowledge to infer personal user information, such as identity, location, activities, social relations, and profile information. The underlying logic is:

Background knowledge + Authorized information → Social inference opportunity

Background knowledge is defined as any piece of information that is not directly revealed to the users but is available in the outside world or in a system and can be used to infer the attribute in question. It can also be understood as a mental model or world model that provides rules of how to link information together. Such background knowledge is often acquired from real world through experience and observation. Not only may inference require the use of background knowledge, but also the information being inferred (e.g. users' identity at physical appearance granularity) may not be stored in the application database.

In the following sections we will present a more detailed overview of the various ways social inferences can happen and provide examples of real-world incidents as well as user studies which illustrate the impacts and seriousness of social inferences.

2.1. Types of social inferences

First, it is important to understand the variety of ways in which social inferences can occur. In different contexts, different background knowledge or authorized information can yield to a social inference. The way a social inference is made differs depending on:

(i) *Who* makes the inference: person or system;

(ii) *How* the authorized information/background knowledge is acquired (information source);

    (a) over time (historical) or instantaneous and
    (b) from system/web or real world.

(iii) *What* kind of information is used (authorized information/background knowledge) and inferred? (dynamic vs. static, e.g. identity, location, personal information, physical appearance, and social relations).

We refer to social inferences made by people as *human social inferences* and social inferences made by systems as *automated social inferences*. While both types follow the same logic and can be a threat to users' privacy, we focus on the first type, social inferences made by people, in this paper.

The second distinguishing feature of social inferences is the information used to make the inference and how it is acquired. Background knowledge and authorized information might be available only at a certain point in time allowing for an *instantaneous* social inference. On the other hand, when users build up background knowledge over time, e.g. as users interact with applications or observe the world around them over time, which enables a social inference, we refer to it as *historical* social inference.

Both authorized information and background knowledge can come from various different sources, e.g.

(i) different systems (e.g. through the web);

(ii) real-world observations (e.g. who is nearby at the moment); and

(iii) different time periods, accumulated knowledge resulting from aggregating previously revealed information.

For example, in mobile social applications, social inferences can be the result of either accessing location-based information or the result of social communications, or both. Table 1 provides an overview of different types of background knowledge and authorized information used in the various ways social inferences can occur.

While the focus of this paper is on human social inferences, we briefly describe automated social inferences, which are inferences a system can make about a user based on the system's model of the world. This model consists of explicit and authorized data revealed by the user linked to automatically collected background knowledge (e.g. from other sources like the web, from sensors about the user's context) following certain rules and associations. In this way, automated social inferences can be seen as occurring when systems extend their knowledge by emulating human reasoning, which is why such methods are well known in the field of artificial intelligence, machine learning, and the semantic web. Automated social inferences can be used for both personalization and the collection of unauthorized personal information. A widely used method for inferring identity information is to ask visitors to a website for their date of birth, gender, and zip code, supposedly to allow for personalized services. This information is then used to infer a user's identity typically for additional marketing purposes. Researchers started investigating risks related to automated social inferences made by malicious websites which could find out what groups a user belongs to, and use that information to identify the users [9]. By 'capturing' people's social networking groups from their browser with a trick known as history stealing and then cross-referencing these groups, a user's social-network profile—and therefore his real-life identity—could be inferred 42% of the time. This means that an otherwise anonymous

web user could be identified correctly by a malicious site simply because the user visited that site.

While automated social inferences also pose a serious threat to users' privacy, this paper focuses on the risks and challenges associated with human social inferences. Human social inferences are made based on people's model of the world (physical, online, social) combined with what a system reveals about the user. Background knowledge here refers to a user's world model consisting of learned, observed, experienced knowledge combined with rules and relations (e.g. if the person is not a boy, it is very likely that she is a girl; or I can find his home address through a Google search). Due to the limitations of people's memory and associated assessment of social inference risks it is important that we distinguish between instantaneous and historical inferences. An example of an instantaneous social inference is when Alice's cell phone shows her that there is a romantic match nearby, Bob. Since Alice sees only one individual with a similar cell phone nearby, Alice infers this must be Bob, thus increasing her chance of identifying him. An example of a historical social inference is when an anonymous nickname is repeatedly shown through a mobile social computing application as being on the first floor of a gym, where the gym assistant normally sits. Despite a wide variety of other nicknames appearing at the same location, given time this association allows other users to infer that the particular online nick name of the gym assistant.

2.2. Examples and implications

The risk of social computing enabled social inferences is growing both with increasing popularity of social media and with the advent of mobile and location-aware social computing. The subtle nature of the social inference risk is illustrated by an incident that occurred during our deployment of a location-aware campus-based Wiki. *CampusWiki* [10] allowed students to create and edit location-linked content. Editors of pages on CampusWiki could be anonymous or identified, and hide or reveal their physical location. During the first semester CampusWiki was deployed, a student added unpleasant comments about

Table 1. Different ways social inferences can occur.

	Instantaneous	Historical
Human	*Background knowledge*—what user knows about his/her current situation through observing and interacting with the world or the web. *Authorized information*—temporary revealed information by the system, e.g. current location of another user.	*Background knowledge*—what user has learned or acquired over time from experience and observations of the world around or from the web. *Authorized information*—information revealed by the systems over time and collected by user.
Automated	*Background knowledge*—temporarily available information from other sources, e.g. web. *Authorized information*—temporary authorized information, real-time shared information, e.g. current location of user.	*Background knowledge*—algorithms to link or cross-reference, e.g. machine learning. *Authorized information*—system-collected user information over time, e.g. buying behavior, cookies from browsing history contextual data collected (GPS and other sensor data).

a professor. In the process the student kept his name hidden, but revealed the time of his edits, and his location at the time. However, the professor was able to infer his identity by realizing that the comments were added in his classroom when he was teaching. Since only two students were using a laptop during the class in question, he was able to identify the student editor. In this case, the inferrer (professor) used knowledge obtained from physical observations (background knowledge) to ascertain who the student in question was. The result was a confrontation, which led to the student dropping the course.

Unfortunately, the risk of social inferences is underestimated not only by individual users but also by system designers. The release of the Google's BUZZ application raised consumer concerns about how social computing applications protect their users' privacy. Initially, Google automatically allowed BUZZ users to see which other users a user had been frequently chatting or emailing. It soon became obvious that having your communication partners revealed in an uncontrolled fashion could result in leaking of quite sensitive and private information about personal relations and/or activities. The Huffington Post discussed problems of the application which could, e.g. let your current employer know you are engaged in conversations with a competitor [11], which in turn could be used to make an unwanted inference about plans for switching jobs. Google's step of making the contact list sharing in BUZZ an opt-in feature was a good initial step, but users still need more help managing the risk of social inferences as they do not have a thorough understanding of the possible consequence of their actions. These examples suggest that the risk of unwanted social inferences is serious and could jeopardize one's education, career, or personal life.

To explore people's understanding of the privacy threats associated with social computing, we carried out a survey of students at our North Eastern, urban university campus (107 subjects from 17 different majors) [12]. The following seven categories of threats to user privacy in social computing system were introduced to the subjects together with example scenarios:

(i) *Inappropriate Use by Administrators*: The system admin sells personal data without permission [13].

(ii) *Legal Obligations*: The system admin is forced by an organization such as the police to reveal personal data [13].

(iii) *Inadequate Security*: The server is not protected against intrusions or wireless transmission through the air is not secured [14].

(iv) *Designed Invasion* (due to poorly designed features): A cell phone application that reveals location to friends, but does this without informing the user or providing control of this feature [15, 16, 17].

(v) *Instantaneous Social Inferences*: Alice's cell phone shows her that there is a romantic match nearby, Bob. Since Alice sees only one individual with a similar cell phone nearby, Alice infers this must be Bob, thus increasing her chance of identifying him.

(vi) *Historical Social Inferences*: Bob is so often in Alice's office. Their relationship must be romance.

(vii) *Social Leveraging of Privileged Data*: David can't access my location, but Jane can. David asks Jane my location.

Then subjects were asked to rate their awareness of each threat as well as their privacy concerns depending on the type of information, such as identity, location, status, profile information, and social ties. Respondents' answers highlighted the expected issue that people have comparatively little awareness or concerns about the risk of social inferences. Subjects expressed less awareness over inferences (categories v, vi, and vii, where categories v and vi represent the social inference problem) and more awareness of the first four categories (Friedman, $\chi^2 = 299$, $df = 6$, $n = 102$, $p < 0.001$). Combining the inference categories into one new variable and the first three categories into another one also shows a statistically significant difference between them (Wilcoxon's Signed Rank $u = -7.91$, $n = 102$, $p < 0.001$). Furthermore, subjects were generally more concerned over the threats they were more aware of. They are most concerned over being hacked, which is not surprising considering the ability of hacking stories to make the news headlines. *Inappropriate Use by Administrators* (i) and *Designed Invasion* (iv) come next and the other categories were least worrisome for them (Friedman, $\chi^2 = 64.4$, $df = 2$, $n = 99$, $p < 0.001$).

To determine if people's general lack of concern about social inferences is due to their ability to manage such risk, or due to a lack of understanding of the extent of such risks, we conducted an experiment [12, 18]. Two hundred ninety-two individuals (146 pairs) engaged in anonymous chat with each other. Subjects were asked questions about their desired level of anonymity, if they think they had maintained their desired level of anonymity, and if they could identify who they chatted with. The level of desired anonymity varied greatly, however 72% of the subjects who had anonymity concerns did not want to be exactly identified by their name or face and 6.3% of them did not want to be narrowed down to two people or less. It was surprising to us just how poorly subjects were able to assess what their chat partner knew about their identity. Subject's estimated-degree-of-anonymity was smaller than maintained-degree-of-anonymity in 20% of the cases, which means at least 20% of the subjects revealed identifiers that put them at the risk of *unwanted* identity inference. This illustrates that while people are often able to make social inferences, they are unable to estimate social inference risks and are therefore unable to maintain their desired degree of anonymity.

The above examples, survey, and experiment illustrate the mismatch between real privacy risks and user perceptions and suggest that at present users of social computing applications are unable to routinely maintain their desired and expected level of personal privacy. Even when users are in complete control of the information they reveal they are not able to maintain their desired degree of anonymity because individuals are unable to correctly judge inference risks. The research community is yet to come to grips with the unique privacy challenges associated with social inferences. This calls for a more thorough examination of the social inference problem as a privacy threat in order to build effective social inference protection systems.

To address this fundamental gap in the literature, below we present a social inference risk prediction framework and associated confirmatory studies, and then outline some of the major research challenges facing those that wish to build effective social inference protection systems.

3. Determining the social inference risk

In order to protect individuals from unwanted social inferences, we need to be able to measure the extent of the social inference risk in various contexts. This determination can be achieved through the steps outlined in our social inference risk prediction framework. We start this section with an overview of previous attempts to predict social inference risk. Then we will introduce our social inference prediction framework and present findings from validating studies.

3.1. Previous work on the inference problem

To date, most of the research into inferences has addressed the general inference problem in databases, which focuses on the problem of detecting and removing unwanted inference channels. An inference channel in a database is a means by which one can deduce unauthorized data from authorized data. In order to detect an inference channel in a database, the inference risk has to be predicted. Two types of techniques have been previously proposed. One makes use of semantic data modeling methods to predict chances of inferences and locate inference channels in the database design, in order to redesign the database for the removal of these channels [19]. The other one evaluates database queries to understand whether they lead to unauthorized inferences [20]. These techniques have been studied for statistical databases [20], multilevel secure databases [21, 22], and general-purpose databases [8, 23].

In addition, deductive inferences can also result from analysis of data sets generated for data mining. Although such data sets are usually cleaned up from sensitive information, common data mining techniques could lead to leakage of some previously eliminated sensitive information. Therefore, a few researchers have attempted to prevent the inferences in data sets generated for data mining, such as medical records [7].

However, as noted in the Section 1, social inferences are often made about personal user information that is not stored in the application database and leveraging background knowledge that is also not stored in the database. As a result, systems cannot protect users from social inferences by applying traditional database inference protection techniques. An alternative way is to use an entropy-based approach. An early attempt to use entropy measurements in anonymity protection came from Serjantov and Danezis [24]. They suggested that the anonymity level of networking nodes (transmission and routing systems) could be measured using an information theory approach, i.e. entropy measurements. They proposed measuring the degree of anonymity of geographically fixed nodes (such as desktops) assuming that network attackers have partial information about the topology of the network (which can be considered the attacker's background knowledge).

Denning and Morgenstern, pioneers in calculating the partial inference risk, employed classical information theory to measure the inference chance [3, 25]. Given two data items x and y, let $H(y)$ denote the entropy of y and $H_x(y)$ denote the entropy of y given x, where entropy is as defined in information theory. Then, the reduction in uncertainty of y given x is defined as follows:

$$Infer(x \to y) = \frac{H(y) - H_x(y)}{H(y)}.$$

The value of $Infer (x \to y)$ is between 0 and 1, representing how likely it is to derive y given x. If the value is 1, then y can be definitely inferred given x. Denning and Morgenstern did not know how to use this formulation in real situations and they mention the serious drawbacks of using this technique [3]. Firstly, it is difficult, if not impossible, to determine the value of $H_x(y)$; secondly, the computational complexity that is required to draw the inference is ignored [3]. Nevertheless, this formulation has the advantage of presenting the probabilistic nature of inference (i.e. inference is a relative not an absolute concept).

Another approach addressing the problem of identity inferences based on usernames was proposed by Lemos [26]. They produce an analytical model that estimates the uniqueness of a username and then assign a probability that a single username from two different online services refers to the same user. Probability estimates are calculated using language models and Markov chain techniques. Their results show that entropy measures of usernames can be used to link accounts and to identify users.

The above overview shows that previous entropy-based research into predicting the impact of background knowledge on inference risks has made important steps toward privacy protection and provides significant insights, such as the use of entropy-based modeling approaches. However, this research has failed to take into account a number of key considerations. The above methods are based on the assumption that all data used by the inferrer are inside a self-contained system. Therefore, the presented solutions can only protect from a narrow range of social inferences because, as seen above, social inference opportunities are enabled through use of information from outside the system in question (background information). In addition, the proposed solutions assume that the information that is at risk of being inferred is stored inside the system in question, which is, e.g. usually not the case for users' identity at physical appearance granularity. Therefore, if we are to help users protect themselves from unwanted social inferences, mechanisms will need to be developed that take into account inferrer's potential background knowledge. They also do not deal with the dynamic nature of personal and contextual information in social computing applications and they make unverified assumptions about the inferrer's background knowledge. We address these drawbacks in our framework by

(i) considering the probabilistic nature of social inferences by using an attribute's information entropy to measure the level of its uncertainty;

(ii) taking the inferrer's background knowledge and historical data into account by calculating the conditional entropies conditioned on both revealed data and background knowledge; and

(iii) dynamically updating the level of entropy.

3.2. Social inference risk prediction framework

We developed a framework [27, 28] to model and reliably predict social inference risk. As seen before, social inferences happen as a result of low information entropy. The framework is based on this logic: as individuals or applications collect more information about a user, such as his/her current situation, our uncertainty about other attributes, such as his/her identity, may be reduced, thus increasing the opportunity of a social inference. This uncertainty can be measured by *information entropy*. *Information* [29], as used in information theory for telecommunications, is a measure of the decrease of uncertainty in a signal value at the receiver site. Here we use the fact that the more uncertain or random an event (outcome) is, the higher the *entropy* it possesses. If an event is

very likely or very unlikely to happen, it will not be highly random and will have low entropy. Therefore, entropy is influenced by the probability of possible outcomes. It also depends on the number of possible events, because more possible outcomes make the result more uncertain. In our context the probability of an event is the probability that an attribute (such as a user's name) takes a specific value. As the inferrer collects more information, the number of entities that match her/his collected information decreases, resulting in fewer possible values for the attribute and lower information entropy.

The scenario below illustrates the logic of our framework. It describes the actual behavior of a pair of subjects in our experimental examination of the social inference risk prediction framework [18]:

Bob engages in an online chat with Alice. At the start of communication, Bob does not know anything about his chat partner. He is not told the name of the chat partner or anything else about her, so all users are equally likely to be his partner. After they start chatting, Alice's language and chat style help Bob guess her gender and that she is Hispanic (and Alice confirms his guess during the course of conversation). After a while, Alice reveals that she plays for the university's women's soccer team. Bob, who has prior knowledge of this soccer team, knows that it has only one Hispanic member. This allows Bob to then infer Alice's identity.

Here, as Bob combines his background knowledge of the female Hispanic soccer players on campus with what Alice reveals, his uncertainty about his chat partner's identity decreases, thus increasing the opportunity of a social inference. This uncertainty can be measured by *information entropy*. In the case of the above scenario Bob's background knowledge of the ethnic makeup of soccer players on his campus is all that is necessary for an identifying social inference. In order to calculate the information entropy of an attribute, the background knowledge of a user has to be modeled. This is needed in order to identify:

(i) What attributes, if revealed, can help the inferrer to reduce the identity entropy of a user and how they change conditional probabilities.

(ii) What attributes, even if not revealed, can help the inferrer reduce the identity entropy of a user and how they change conditional probabilities (such as guessing Alice's gender from her chat style and using that to infer her identity).

However, as Jajodia and Meadows [21] say, 'we have no way of controlling what data is learned outside of the database, and our abilities to predict it will be limited'. Thus, even the best model can give us only an approximate idea of how safe a database is from illegal inferences. Nevertheless, the results of our studies suggest that considering the context and community of users enables systems to effectively model background knowledge.

Background knowledge can be estimated using the following different methods listed here with increasing levels of accuracy [11, 28].

(i) *Method 1*. Simply assume that the inferrer can link the information in the application database to the outside world, thus being able to estimate the number of matching users and their probabilities based on the existing database. *Weakness*: Some of the attributes in the database are not usually known to the inferrer while some parts of the inferrer's background knowledge may not exist in the database.

(ii) *Method 2*. Hypothesize about the inferrer's likely background knowledge taking the context of the application into consideration.

(iii) *Method 3*. Utilize the data from user studies designed to capture the users' background knowledge. The advantage of this method is a reliable modeling of background knowledge.

(iv) *Method 4*. Extension of the latter two methods with application usage data that allow for continuous monitoring of an inferrer's knowledge.

Two user studies provided comparative value and practicality of the second and third method. The results suggested that Method 2 was almost as accurate as Method 3 in the realm of computer-mediated communication and proximity-based applications [11, 28]. This means that context and community of users allows to effectively model background knowledge for a specific context.

After estimating all the significant information (including background knowledge) available to the inferrer, Q, the conditional instantaneous information entropy of attribute Φ, is defined as

$$H_c = H(\Phi|Q) = \sum_{i=1}^{V} P_c(i) \cdot \log_2 P_c(i),$$

where V is the number of possible values for attribute Φ, $P_c(i)$ is the probability that the ith possible value is thought to be the true value by the inferrer. $P_c(i)$ is the posterior probability of each value given Q. Q includes the inferrer's background knowledge as well as the information currently being revealed by the system. In the case of Bob and Alice mentioned above, at the beginning of the chat, V equals the number of students and $P_c(i)$ is uniformly distributed over V. As the revelations continue, V decreases and also $P_c(i)$ deviates from a uniform distribution until entropy is lower than a risky threshold. Historical information entropy is defined in the same way, but by including previously revealed information in Q.

More details on calculating the conditional probabilities and information entropy can be found in [11, 28] where the authors also set thresholds for information entropy based on users' preferences. If the information entropy is lower than its threshold, there is a high inference risk. Therefore, instantaneous and historical inference functions are based on instantaneous and historical information entropies [11, 28].

3.3. Testing the risk prediction framework

To the best of our knowledge, the above risk prediction framework is the only method proposed to predict the social inference risk. The following empirical studies showed that it is able to strongly predict the risk of social inferences [11, 18, 28].

The information entropy modeling approach was tested in two different areas of social computing:

(i) for anonymous computer-mediated communication (Anon-CMC), through a laboratory chat experiment between unknown chat partners [18] described in part earlier in this paper; and

(ii) for proximity-based social applications, through a mobile phone field study that lasted for 4 weeks and explored patterns of co-location and anonymity of the subjects [11, 28].

Moreover, large-scale simulations were done for both areas to verify the entropy-based social inference risk modeling framework and to investigate the problem and appropriate actions on a larger scale for various situations.

The laboratory experiment involved 532 collected user profiles, out of which 292 subjects completed a chat experiment and the post-experiment survey [18]. Subjects participated in a study consisting of three phases:

(i) Phase I—*Online Personal Profile Entry*: Subjects were filling out an extensive online profile (67 individual profile items clustered into five broad categories: basic information, personal information, education information, contact information, and interests).

(ii) Phase II—*Introductory Chat Experiment*: Subjects were chatting with a randomly assigned unknown chat partner using a custom-developed software application designed to aid in communication and exchange of personal profile information. The user interface (UI) of the introductory chat experiment is shown in Figure 1. During the chat, subjects were able to see their own profile on the left side of the screen, a chat box for typing in the center, and their chat partner's profile (all information hidden by default) on the right side. Throughout the conversation subjects could decide to reveal parts of the profile or request their

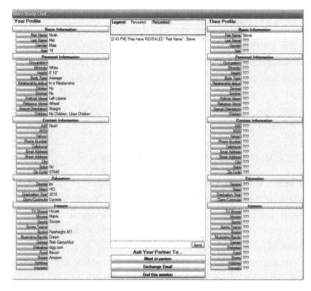

Figure 1. User interface of introductory chat study between anonymous chat partners.

chat partner to reveal a certain field. Revealed fields were highlighted yellow while requested fields were highlighted green.

(iii) Phase III—*Post-Chat Survey*: Subjects were asked if they could guess their chat partner's identity, or other attributes (physical characteristics), and how they made the guesses, as well as how anonymous they wanted to be and what they revealed about themselves.

Phases I and III served to inform us as to how people perceived and understood social inference (detailed results can be found in [18]) while Phase II allowed us to test the entropy-based risk prediction and validate our framework. Many variables were measured from this laboratory experiment such as the duration of the chat, the number of revealed profile items, subjects' intended level of anonymity, and subjects' demographics. Results of a binary logistic regression showed that among all measured variables, information entropy was the only statistically significant predictor of the inference risk (Wald's $\chi^2 = 6.018$, $\exp(\beta) = 0.705$, $p = 0.014$). Background knowledge of the users was investigated and taken into account in calculating the information entropy.

The proximity-based social application field study was carried out in two stages. The first stage lasted for 3 weeks and 180 subjects participated in it. The second stage lasted for 4 weeks. One hundred sixty-five subjects started the field study, out of which 129 subjects completed 4 weeks of the study. The subjects used a proximity-based social application that showed user the nicknames of users in their vicinity (as shown in Figure 2). Every time subjects changed their location and stayed in a new location for 5 min or when they had not answered a questionnaire

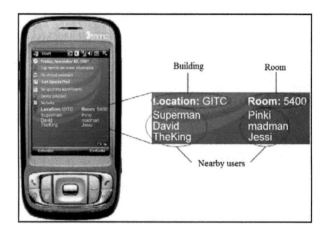

Figure 2. Proximity-based social application study.

for at least 2 hours, they answered pop-up questionnaires asking what they could guess about the identity of nickname owners and if they could map them to people in their vicinity.

Various place-, time-, proximity-, and subject-related parameters were measured. A binary stepwise logistic regression analysis was performed on the identity-inference-incident (dependent variable) and all the independent variables listed above. The only variables left in the analysis were the instantaneous inference function (Wald's $\chi^2 = 5.818$, $\exp(\beta) = 0.970$, $p = 0.012$) and historical inferences function (Wald's $\chi^2 = 53.001$, $\exp(\beta) = 1.084$, $p < 0.001$). Users' background knowledge in this context was investigated and modeled for calculating information entropy [11, 12]. In both studies, entropy-based inference modeling was again the strongest predictor of social inference opportunities.

Experimental results show that social inferences are not rare and that they are more common in CMC than in proximity-based application. We used the experimental data from the above studies to investigate the extent of the risk of identity inference for both applications on a larger scale. Parameters for the simulation models, such as the diversity of profile items and their statistical distribution as well as the probability of revealing profile items and statistical distribution of nearby users, were derived from the experimental data from previously explained user studies to approximate real-world deployments. Additional information such as the number of courses, statistical distribution of the number of students in a class, and enrolment statistics were obtained from university admission statistics. Entropy thresholds were calculated based on the desired degree of anonymity (a desired degree of anonymity of u means that the users wished to be indistinguishable from $u - 1$ other users).

For the Anon-CMC simulation online interactions of the users were simulated and information entropy was calculated for each simulated chat based on their revealed profile information. Results shown in Figure 3 indicate

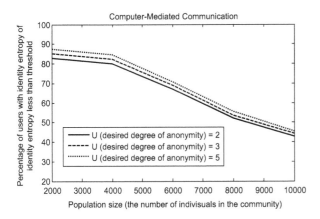

Figure 3. Risk of identity inference for computer-mediated communication.

that users reveal information that 50% of the time could lead to the invasion of their desired degree of anonymity (failure to maintain their desired degree of anonymity), which shows that identity inferences can be quite prevalent in CMC.

Figure 4 shows the probability that a user is at the risk of instantaneous identity inference in a proximity-based application. The *y*-axis shows the percentage of users whose identity entropy was lower than its threshold. Entropy threshold was calculated based on their desired degree of anonymity, Φ. The *x*-axis represents the desired degree of anonymity. Each curve depicts the risk for a different mean of nearby population density. As expected, more crowded environments have a lower chance of being at the identity-inference risk. Simulation of the risk of historical inferences and experimental results show that for a given population density, historical inferences happen less frequently than instantaneous inferences. Therefore, identity-inference risks happen less frequently for proximity-based applications than for CMC.

To conclude, these experiments and simulations verified the entropy-based social inference risk prediction framework and showed that we are able to predict social

inference risk. In order to protect individuals from unwanted social inferences, systems will need to be deployed that systematically reduce this risk.

In the next section we propose several features of social inference protection systems and point out open key challenges for future research.

4. Social inference protection systems

Historically, inference protection has been thought of as an access control issue, where the main challenge is to make sure that potential privacy invaders cannot get the results of dangerous queries that enable inferences. As we have shown social inferences cannot be protected in this way because the relevant data are not stored in the user-application database. As a result, new enhanced techniques and systems will need to be developed if we are to protect users from unwanted social inferences. As shown above, we are able to reliably predict the social inference risk using an entropy-based approach, but this alone provides little protection for the user. Instead, social inference protection systems need to be developed as shown in Figure 5 that:

(i) determine users' privacy preferences;

(ii) monitor users and their environments;

(iii) use stored and contextual data collected to calculate users' current social inference risks; and

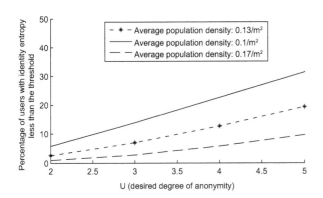

Figure 4. Risk of identity inference for proximity-based applications.

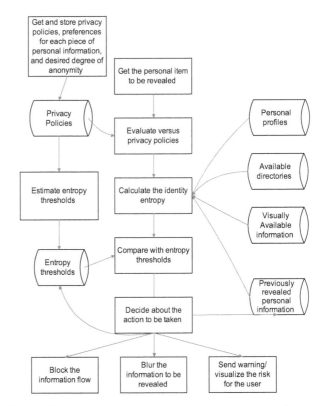

Figure 5. Components and processes in a social inference protection system instantiation.

(iv) utilize this knowledge to reduce the dangers by either

  (a) blocking information exchange;
  (b) blurring information exchanged; or
  (c) providing visualizations/warnings to users that support their taking appropriate actions.

Unfortunately, there is limited knowledge about how to build such a system. The key areas where further research is needed can be divided into risk prediction and risk management challenges.

4.1. Challenges in risk prediction

Information entropy as used in our presented framework showed to be a reliable predictor for social inference risk. The next steps to be taken are to explore other sources of background knowledge and improve modeling of background knowledge as well as historical information.

Modeling background knowledge. Background knowledge can be acquired from a variety of sources. For example, users can socially leverage privileged data and obtain knowledge from their friends and social ties that could inherently become part of their background knowledge. Such sources were insignificant in the above studies. However, in the case of finding such knowledge or other distributed types of background knowledge, such information sources have to be appropriately rated based on their reliability and access probability. Such effects should be parameterized in order to tailor the conditional probability equations used to estimate the information entropy. Studies of background knowledge can be merged with initial studies of the application, such as usability studies, so that the estimation can be obtained with a low cost. Finding efficient and inexpensive ways of background knowledge modeling for various applications of ubiquitous and social computing is a challenge yet to be addressed.

Modeling the historical information. Moreover, modeling of historical information has to be improved. Historical information enables social inferences, thus achieving a precise calculation of information entropy needs a model of the information previously presented to a potential inferrer. In the few solutions suggested for the information aggregation problems in databases, historical information of an adversary is usually considered to include all of the previously revealed information. However, a study that focused on the modeling of historical data for social inference risk prediction shows that the exact proximity history of the users determines the inference risk and that the optimum history length was 2 weeks [30]. In fact, the users do not have a perfect memory of their past visual observations.

It should be noted that previous work had a number of limitations. It derived the best values of time-weighting parameters and the history length from a simple proximity-based social application that only reveals nicknames. Furthermore, this study was carried out for 4 weeks, whereas a real application may be used for years. Gaining data over a longer period of time brings up new issues and calls for improved models. Therefore, researchers have to look more into this problem of historical information modeling.

Computational complexity. Another practical challenge in implementing social inference prediction algorithms is the computation complexity. The framework explained in this section can estimate the level of anonymity in any situation where personal attributes are shared, especially in social computing. However, the computational complexity of calculating parameters such as V (the number of relevant entities) and the probability of each one might raise concerns over the practicality of building a social inference protection system for synchronous communications. In synchronous online communications, a social inference protection system should be able to estimate the risk in acceptable time. An algorithm with acceptable delay and computational complexity was proposed in [27] that used basic properties of information entropy for this purpose. The aim of this work was not to find an optimal solution to this problem, but to find an algorithm that reduces the complexity substantially, as compared to the brute-force algorithm. It also considered the worst case where one server was used for anonymity estimation. Nevertheless, in an application with a very large number of users, many linkable attributes, and highly dynamic profiles, distributed servers or even more efficient algorithms may be needed. In distributed systems, each server may not be fully aware of a user's history or other users' profiles. In that case, decision making under incomplete knowledge may be inevitable and risk prediction algorithms need to make an optimal decision based on the locally known data. For example, the user's previous revelations or the values of certain attributes of other users will also have a probabilistic nature. Entropy thresholds may be fuzzy rather than exact preset values. Therefore, future research into this problem is needed.

4.2. Challenges in risk management

After a social inference protection system detects a situation with high inference risk (with entropy below a certain threshold), the system has to take proper action to reduce this risk. This section presents two different levels of system action which could be taken:

  (i) automated risk management; and

  (ii) semi-automated risk management.

Moreover, UI approached for risk management features in social inference protection systems is proposed and discussed.

Automated inference protection systems directly intervene in risky information exchanges. Users, designers, or system administrators could set the desired anonymity level on a privacy control setting page and the system would then act accordingly to prevent users from unwanted social inferences. Possible methods to be used in such systems are, e.g. blocking the exchange of specific items of information, or blurring the exchanged information automatically when the predicted social inference risk is too high. Lowering the granularity of revealed information, e.g. revealing the location at floor precision instead of room precision, or showing an anonymous name instead of a nickname can lower the social inference risk. However, the practicality and comparative utility of blurring or blocking the information exchanged (e.g. providing a user's home state as opposed to home city or age range as opposed to birth date) need to be further examined.

However, the main purpose of social computing is to allow people communicating with each other and therefore automated protection by simply blocking the information exchange might defeat this purpose. Our simulation results imply that automatic management of the risk, such as blocking the information flow, can severely degrade the utility of CMC applications, because such interferences with the flow of communication can happen quite frequently. Instead, more sophisticated and dynamic solutions for risk management are needed.

The semi-automated approach detects the risk automatically, but the action is left to the user. Instead of specifying general privacy preferences, users are informed about their social inference risk and given control about the reduction method at all time. To enable users to take appropriate risk reduction actions, they need to be made aware of the risk.

User interfaces need to present users with understandable social inference risk visualizations and controls to allow them adjust their information sharing accordingly. Many social computing applications already provide UIs intended to inform users as to the personal information they are sharing. They aim at supporting user impression and privacy management, providing users with the ability to manage their personal privacy, as a communication aid, and to inform users about the reasoning behind a software personalization. Generally, users are able to view how their profile is seen by other users. However, current privacy management interfaces fail to provide users with an understanding of the privacy risks they face or suitable options for truly controlling the information being shared. Simple rule-based privacy settings do not cope well with the dynamic and context-dependent nature of people's privacy preferences and information needs. Users do not wish to constantly set rules to manage what they are sharing, and at the same time they do not want to have sensitive information put in danger as a result of sharing fairly innocuous information. Moreover, a user may need to know not only what is directly being shared with other users but also what aspects of their profile can be inferred.

Semi-automated risk management interfaces need to inform users about their current social inference risk (risk visualizations) and allow users to control their privacy preferences accordingly (control interfaces). Risk visualizations aim at supporting users' awareness of social inference risks by providing a status of the current social inference risk the user is exposed to, while control interfaces provide users the means to adjust both their general desired anonymity level as well as their current personal information sharing based on potential social inference risks rather than static information sharing rules.

The systems can, for example, show users how uniquely they have specified themselves so far, or send a warning message when revealing a piece of information would enable their partner to invade their desired degree of anonymity.

Alternative UI approaches to inform users about their current social inference risk can include

(i) *pop-up warnings* during application use, which highlight the risks associated with taking the decision to reveal particular items of info- rmation;

(ii) *risk status lookup* UI where users can check now and then their overall current social inference risk; and

(iii) *awareness displays* that permanently show users their social inference risk in an unobtrusive way.

These alternative approaches need to be further explored in terms of their utility and usability. User studies need to investigate their usefulness and effectiveness in various contexts to understand how to best design systems that truly help the user manage their social inference risks.

To decide whether semi-automatic or fully automatic methods should be used in social computing application, the seriousness and frequency of identity inferences in the domains of computer-mediated communication and proximity-based social applications were explored in a previous study [28]. The results suggested that automatic control of information exchanges in computer-mediated interpersonal communication can overly interrupt the information exchange because social computing applications are designed to exchange information. Automated control of information exchange can degrade system usability or be frustrating for the user. Therefore, the semi-automated methods seem to be more appropriate solutions to reduce social inference risk. However, the automated method appeared to work fine in proximity-based social applications.

5. Discussion and conclusion

Social computing applications are becoming an essential part of our lives and as a result are fundamentally changing the ways in which we need to think about privacy and privacy management. We illustrated that the serious threat social inferences pose to user privacy is real and that they happen frequently. Users are often unaware of the possible negative consequences of their personal information sharing. Furthermore, many social computing applications currently misrepresent the privacy protection they provide to users by implying that a user's anonymity can be effectively protected by simply allowing them to select not to publicly share a subset of personal information.

The comprehensive overview of the social inference problem in this paper illustrated that the social inferences problem is a serious threat to user privacy in social computing applications. Unfortunately, this has not resulted in researchers looking holistically at the social inference problem. Nevertheless, the social inference problem becomes more and more prominent and seems to be expanding along with the ever-increasing adoption of mobile social computing systems that merge online social interactions with context-aware computing. The problem is further complicated by social computing application users sharing different personal information with different potential inferrers, and by sensitivity of user information as well as user's privacy preferences often being highly dynamic depending on changing user location and context. Users may be willing to compromise their privacy settings to have more meaningful and productive communication. Social inference protection systems have to take into account the need for adjustments of privacy preferences based on context, social inference risks, and user needs.

Our empirically validated social inference risk prediction framework suggests mechanisms by which many of the current inadequacies could be addressed. We believe that this is a significant first step toward providing individuals with tools for managing their social inference risks so that privacy needs are better met and more importantly people's awareness of the possible consequences of their information sharing choices made apparent. However, our knowledge regarding the social inference problem in other circumstances than the ones examined in this paper is lacking. We see future research progressing along several trajectories. First, we expect to see more work on background knowledge estimation and modeling. Second, there is potential for a thorough empirical examination of how to model historical information to better predict social inference risks. Third, entropy estimation algorithms need to be optimized for computational complexity. Finally, research into alternative visualization approaches to inform users about social inference risks is needed that can provide design guidelines.

References

[1] HOADLEY, M.C., XU, H., LEE, J. and ROSSON, M.B. (2009) Privacy as information access and illusory control: the case of the facebook news feed privacy outcry. *Electron. Com. Res. Appl. (Special Issue on Social Networks and Web 2.0)* **9**(1): 50–60.

[2] MIFFLIN, H.e. (2004) *The American Heritage Dictionary of the English Language* (New York, NY: Houghton Mifflin).

[3] DENNING, D.E. and MORGENSTERN, M. (1986) Military database technology study: AI techniques for security and reliability. *SRI Technical Report.*

[4] MACHANAVAJJHALA, A., GEHRKE, J. and KIFER, D. (2006) l-diversity: privacy beyond k-anonymity. In *Proceedings of the 22nd IEEE International Conference on Data Engineering (ICDE 2006).*

[5] O'LEARY, D.E. (1995) Some privacy issues in knowledge discovery: The OECD personal privacy guidelines. *IEEE Expert: Intell. Syst. Appl.* **10**(2): 48–52.

[6] SWEENEY, L. (2002) Achieving k-anonymity privacy protection using generalization and suppression. *Int. J. Uncertainty Fuzziness Knowledge Based Syst.* **10**(5): 571–588.

[7] ZHAN, J. and MATWIN, S. (2006) A crypto-based approach to privacy-preserving collaborative data mining. In *Sixth IEEE International Conference on Data Mining Workshops*, 546–550.

[8] BRODSKY, A., FARKAS, C. and JAJODIA, S. (2000) Secure databases: constraints, inference channels, and monitoring disclosures. *IEEE Trans. Knowl. Data Eng.* **2**(6): 900–919.

[9] WONDERACEK, G., HOLZ, T., KIRDA, E. and KRUEGEL, C. (2010) A practical attack to de-anonymize social network users. In *IEEE Symposium on Security and Privacy.*

[10] SCHULER, R.P., LAWS, N., BAJAJ, S., GRANDHI, S.A. and JONES, Q. (2007) Finding your way with CampusWiki: a location-aware Wiki to support community building. In *The ACMs Conference on Human Factors in Computing Systems CHI2007* (San Jose, CA).

[11] MAGID, L. (2010) EPIC complains while Google tries to fix buzz privacy snafus. *The Huffungton Post.*

[12] MOTAHARI, S., MANIKOPOULOS, C., HILTZ, R. and JONES, Q. (2007) Seven privacy worries in ubiquitous social computing. In *ACM International Conference Proceeding Series; Proceedings of the 3rd Symposium on Usable Privacy and Security* (ACM), 171–172.

[13] LANGHEINRICH, M. (2001) Privacy by design—principles of privacy-aware ubiquitous systems. In *Third International Conference on Ubiquitous Computing (UbiComp 2001)* (ACM), 273–291.

[14] STALLINGS, W. (1999) *Cryptography and Network Security Principles and Practices* (New Jerssey, NJ: Pearson Prentice Hall).

[15] CORNWELL, J., FETTE, I., HSIEH, G., PRABAKER, M., RAO, J., TANG, K., VANIEA, K., *et al.* (2007) User-controllable security and privacy for pervasive computing. In *Proceedings of the 8th IEEE Workshop on Mobile Computing Systems & Applications.*

[16] HONG, J.I. and LANDAY, J.A. (2004) An architecture for privacy-sensitive ubiquitous computing. In *2nd International Conference on Mobile Systems, Applications, and Services* (ACM), 177–189.

[17] SAMARATI, P. and DE CAPITANI di VIMERCATI, S. (2001) Access control: policies, models, and mechanisms. In FOCARDI, R. and GORRIERI, R. [eds.] *Foundations of Security Analysis and Design* (Berlin, Heidelberg, New York: Springer-Verlag), 137–196.

[18] MOTAHARI, S., ZIAVRAS, S., SCHULAR, R. and JONES, Q. (2008) Identity inference as a privacy risk in computer-mediated communication. *IEEE Hawaii International Conference on System Sciences (HICSS-42)*, 1–23.

[19] CHEN, Y. and CHU, W. (2010) Database security protection via inference detection. In *IEEE International Conference on Intelligence and Security Informatics*.

[20] LUNT, T.F. (1991) Current issues in statistical database security. *IFIP Transactions, Results of the IFIP WG 11.3 Workshop on Database Security V: Status and Prospects A-6*, 381–385.

[21] JAJODIA, S. and MEADOWS, C. (1995) *Inference Problems in Multilevel Secure Database Management Systems* (Los Alamitos, CA: IEEE Computer Society Press).

[22] STACHOUR, P.D. and THURAISINGHAM, B. (1990) Design of LDV: a multilevel secure relational database management. *IEEE Trans. Knowl. Data Eng.* 2(2): 190–209.

[23] DAWSON, S., CAPITANI, S.D. and SAMARATI, d.V.P. (1999) Specification and enforcement of classification and inference constraints. In *IEEE Symposium on Security and Privacy*, 181–195.

[24] SERJANTOV, A. and DANEZIS, G. (2002) Towards an information theoretic metric for anonymity. In *Proceedings of Privacy Enhancing Technologies Workshop (PET 2002)* (ACM).

[25] MORGENSTERN, M. (1987) Security and inference in multilevel database and knowledge-base systems. In *Proceedings of the 1987 ACM SIGMOD International Conference on Management of Data* (ACM).

[26] LEMOS, R. (2011) How Your Username May Betray You, http://www.technologyreview.com/web/32326/?p1=MstRcnt&a=f, 2011.

[27] MOTAHARI, S., ZIAVRAS, S. and JONES, Q. (2010) Online anonymity protection in computer-mediated communication. *IEEE Trans. Inf. Forensics Secur.* 5(3): 570–580.

[28] MOTAHARI, S., ZIAVRAS, S., NAAMAN, M., ISMAIL, M. and JONES, Q. (2009) Social inference risk modeling in mobile and social applications. *IEEE International Conference on Information Privacy, Security, Risk and Trust (PASSAT)*, 125–132.

[29] SHANNON, C.E. (1950) Prediction and entropy of printed English. *Bell Syst. Tech. J.* 30: 50–64.

[30] MOTAHARI, S. (2010) *Inference Prevention in Ubiquitous Social Computing.* (Newark, NJ: Electrical and Computer Engineering Department, Institute of Technology).

16

Exploring Relay Cooperation for Secure and Reliable Transmission in Two-Hop Wireless Networks

Yulong Shen[1],[2],*and Yuanyu Zhang[1],[3]

[1]School of Computer Science and Technology, Xidian University, China
[2]State Key Lab. of Integrated Service Network, Xi'an, Shaanxi, China
[3]School of Systems Information Science, Future University Hakodate, Japan

Abstract

This work considers the problem of secure and reliable information transmission via relay cooperation in two-hop relay wireless networks without the information of both eavesdropper channels and locations. While previous work on this problem mainly studied infinite networks and their asymptotic behavior and scaling law results, this papers focuses on a more practical network with finite number of system nodes and explores the corresponding exact result on the number of eavesdroppers one network can tolerate to ensure desired secrecy and reliability. We first study the scenario where path-loss is equal between all pairs of nodes and consider two transmission protocols there, one adopts an optimal but complex relay selection process with less load balance capacity while the other adopts a random but simple relay selection process with good load balance capacity. Theoretical analysis and numerical results are then provided to determine the maximum number of eavesdroppers one network can tolerate to ensure a desired performance in terms of the secrecy outage probability and transmission outage probability. We further extend our study to the more general scenario where path-loss between each pair of nodes also depends on the distance between them, for which a new transmission protocol with both preferable relay selection and good load balance as well as the corresponding theoretical analysis and numerical results are presented.

Keywords: Two-Hop Wireless Networks, Cooperative Relay, Physical Layer Security, Transmission Outage, Secrecy Outage.

1. Introduction

Two-hop ad hoc wireless networks, where each packet travels at most two hops (source-relay-destination) to reach its destination, have been a class of basic and important networking scenarios [1]. Actually, the analysis of basic two-hop relay networks serves as the foundation for performance study of general multi-hop networks. Due to the promising applications of ad hoc wireless networks in many important scenarios (like battlefield networks, vehicle networks, disaster recovery networks), the consideration of secrecy (and also reliability) in such networks is of great importance

for ensuring the high confidentiality requirements of these applications.

Traditionally, the information security is provided by adopting the cryptography approach, where a plain message is encrypted through a cryptographic algorithm that is hard to break (decrypt) in practice by any adversary without the key. The cryptography is acceptable for general applications with standard security requirement (like education system and public networks). Based on the cryptography, H. Wang et al. proposed a rule-based framework to identify and address issues of sharing for the global education system in [2], studied a problem of protecting privacy of individuals in large public survey rating data in [3] and proposed a privacy-aware access control model in web service environments in [4]. While these methods may not be sufficient for applications with a requirement

*Corresponding author. Email: ylshen@mail.xidian.edu.cn

of strong form of security (like military networks and emergency networks). This is because that the cryptographic approach can hardly achieve everlasting secrecy, since the adversary can record the transmitted messages and try any way to break them [5]. That is why there is an increasing interest in applying signaling scheme in physical layer to provide a strong form of security, where a degraded signal at an eavesdropper is always ensured such that the original data can be hardly recovered regardless of how the signal is processed at the eavesdropper. We consider applying physical layer method to achieve secure and reliable information transmission in the two-hop wireless networks. By now, a lot of research works have been dedicated to the study of physical layer security based on cooperative relays and artificial noise, and these works can be roughly classified into two categories depending on whether the information of eavesdroppers channels and locations is known or not.

For the case that the information of eavesdroppers channels and locations is available, to achieve the goal of maximizing the secrecy rates while minimizing the total transmit power, a few cooperative transmission schemes have been proposed in [6][7][8], and for two-hop wireless networks the optimal transmission strategies were presented in [9][10]. With respect to small networks, cooperative jamming with multiple relays and multiple eavesdroppers and knowledge of channels and locations was considered in [11][12]. Even if only local channel information rather than global channel state information is known, it was proved that the near-optimal secrecy rate can achieved by cooperative jamming schemes [13][14]. Except channel information, the relative locations were also considered for optimizing cooperative jamming and power allocation to disrupt an eavesdropper with known location [15][16]. In addition, L. Lai et al. established the utility of user cooperation in facilitating secure wireless communications and proposed cooperation strategies in the additive White Gaussian Noise (AWGN) channel [17], R. Negi et al. showed how artificially generated noise can be added to the information bearing signal to achieve secrecy in the multiple and single antenna scenario under the constraint on total power transmitted by all nodes [18]. The physical layer security issue in a two-way untrusted relay system was also investigated with friendly jammers in [19][20]. The cooperative communication in mobile ad hoc networks was discussed in [21]. Effective criteria for relay and jamming node selection were developed to ensure nonzero secrecy rate in case of given sufficient relays in [22]. In practice, however, it is difficult to gain the information of eavesdropper channels and locations, since the eavesdroppers always try to hide their identity information as much as possible. To alleviate such a requirement on eavesdroppers information, some recent works explored the implementation of secure and reliable information transmission in wireless networks without the information of both eavesdropper channels and locations.

For the case that the information of eavesdropper channels and locations is unknown, the works in [23][24] considered the secrecy for two-hop wireless networks, the works in [25][26][27] considered the secrecy for large wireless networks, and the further work in [28] considered the energy efficiency cooperative jamming strategies. These works considered how cooperative jamming by friendly nodes can impact the security of the network and compared it with a straight-forward approach based on multi-user diversity. They also proposed some protocols to embed cooperative jamming techniques for protecting single links into a large multi-hop network and explored network scaling results on the number of eavesdroppers one network can tolerate. A.Sheikholeslami et al. explored the interference from multiple cooperative sessions to confuse the eavesdroppers in a large wireless network [29]. The cooperative relay scheme for the broadcast channel was further investigated in [30][31]. It is notable, however, that these works mainly focus on exploring the scaling law results in terms of the number of eavesdroppers one network can tolerate as the number of system nodes there tends to infinity. Although the scaling law results are helpful for us to understand the general asymptotic network behavior, they tell us a little about the actual and exact number of eavesdroppers one network can tolerate. In practice, however, such exact results are of great interest for network designers.

This paper focuses on applying the relay cooperation to achieve secure and reliable information transmission in a more practical finite two-hop wireless network without the knowledge of both eavesdropper channels and locations. The main contributions of this paper as follows.

- For achieving secure and reliable information transmission in a more practical two-hop wireless network with finite number of system nodes and equal path-loss between all pairs of nodes, we consider the application of the cooperative protocol proposed in [24] with an optimal and complex relay selection process but less load balance capacity, and also propose to use a new cooperative protocol with a simple and random relay selection process but good load balance capacity.

- Rather than exploring the asymptotic behavior and scaling law results, we provide theoretic analysis for above two cooperative protocols to determine the corresponding exact results on the number of eavesdroppers one network can tolerate to meet a specified requirement in terms of the maximum secrecy outage probability and the maximum transmission outage probability allowed.

- We further extend our study to the more general and practical scenario where the path-loss between each pair of nodes also depends on their relative locations, for which we propose a new transmission protocol with both preferable relay selection and good load balance and also present the corresponding theoretical analysis under this new protocol.

The remainder of the paper is organized as follows. Section 2 presents system models and also introduces transmission outage and secrecy outage for the analysis of transmission protocols. Section 3 considers two transmission protocols for the scenario of equal path-loss between all pairs of nodes and provides the corresponding theoretical analysis and numerical results. Section 4 further presents a new transmission protocol and its theoretical analysis to address distance-dependent path-loss issue. Section 5 presents the analysis on load balance. The numerical results are in Section 6 and Section 7 concludes this paper.

2. System Models

2.1. Network Model

As illustrated in Fig.1 that we consider a network scenario where a source node S wishes to communicate securely with its destination node D with the help of multiple relay nodes $R_1, R_2, \cdots, R_n$. In addition to these normal system nodes, there are also m eavesdroppers $E_1, E_2, \cdots, E_m$ that are independent and also uniformly distributed in the network. Our goal here is to ensure the secure and reliable information transmission from source S to destination D under the condition that no real time information is available about both eavesdropper channels and locations.

2.2. Transmission Model

Consider the transmission from a transmitter A to a receiver B, and denote by $x_i^{(A)}$ the ith symbol transmitted by A and denote by $y_i^{(B)}$ the ith signal received by B. We assume that all nodes transmit with the same power E_s, path-loss between all pairs of nodes is independent, and the frequency-nonselective multi-path fading from A to B is a complex zero-mean Gaussian random variable. The fading is assumed constant during K message transmissions, called flat interval hereafter, and varies randomly and independently from interval to interval. Under the condition that all nodes in a group of nodes, $\mathcal{R}$, are generating noises, the ith signal received at node B from node A is determined as:

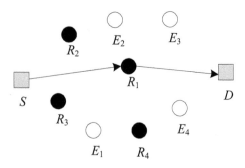

Figure 1. System scenario: Source S wishes to communicate securely with destination D with the assistance of finite relays $R_1, R_2, \cdots, R_n$ (n=4 in the figure) in the presence of passive eavesdroppers $E_1, E_2, \cdots, E_m$ (m=4 in the figure). Cooperative relay scheme is used in the two-hop transmission. A assistant node is selected randomly as relay (R_1 in the figure).

$$y_i^{(B)} = \frac{h_{A,B}}{d_{A,B}^{\alpha/2}} \sqrt{E_s} x_i^{(A)} + \sum_{A_i \in \mathcal{R}} \frac{h_{A_i,B}}{d_{A_i,B}^{\alpha/2}} \sqrt{E_s} x_i^{(A_i)} + n_i^{(B)}$$

where $d_{A,B}$ is the distance between nodes A and B, $\alpha \geq 2$ is the path-loss exponent. The noise $\left\{ n_i^{(B)} \right\}$ at receiver B is assumed to be i.i.d complex Gaussian random variables with $E\left[\left| n_i^{(B)} \right|^2 \right] = N_0$, and $\left| h_{A,B} \right|^2$ is exponentially distributed with mean $E\left[\left| h_{A,B} \right|^2 \right]$. Without loss of generality, we assume that $E\left[\left| h_{A,B} \right|^2 \right] = 1$. The SINR $C_{A,B}$ from A to B is then given by

$$C_{A,B} = \frac{E_s \left| h_{A,B} \right|^2 d_{A,B}^{-\alpha}}{\sum_{A_i \in \mathcal{R}} E_s \left| h_{A_i,B} \right|^2 d_{A_i,B}^{-\alpha} + N_0/2}$$

In the wireless transmission, the receiver can decode the message if and only if the received signal quality is better than a special threshold. In this paper, for a legitimate node and an eavesdropper, we use two separate SINR thresholds γ_R and γ_E to define the minimum SINR required to recover the transmitted messages, respectively. Therefore, a receiver (relay or destination) is able to decode a packet if and only if the SINR at receiver is greater than γ_R, while the transmitted message is secure if and only if the SINR at each eavesdropper is less than γ_E. In this paper, γ_R and γ_E depend on the sensitivity of the receiver. When the node is manufactured, the SINR threshold is determined. In the subsequent analysis on the numerical results, (γ_R, γ_E) is fixed as $(0.5, 0.5)$ for Protocol 1 and 2, and $(0.5, 15)$ for Protocol 3. If the eavesdroppers are original system nodes which are captured by the adversary, the SINR threshold γ_E can be identical to γ_R. If the eavesdroppers are illegal nodes distributed by the adversary, γ_E is not relative to γ_R.

2.3. Transmission Outage and Secrecy Outage

For a transmission from the source S to destination D, we call transmission outage happens if D can not decode the transmitted packet, i.e., D received the packet with SINR less than the predefined threshold γ_R. The transmission outage probability, denoted as $P_{out}^{(T)}$, is then defined as the probability that transmission outage from S to D happens.

We predefined a upper bound ε_t on $P_{out}^{(T)}$, which is an extremely small quantity and if the $P_{out}^{(T)}$ is less than ε_t, the communication is regarded as reliable. ε_t is determined according to the transmission outage the application can tolerate. For the video transmission, it can be tolerated that some messages are lost, thus ε_t can take a larger value i.e. 10^{-1}. For the applications in which the messages can not be lost, ε_t can take a smaller value i.e. 10^{-2}.

We call the communication between S and D is reliable if $P_{out}^{(T)} \le \varepsilon_t$. Notice that for the transmissions from S to the selected relay R_{j^*} and from R_{j^*} to D, the corresponding transmission outage can be defined in the similar way as that of from S to D. We use $O_{S \to R_{j^*}}^{(T)}$ and $O_{R_{j^*} \to D}^{(T)}$ to denote the events that transmission outage from source S to R_{j^*} happens and transmission outage from relay R_{j^*} to D happens, respectively. Due to the link independence assumption, we have

$$P_{out}^{(T)} = P\left(O_{S \to R_{j^*}}^{(T)}\right) + P\left(O_{R_{j^*} \to D}^{(T)}\right) - P\left(O_{S \to R_{j^*}}^{(T)}\right) \cdot P\left(O_{R_{j^*} \to D}^{(T)}\right)$$

Regarding the secrecy outage, we call secrecy outage happens for a transmission from S to D if at least one eavesdropper can recover the transmitted packets during the process of this two-hop transmission, i.e., at least one eavesdropper received the packet with SINR larger than the predefined threshold γ_E. The secrecy outage probability, denoted as $P_{out}^{(S)}$, is then defined as the probability that secrecy outage happens during the transmission from S to D.

We predefined a upper bound ε_s on $P_{out}^{(S)}$, which is an extremely small quantity and if the $P_{out}^{(S)}$ is less than ε_s, the communication is regarded as secure. ε_s is determined by the amount of messages obtained by the eavesdroppers can be tolerated which represents a level of security. If the higher secure level is required, ε_s can take a smaller value i.e. 10^{-1}. Otherwise, ε_s can take a larger value i.e. 0.2, which means it is able to tolerate 20% messages obtained by the eavesdroppers.

We call the communication between S and D is secure if $P_{out}^{(S)} \le \varepsilon_s$. Notice that for the transmissions from S to the selected relay R_{j^*} and from R_{j^*} to D, the corresponding secrecy outage can be defined in the similar way as that of from S to D. We use $O_{S \to R_{j^*}}^{(S)}$ and $O_{R_{j^*} \to D}^{(S)}$ to denote the events that secrecy outage from source S to R_{j^*} happens and secrecy outage from relay R_{j^*} to D happens, respectively. Again, due to the link independence assumption, we have

$$P_{out}^{(S)} = P\left(O_{S \to R_{j^*}}^{(S)}\right) + P\left(O_{R_{j^*} \to D}^{(S)}\right) - P\left(O_{S \to R_{j^*}}^{(S)}\right) \cdot P\left(O_{R_{j^*} \to D}^{(S)}\right)$$

3. Secure and Reliable Transmission under Equal Path-Loss

In this section, we consider the case where the path-loss is equal between all pairs of nodes in the system (i.e., we set $d_{A,B} = 1$ for all $A \ne B$). We first introduce two transmission protocols considered for such scenario, and then provide theoretical analysis to determine the numbers of eavesdroppers one network can tolerate under these protocols.

3.1. Transmission Protocols

The first protocol we consider (hereafter called Protocol 1) is the one proposed in [24], in which the optimal relay node with the best link condition to both source and destination is always selected for information relaying. Notice that although the Protocol 1 can guarantee the optimal relay node selection, it suffers from several problems. Protocol 1 involves a complicated process of optimal relay selection, which is not very suitable for the distributed wireless networks, in particular when the number of possible relay nodes is huge. More importantly, since the channel state is relatively constant during a fixed time period, some relay nodes with good link conditions are always preferred for information relaying, resulting in a severe load balance problem and a quick node energy depletion in energy-limited wireless environment.

Based on above observations, we propose to use a simple and random relay selection rather than the optimal relay selection in Protocol 1 to achieve a better load balance among relay nodes in terms of the energy consumption. Notice that the energy consumption hereafter indicates the energy consumed by each relay for relaying message to the destination for the source. By modifying the Protocol 1, the new transmission protocol (hereafter called Protocol 2) works as follows.

1) **Relay selection:** A relay node, indexed by j^*, is randomly selected from all candidate relay nodes $R_j, j = 1, 2, \cdots, n$.

2) **Channel measurement:** The selected relay R_{j^*} broadcasts a pilot signal to allow each of other

relays to measure the channel from R_{j^*} to itself. Each of the other relays $R_j, j = 1, 2, \cdots, n, j \neq j^*$ then knows the corresponding value of $h_{R_j,R_{j^*}}$. Similarly, the destination D broadcasts a pilot signal to allow each of other relays to measure the channel from D to itself. Each of the other relays $R_j, j = 1, 2, \cdots, n, j \neq j^*$ then knows the corresponding value of $h_{R_j,D}$.

3) **Two-hop transmission:** The source S transmits the messages to the selected relay R_{j^*}, and concurrently, the relay nodes with indexes in $\mathcal{R}_1 = \left\{ j \neq j^* : \left| h_{R_j,R_{j^*}} \right|^2 < \tau \right\}$ transmit noise to generate interference at eavesdroppers. The relay R_{j^*} then transmits the messages to destination D, and concurrently, the relay nodes with indexes in $\mathcal{R}_2 = \left\{ j \neq j^* : |h_{R_j,D}|^2 < \tau \right\}$ transmit noise to generate interference at eavesdroppers.

Remark 1: The parameter τ involved in the Protocol 1 and Protocol 2 serves as the threshold on fading, based on which the set of noise generating relay nodes can be identified. Notice that a too large τ may disable legitimate transmission, while a too small τ may not be sufficient for interrupting all eavesdroppers. Thus, the parameter τ should be set properly to ensure both secrecy requirement and reliability requirement.

Remark 2: The two protocols considered here have their own advantages and disadvantages and thus are suitable for different network scenarios. For the protocol 1, it can achieve a better performance in terms of the number of eavesdroppers can be tolerated (see Theorem 1). However, it involves a complex relay selection process, and more importantly, it results in an unbalanced load and energy consumption distribution among systems nodes. Thus, such protocol is suitable for small scale wireless network with sufficient energy supply rather than large and energy-limited wireless networks (like wireless sensor networks). Regarding the Protocol 2, although it can tolerate less number eavesdroppers in comparison with the Protocol 1 (see Theorem 2), it involves a very simple random relay selection process to achieve a good load and energy consumption distribution among system nodes. Thus, this protocol is more suitable for large scale wireless network environment with stringent energy consumption constraint.

3.2. Analysis of Protocol 1

We now analyze that under the Protocol 1 the number of eavesdroppers one network can tolerate subject to specified requirements on transmission outage and secrecy outage. We first establish the following two lemmas regarding some basic properties of $P_{out}^{(T)}$, $P_{out}^{(S)}$

and τ, which will help us to derive the main result in Theorem 1.

Lemma 1: Consider the network scenario of Fig.1 with equal path-loss between all pairs of nodes, under the Protocol 1 the transmission outage probability $P_{out}^{(T)}$ and secrecy outage probability $P_{out}^{(S)}$ there satisfy the following conditions.

$$P_{out}^{(T)} \leq 2\left[1 - e^{-2\gamma_R(n-1)(1-e^{-\tau})\tau}\right]^n - \left[1 - e^{-2\gamma_R(n-1)(1-e^{-\tau})\tau}\right]^{2n}$$

$$P_{out}^{(S)} \leq 2m \cdot \left(\frac{1}{1+\gamma_E}\right)^{(n-1)(1-e^{-\tau})} - \left[m \cdot \left(\frac{1}{1+\gamma_E}\right)^{(n-1)(1-e^{-\tau})}\right]^2$$

The proof of Lemma 1 can be found in the Appendix A.

Lemma 2: Consider the network scenario of Fig.1 with equal path-loss between all pairs of nodes, to ensure $P_{out}^{(T)} \leq \varepsilon_t$ and $P_{out}^{(S)} \leq \varepsilon_s$ under the Protocol 1, the parameter τ must satisfy the following condition.

$$\tau \in \left[-\log\left[1 + \frac{\log\left(\frac{1-\sqrt{1-\varepsilon_s}}{m}\right)}{(n-1)\log(1+\gamma_E)}\right], \sqrt{\frac{-\log\left[1 - \left(1 - \sqrt{1-\varepsilon_t}\right)^{\frac{1}{n}}\right]}{2\gamma_R(n-1)}} \right]$$

The proof of Lemma 2 can be found in the Appendix B.

Based on the results of Lemma 2, we now can establish the following theorem regarding the performance of Protocol 1.

Theorem 1. Consider the network scenario of Fig.1 with equal path-loss between all pairs of nodes. To guarantee $P_{out}^{(T)} \leq \varepsilon_t$ and $P_{out}^{(S)} \leq \varepsilon_s$ under the Protocol 1, the number of eavesdroppers m one network can tolerate must satisfy the following condition.

$$m \leq \left(1 - \sqrt{1-\varepsilon_s}\right) \cdot (1+\gamma_E)^{\sqrt{\frac{-(n-1)\log\left[1-\left(1-\sqrt{1-\varepsilon_t}\right)^{\frac{1}{n}}\right]}{2\gamma_R}}}$$

The proof of Theorem 1 can be found in the Appendix C.

3.3. Analysis of Protocol 2

Similar to the analysis of Protocol 1, we first establish the following two lemmas regarding some basic properties of $P_{out}^{(T)}$, $P_{out}^{(S)}$ and τ under the Protocol 2.

Lemma 3: Consider the network scenario of Fig.1 with equal path-loss between all pairs of nodes, the transmission outage probability $P_{out}^{(T)}$ and secrecy outage probability $P_{out}^{(S)}$ under the Protocol 2 satisfy the following conditions.

$$P_{out}^{(T)} \le 2\left[1 - e^{-\gamma_R(n-1)(1-e^{-\tau})\tau}\right] - \left[1 - e^{-\gamma_R(n-1)(1-e^{-\tau})\tau}\right]^2$$

$$P_{out}^{(S)} \le 2m \cdot \left(\frac{1}{1+\gamma_E}\right)^{(n-1)(1-e^{-\tau})} - \left[m \cdot \left(\frac{1}{1+\gamma_E}\right)^{(n-1)(1-e^{-\tau})}\right]^2$$

The proof of Lemma 3 can be found in the Appendix D.

Lemma 4: Consider the network scenario of Fig.1 with equal path-loss between all pairs of nodes, to ensure $P_{out}^{(T)} \le \varepsilon_t$ and $P_{out}^{(S)} \le \varepsilon_s$ under the Protocol 2, the parameter τ must satisfy the following condition.

$$\tau \in \left[-\log\left[1 + \frac{\log\left(\frac{1-\sqrt{1-\varepsilon_s}}{m}\right)}{(n-1)\log(1+\gamma_E)}\right], \sqrt{\frac{-\log(1-\varepsilon_t)}{2\gamma_R(n-1)}}\right]$$

The proof of Lemma 4 can be found in the Appendix E.

Theorem 2. Consider the network scenario of Fig.1 with equal path-loss between all pairs of nodes. To guarantee $P_{out}^{(T)} \le \varepsilon_t$ and $P_{out}^{(S)} \le \varepsilon_s$ based on the Protocol 2, the number of eavesdroppers m the network can tolerate must satisfy the following condition.

$$m \le \left(1 - \sqrt{1-\varepsilon_s}\right) \cdot (1+\gamma_E)^{\sqrt{\frac{-(n-1)\log(1-\varepsilon_t)}{2\gamma_R}}}$$

The proof of Theorem 2 can be found in the Appendix F.

Remark 3: It can be observed that the right hand side in Theorem 1 is larger than that of Theorem 2, when all the given parameters are same for Theorem 1 and 2. This means that more eavesdroppers can be tolerated in the network applying Protocol 1 than Protocol 2, given the same system settings and reliability and security requirements.

4. Secure and Reliable Transmission under Distance-Dependent Path-Loss

In this section, we consider the more general scenario where the path-loss between each pair of nodes also depends on the distance between them. We first introduce the coordinate system adopted in our discussion, and then propose a flexible transmission protocol to achieve both the preferable relay selection and good load balance under such distance-dependent path-loss scenario. The related theoretic analysis is further provided to determine the number of eavesdroppers one network can tolerate by adopting this protocol.

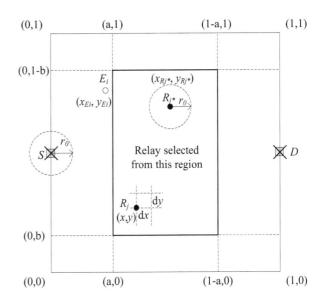

Figure 2. Coordinate system for the scenario where path-loss between pairs of nodes is based on their relative locations.

4.1. Coordinate System

To address the distance-dependent path-loss, we consider a two-hop relay wireless network deployed in a square of unit area and defined by the the coordinate system shown in Fig.2, where the source S located at coordinate $(0, 0.5)$ wishes to establish two-hop transmission with destination D located at coordinate $(1, 0.5)$. Since only one $S - D$ pair is considered in this paper, we can define a coordinate axis system with x-axis parallel to the segment $S - D$. In addition to the source S and destination D, we assume that there are n cooperative relays and m eavesdroppers of unknown channels and locations independently and uniformly distributed in the network area. This scenario is similar to ones in [23][24][29].

4.2. Transmission Protocol

Notice that under the distance-dependent path-loss scenario, the further the distance between a transmitter and a receiver, the weaker the signal received at the receiver. Thus, the system nodes located in the middle region between source S and destination D are preferable relays. Based on such observation, we propose here a general and practical protocol (hereafter called Protocol 3) to ensure both the preferable relay selection and good load balance for distance-dependent path-loss scenario, which works as follow.

1) *Relay selection*: We first define a relay selection region $[a, 1 - a] \times [b, 1 - b]$ between source S and destination D, where a and b are two parameters determining the area of the selection region. Since

this region is located in the normalized network, we can scale it such that $0 \leq a \leq 0.5, 0 \leq b \leq 0.5$. A relay node, indexed by j^*, is then selected randomly from relays falling within the relay selection region.

2) **Channel measurement:** Each of the other relays measures the channel from the selected relay R_{j^*} and destination D by accepting the pilot signal from R_{j^*} and D for determining the noise generation nodes.

3) **Two-hop transmission:** The source S and the selected relay R_{j^*} transmit the messages in two-hop transmission. Concurrently, the relay nodes with indexes in $\mathcal{R}_1 = \left\{ j \neq j^* : |h_{R_j,R_{j^*}}|^2 < \tau \right\}$ in the first hop and the relay nodes with indexes in $\mathcal{R}_2 = \left\{ j \neq j^* : |h_{R_j,D}|^2 < \tau \right\}$ in the second hop transmit noise respectively to help transmission.

Remark 4: In the Protocol 3, a trade off between the preferable relay selection and better load balance can be controlled through the parameters a and b, which determine the relay selection region. As to be shown in Theorem 3 that by adopting a small value for both a and b (i.e., a larger relay selection region), a better load balance capacity can be achieved at the cost of a smaller number of eavesdroppers one network can tolerate.

4.3. Analysis of Protocol 3

To address the near eavesdropper problem and also to simply the analysis for the Protocol 3, we assume that there exits a constant $r_0 > 0$ such that any eavesdropper falling within a circle area with radius r_0 and center S or R_{j^*} can eavesdrop the transmitted messages successfully with probability 1, while any eavesdropper beyond such area can only successfully eavesdropper the transmitted messages with a probability less than 1. Based on such a simplification, we can establish the following two lemmas regarding some basic properties of $P_{out}^{(T)}$, $P_{out}^{(S)}$ and τ under this protocol.

Lemma 5: Consider the network scenario of Fig.2, under the Protocol 3 the transmission outage probability $P_{out}^{(T)}$ and secrecy outage probability $P_{out}^{(S)}$ there satisfy the following conditions.

$$P_{out}^{(T)} \leq \left[1 - e^{-\frac{\gamma_R \tau (n-1)(1-e^{-\tau})}{\phi^{-\alpha}}(\varphi_1+\varphi_2)} \right] (1 - \vartheta) + 1 \cdot \vartheta$$

$$P_{out}^{(S)} \leq 2m \left[\pi r_0^2 + \left(\frac{1}{1 + \gamma_E \psi r_0^{\alpha}} \right)^{(n-1)(1-e^{-\tau})} \left(1 - \pi r_0^2 \right) \right] - \left[m \left(\pi r_0^2 + \left(\frac{1}{1 + \gamma_E \psi r_0^{\alpha}} \right)^{(n-1)(1-e^{-\tau})} \left(1 - \pi r_0^2 \right) \right) \right]^2$$

here,

$$\vartheta = \left[1 - (1 - 2a)(1 - 2b) \right]^n$$

$$\varphi_1 = \int_0^1 \int_0^1 \frac{1}{\left[(x - 0.5)^2 + (y - 0.5)^2 \right]^{\frac{\alpha}{2}}} dx dy$$

$$\varphi_2 = \int_0^1 \int_0^1 \frac{1}{\left[(x - 1)^2 + (y - 0.5)^2 \right]^{\frac{\alpha}{2}}} dx dy$$

$$\phi = \sqrt{(1 - a)^2 + (0.5 - b)^2}$$

$$\psi = \int_0^1 \int_0^1 \frac{1}{(x^2 + y^2)^{\frac{\alpha}{2}}} dx dy$$

The proof of the Lemma 5 can be found in the Appendix G.

Lemma 6: Consider the network scenario of Fig.2, to ensure $P_{out}^{(T)} \leq \varepsilon_t$ and $P_{out}^{(S)} \leq \varepsilon_s$ by applying the Protocol 3, the parameter τ must satisfy the following condition.

$$\tau \in \left[-\log \left[1 + \frac{\log \left(\frac{\frac{1-\sqrt{1-\varepsilon_s}}{m} - \pi r_0^2}{1-\pi r_0^2} \right)}{(n-1)\log(1 + \gamma_E \psi r_0^{\alpha})} \right], \sqrt{\frac{-\log\left(\frac{1-\varepsilon_t}{1-\vartheta}\right)\phi^{-\alpha}}{\gamma_R (n-1)(\varphi_1 + \varphi_2)}} \right]$$

here, ϑ, φ_1, φ_2, ϕ and ψ are defined in the same way as that in Lemma 5.

The proof of the Lemma 6 can be found in the Appendix H.

Based on the results of Lemma 6, we now can establish the following theorem about the performance of Protocol 3.

Theorem 3. Consider the network scenario of Fig.2. To guarantee $P_{out}^{(T)} \leq \varepsilon_t$ and $P_{out}^{(S)} \leq \varepsilon_s$ based on the Protocol 3, the number of eavesdroppers m the network can tolerate must satisfy the following condition.

$$m \leq \frac{1 - \sqrt{1 - \varepsilon_s}}{\pi r_0^2 + (1 - \pi r_0^2)\omega}$$

here,

$$\omega = (1 + \gamma_E \psi r_0^{\alpha})^{-\sqrt{\frac{-(n-1)\log\left(\frac{1-\varepsilon_t}{1-\vartheta}\right)}{\gamma_R(\varphi_1+\varphi_2)\phi^{\alpha}}}}$$

$\vartheta, \varphi_1, \varphi_2, \phi$ and ψ are defined in the same way as that in Lemma 5.

The proof of the Lemma 6 can be found in the Appendix I.

Remark 5: From the practical point of view, the two-hop transmissions are not independent as the common relay is used. In practical, the distance between source S and the selected relay R_{j*} is correlative with the distance between the selected relay R_{j*} and destination D. However, the independence assumption has no any affects on our results, which is described as follow. 1) In the case of equal path-loss (described in the section 3), the channel state information is exponentially distributed, which is independent of distance between transmitter and receiver. Thus, in this case, this assumption holds. 2) In the case of distance-dependent path-loss (described in this section), in each hop, the worst case about distance between source S and the selected relay R_{j*} and between the selected relay R_{j*} and destination D is considered, and a minimum value of maximum number of eavesdroppers one network can tolerate is obtained. Thus this assumption has no affect on our results.

5. Analysis on Load Balance

Load balance is used to characterize how the energy consumptions among n system relays are balanced. For the three protocols considered in this paper, we denote the corresponding load balance by $L^{(1)}, L^{(2)}$ and $L^{(3)}$. Recall that each of the channels stays constant during K message transmissions (i.e., flat interval). We assume that total N messages are transmitted are from the source S to the destination D and the unit energy consumption of one message transmission is I_0 for the message relay. In addition, we define the expectation of the energy consumption for each relay $R_i, i = 1, 2, \cdots, n$ under protocol 1,2 and 3 by $C_i^{(1)}, C_i^{(2)}$ and $C_i^{(3)}$ and variance by $\sigma_i^{(1)}$, $\sigma_i^{(2)}$ and $\sigma_i^{(3)}$, respectively. In order to analyze the load balance performance of different protocols, we simply characterize it as follows

$$L^{(A)} \propto \frac{\max_i C_i^{(A)} - \min_i C_i^{(A)}}{\sum_{i=1}^n C_i^{(A)}} \propto \frac{\sigma_i^{(A)}}{E[C_i^{(A)}]}$$

, where $A = 1, 2$ and 3. Notice that the larger the $L^{(A)}$ is, the worse the energy consumption is balanced among n relays.

5.1. Analysis of Protocol 1

As the flat interval is K message transmissions, there are $\frac{N}{K}$ intervals in each of which all the channels stay constant. We define the number of such intervals in which relay $R_i, i = 1, 2, \cdots, n$ is selected as the message relays by Y_i. Notice that a relay is always selected as

the message relay during the same interval. Hence, the energy consumption of relay R_i is $C_i^{(1)} = I_0 K Y_i$. Defining the event that relay R_i is selected as the message relay in one interval by A_i, we have

$$P(A_i) = \int_0^\infty P(A_i \mid x) f(x) dx$$

$$= \int_0^\infty \prod_{j=1, j\neq i}^n P(min(|h_{S,R_j}|^2, |h_{R_j,D}|^2) \le x) f(x) dx$$

$$= \int_0^\infty 2(1 - e^{-2x})^{n-1} e^{-2x} dx$$

$$= \frac{1}{n}$$

where $f(x) = 2e^{-2x}$ for $x > 0$ is the pdf of $min(|h_{S,R_i}|^2, |h_{R_i,D}|^2)$. From the above, we know that Y_i is a binomial random variable with parameters $\frac{N}{K}$ and $\frac{1}{n}$. Therefore, the expected energy consumption of each relay R_i is

$$E[C_i^{(1)}] = I_0 K E[Y_i] = \frac{I_0 N}{n}$$

and the variance is

$$\sigma_i^{(1)} = I_0^2 K^2 \sigma_{Y_i} = K \frac{I_0^2 N(n-1)}{n^2}$$

Consequently,

$$L^{(1)} \propto \frac{\sigma_i^{(1)}}{E[C_i^{(1)}]} = K \frac{I_0(n-1)}{n}$$

5.2. Analysis of Protocol 2

For each relay $R_i, i = 1, 2, \cdots, n$, define the times of being selected as the message relay among N transmissions by X_i. Obviously, $C_i^{(2)} = I_0 X_i$. Since the message relay is randomly selected, the probability that relay R_i is selected as the message relay is $\frac{1}{n}$ for each transmission, resulting in X_i is a binomial random variable with parameters N and $\frac{1}{n}$. Therefore, the expected energy consumption of each relay R_i is

$$E[C_i^{(2)}] = I_0 E[X_i] = \frac{I_0 N}{n}$$

and the variance is

$$\sigma_i^{(2)} = I_0^2 \sigma_{X_i} = \frac{I_0^2 N(n-1)}{n^2}$$

At last,

$$L^{(2)} \propto \frac{\sigma_i^{(2)}}{E[C_i^{(2)}]} = \frac{I_0(n-1)}{n} = \frac{L^{(1)}}{K}$$

Following the above, we can observe that the random relay selection protocol outperforms the optimal relay selection protocol in terms of the load balance performance.

Remark 6: We notice that the expected number of messages suffering from transmission outage and secrecy outage during N transmissions are $K \cdot \frac{N}{K} P_{out}^{(T)} = N P_{out}^{(T)}$ and $K \cdot \frac{N}{K} P_{out}^{(S)} = N P_{out}^{(S)}$ for Protocol 1, and $N P_{out}^{(T)}$ and $N P_{out}^{(S)}$ for Protocol 2. Besides, the variance of the number of message suffering from outages are $K N P_{out}^{(T)} (1 - P_{out}^T)$ and $K N P_{out}^{(S)} (1 - P_{out}^S)$ for Protocol 1, and $N P_{out}^{(T)} (1 - P_{out}^T)$ and $N P_{out}^{(S)} (1 - P_{out}^S)$ for Protocol 2. Here, $P_{out}^{(T)}$ and $P_{out}^{(S)}$ are the transmission outage probability and secrecy outage probability of the corresponding protocol. The derivation is similar to that of the load balance and thus is omitted here. Therefore, it can be observed that the non-stationary channel has no impact on the reliability and security for these protocols, while the variance of the number of messages suffering from outages increase with K for Protocol 1.

5.3. Analysis of Protocol 3

Similarly, for each relay $R_i, i = 1, 2, \cdots, n$, we define the times of being selected as the message relay among N message transmissions by Z_i and the energy consumption by $C_i^{(3)}$, which is a function of parameters a and b. Here, we neglect the case $(a = 0.5, b = 0.5)$ where no relays will be selected as the message relay and thus the transmission could not be conducted. Again, we have $C_i^{(3)} = I_0 Z_i$. According to the relay selection protocol, each relay R_i lies in the relay selection region with probability $p = (1 - 2a)(1 - 2b)$. Furthermore, we define the event that relay R_i is selected as the message relay during one transmission by B_i. Thus, $P(B_i) = \frac{p}{U+1}$ where $U \sim B(n-1, p)$ is the number of relays in the relay selection region in addition to R_i. Applying the law of total probability,

$$P(B_i) = E\left[\frac{p}{U+1}\right]$$

$$= \sum_{k=0}^{n-1} \frac{p}{k+1} \binom{n-1}{k} p^k (1-p)^{n-1-k}$$

$$= \sum_{k=0}^{n-1} \frac{1}{n} \binom{n}{k+1} p^{k+1} (1-p)^{n-1-k}$$

$$= \frac{1}{n} \sum_{j=1}^{n} \binom{n}{j} p^j (1-p)^{n-j}$$

$$= \frac{1 - (1-p)^n}{n}$$

Following the above, Z_i is a binomial random variable with parameters N and $\frac{1-(1-p)^n}{n}$. Therefore, the expectation of the energy consumption for each relay R_i is

$$E\left[C_i^{(3)}\right] = I_0 E[Z_i] = \frac{I_0 N (1 - (1-p)^n)}{n}$$

and the variance is

$$\sigma_i^{(3)} = I_0^2 \sigma_{Z_i} = \frac{I_0^2 N (1 - (1-p)^n)}{n} \left(1 - \frac{1 - (1-p)^n}{n}\right)$$

Thus, the load balance of protocol 3 is

$$L^{(3)} \propto \frac{\sigma_i^{(3)}}{E[C_i^{(3)}]} = I_0 \left(1 - \frac{1 - (1-p)^n}{n}\right)$$

Notice that p decreases as the parameters a or b (i.e., the area of the relay selection region) increases. Therefore, $L^{(3)}$ increases with increasing a and b, implying that enlarging the relay selection region can achieve better load balance performance. It is notable that each relay can be selected as the message relay with probability $1/n$ in the case $(a = 0, b = 0)$, which reduces to the Protocol 2 where the message relay is randomly selected from the whole region. Therefore, the load balance of Protocol 3 in this case is identical to that of Protocol 2.

6. Numerical Results

6.1. The Number of Eavesdroppers one Networks can Tolerate

To compare the number of eavesdroppers can be tolerated by Protocol 1 with that of Protocol 2, we show in Fig.3 and Fig.4 respectively how the number of eavesdroppers can be tolerated varies with the number of system relays n when $\varepsilon_t = 0.1$, $\varepsilon_s = 0.1$, $\gamma_R = 0.5$ and $\gamma_E = 0.5$. From these two figures, we can see clearly that the number of eavesdroppers can be tolerated increases exponentially with the number of system relays n as n increases by a proper step (i.e.,10 in Protocol 1 and 100 in Protocol 2). This is because as the number of relays increase, although the threshold τ decreases in order to guarantee the reliability requirement, the expected number of noise-generating nodes increases, resulting in more interference can be generated to suppress the eavesdroppers. It can also be observed from Fig.3 and Fig.4 that Protocol 1 can tolerate significantly more eavesdroppers than Protocol 2. For instance, when the number of relays n is less than 200, Protocol 2 can hardly tolerate any eavesdroppers while Protocol 1 can tolerate relative large number of eavesdroppers. This is due to the reason that larger value of τ can be adopted in Protocol 1 than Protocol 2 under the same reliability requirement (i.e.,ε_t) and number of relays n and also the number of eavesdroppers can be tolerated is relatively sensitive to the threshold τ.

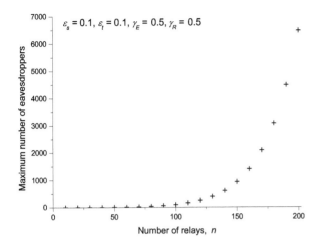

Figure 3. Maximum number of eavesdroppers for Protocol 1 vs. n, when $\varepsilon_t = 0.1$, $\varepsilon_s = 0.1$, $\gamma_R = 0.5$ and $\gamma_E = 0.5$.

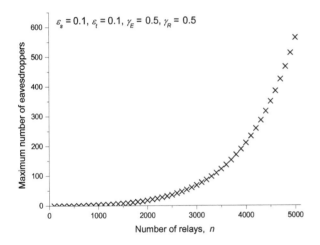

Figure 4. Maximum number of eavesdroppers for Protocol 2 vs. n, when $\varepsilon_t = 0.1$, $\varepsilon_s = 0.1$, $\gamma_R = 0.5$ and $\gamma_E = 0.5$.

Fig.5 shows the maximum number of eavesdroppers one network can tolerate for Protocol 3 with $r_0 = 0.1$, $\varepsilon_t = 0.2$, $\varepsilon_s = 0.2$ and $\gamma_E = 15$. In Fig.5 , the number of eavesdroppers one network can tolerate is small than that of protocol 1 and protocol 2. There are two reasons. On the one hand, For reliable transmission, the worse case that the noise generation nodes can be very near to the receivers (the selected relay or destination) exists and is considered, thus the receivers will be disturbed seriously. On the other hand, the case that the eavesdroppers can be near to the transmitters (source or the selected relay) exists and is considered, especially any eavesdropper falling within a circle area with radius r_0 and center S or R_{j^*} can eavesdrop the

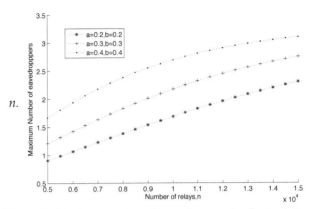

Figure 5. Maximum number of eavesdroppers for Protocol 3 vs.

transmitted messages successfully with probability 1. In the practical networks, these case dose not always happen,thus more eavesdroppers can be tolerated.

6.2. Load Balance among Relays

In order to clearly illustrate the load balance performances of different transmission protocols, simulations are conducted under the settings of $n = 20$, $N = 1000$ and $I_0 = 1$. The comparison between the load balance of optimal relay selection protocol and random relay selection protocol is shown in Fig.6 and we present how the load balance of protocol 3 varies with the area of relay selection region in Fig.7. For ease of simulation, the load balance is represented by the ratio $\frac{\max_i C_i^{(A)} - \min_i C_i^{(A)}}{\sum_{i=1}^{n} C_i^{(A)}}$ defined above.

The Fig.6 indicates clearly that the load balance of random relay selection stays constant with varying flat interval K whereas the load balance of optimal relay selection protocol increase as the flat interval K increases, which implies that the random relay selection protocol can achieve better load balance than the optimal relay selection protocol.

It can be easily observed from Fig.7 that there is a clear threshold (about 1.0 in Fig.7) of the area of relay selection region beyond which the load balance performance cannot be improved by enlarging the area of the relay selection. However, the load balance decreases dramatically from 0.14 to 0.028 (i.e.,the load balance performance is improved) with the area of relay selection region below the threshold. This is due to the reason that the probability $P(B_i)$ increases significantly (from 0.00158 to 0.0439) with the area of the relay selection region when the latter is below the threshold and increases slowly (from 0.0439 to 0.05) when the latter is above the threshold.

A further careful observation of Fig.6, Fig.3 and Fig.4 indicates that Protocol 1 is suitable for small

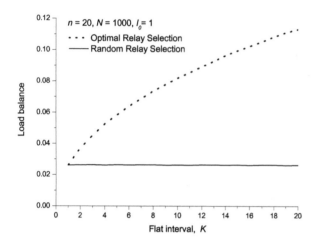

Figure 6. Comparisons between the load balance performance of protocol 1 and 2, when $n = 20, N = 1000$ and $I_0 = 1$.

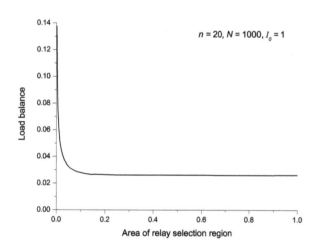

Figure 7. Load balance performance of protocol 3 vs. Area of relay selection region when $n = 20, N = 1000$ and $I_0 = 1$.

networks and Protocol 2 is suitable for large networks with both the considerations of eavesdropper tolerance capability and load balance performance, according with the Remark 2.

Remark 7: Regarding the performances of Protocol 1 and Protocol 2, both of which are proposed in the scenario with equal path loss, we notice that although Protocol 1 has a better eavesdropper-tolerance capability than Protocol 2, it has a worse load balance performance than Protocol 2. Hence, clear tradeoff between the security and load balance performance exists for these two protocols. Therefore, Protocol 1 is suitable for the applications requiring strong security with acceptable load balance performance, while Protocol 2 is suitable for the applications

requiring good load balance with acceptable security requirement. Regarding the performances of Protocol 3 which is designed for the scenario with distance-dependent path loss, we notice that although it has a relatively good load balance, its eavesdropper-tolerance capability is far from acceptable, which limits its application.

7. Conclusion

To achieve reliable and secure information transmission in a two-hop relay wireless network in presence of eavesdroppers with unknown channels and locations, several transmission protocols based on relay cooperation have been considered. In particular, theoretical analysis has been conducted to understand that under each of these protocols how many eavesdroppers one network can tolerate to meet a specified requirement on the maximum allowed secrecy outage probability and transmission outage probability. Our results in this paper indicate that these protocols actually have different performance in terms of eavesdropper-tolerance capacity and load balance capacity among relays, and in general it is possible for us to select a proper transmission protocol according to network scenario such that a desired trade off between the overall eavesdropper-tolerance capacity and load balance among relay nodes can be achieved. As a possible future work, the performances study in more practical network scenarios where the source and destination are distributed within the environment instead of on the border of the network will be of great challenge. Furthermore, it is very interesting to explore the impact of the correlation of the two-hop links on the number of eavesdroppers can be tolerated which is regarded as a promising future work. Besides, conducting analysis on the performance of Protocol 1 in the scenario with distance-dependent path loss will be another interesting future direction. How to improve Protocol 3 to further enhance its eavesdropper-tolerance capability while ensuring an acceptable load balance can also be a promising future research topic.

Appendix A. Proof of Lemma 1

Proof. Notice that $P_{out}^{(T)}$ is determined as

$$P_{out}^{(T)} = P\left(O_{S \to R_{j^*}}^{(T)}\right) + P\left(O_{R_{j^*} \to D}^{(T)}\right) - P\left(O_{S \to R_{j^*}}^{(T)}\right) \cdot P\left(O_{R_{j^*} \to D}^{(T)}\right)$$

Based on the definition of transmission outage probability, we have

$$P\left(O_{S \to R_{j^*}}^{(T)}\right) = P\left(C_{S,R_{j^*}} \leq \gamma_R\right)$$

$$= P\left(\frac{E_s \cdot |h_{S,R_{j^*}}|^2}{\sum_{R_j \in \mathcal{R}_1} E_s \cdot |h_{R_j,R_{j^*}}|^2 + N_0/2} \leq \gamma_R\right)$$

$$\doteq P\left(\frac{|h_{S,R_{j^*}}|^2}{\sum_{R_j \in \mathcal{R}_1} |h_{R_j,R_{j^*}}|^2} \leq \gamma_R\right)$$

Compared to the noise generated by multiple system nodes, the environment noise is negligible and thus is omitted here to simply the analysis. Notice that $\mathcal{R}_1 = \{j \neq j^* : |h_{R_j,R_{j^*}}|^2 < \tau\}$, then

$$P\left(O_{S \to R_{j^*}}^{(T)}\right) \leq P\left(\frac{|h_{S,R_{j^*}}|^2}{|\mathcal{R}_1|\tau} \leq \gamma_R\right)$$

$$= P\left(|h_{S,R_{j^*}}|^2 \leq \gamma_R|\mathcal{R}_1|\tau\right)$$

$$\leq P\left(H^l \leq \gamma_R|\mathcal{R}_1|\tau\right)$$

where $H^l = min\left(|h_{S,R_{j^*}}|^2, |h_{D,R_{j^*}}|^2\right)$ is the largest random variable among the n exponential random variables $min\left(|h_{S,R_j}|^2, |h_{D,R_j}|^2\right)$, $j = 1, 2, \cdots, n$. From reference [32], we can get the distribution function of the $min\left(|h_{S,R_j}|^2, |h_{D,R_j}|^2\right)$ for each relay $R_j, j = 1, 2, \cdots, n$ as following,

$$F_{min\left(|h_{S,R_j}|^2, |h_{D,R_j}|^2\right)}(x) = \begin{cases} 1 - e^{-2x} & x > 0 \\ 0 & x \leq 0 \end{cases}$$

From reference [32], we can also get the distribution function of random variable H^l as following,

$$F_{H^l}(x) = \begin{cases} \left[1 - e^{-2x}\right]^n & x > 0 \\ 0 & x \leq 0 \end{cases}$$

Therefore, we have

$$P\left(O_{S \to R_{j^*}}^{(T)}\right) \leq \left[1 - e^{-2\gamma_R|\mathcal{R}_1|\tau}\right]^n$$

Since there are $n - 1$ other relays except R_{j^*}, the expected number of noise-generation nodes is given by $|\mathcal{R}_1| = (n-1) \cdot P\left(|h_{R_j,R_{j^*}}|^2 < \tau\right) = (n-1) \cdot (1 - e^{-\tau})$. Then we have

$$P\left(O_{S \to R_{j^*}}^{(T)}\right) \leq \left[1 - e^{-2\gamma_R(n-1)(1-e^{-\tau})\tau}\right]^n$$

Employing the same method, we can get

$$P\left(O_{R_{j^*} \to D}^{(T)}\right) \leq \left[1 - e^{-2\gamma_R(n-1)(1-e^{-\tau})\tau}\right]^n$$

Thus, we have

$$P_{out}^{(T)} \leq 2\left[1 - e^{-2\gamma_R(n-1)(1-e^{-\tau})\tau}\right]^n - \left[1 - e^{-2\gamma_R(n-1)(1-e^{-\tau})\tau}\right]^{2n}$$

Similarly, notice that $P_{out}^{(S)}$ is given by

$$P_{out}^{(S)} = P\left(O_{S \to R_{j^*}}^{(S)}\right) + P\left(O_{R_{j^*} \to D}^{(S)}\right) - P\left(O_{S \to R_{j^*}}^{(S)}\right) \cdot P\left(O_{R_{j^*} \to D}^{(S)}\right)$$

According to the definition of secrecy outage probability, we know that

$$P\left(O_{S \to R_{j^*}}^{(S)}\right) = P\left(\bigcup_{i=1}^m \{C_{S,E_i} \geq \gamma_E\}\right)$$

Thus, we have

$$P\left(O_{S \to R_{j^*}}^{(S)}\right) \leq \sum_{i=1}^m P\left(C_{S,E_i} \geq \gamma_E\right)$$

Based on the Markov inequality,

$$P\left(C_{S,E_i} \geq \gamma_E\right) \leq P\left(\frac{E_s \cdot |h_{S,E_i}|^2}{\sum_{R_j \in \mathcal{R}_1} E_s \cdot |h_{R_j,E_i}|^2} \geq \gamma_E\right)$$

$$= E_{\{h_{R_j,E_i}, j=0,1,\cdots,n+mp, j \neq j^*\}, \mathcal{R}_1}\left[P\left(|h_{S,E_i}|^2 > \gamma_E \cdot \sum_{R_j \in \mathcal{R}_1} |h_{R_j,E_i}|^2\right)\right]$$

$$\leq E_{\mathcal{R}_1}\left[\prod_{R_j \in \mathcal{R}_1} E_{h_{R_j,E_i}}\left[e^{-\gamma_E|h_{R_j,E_i}|^2}\right]\right]$$

$$= E_{\mathcal{R}_1}\left[\left(\frac{1}{1+\gamma_E}\right)^{|\mathcal{R}_1|}\right]$$

Therefore,

$$P\left(O_{S \to R_{j^*}}^{(S)}\right) \leq \sum_{i=1}^m \left(\frac{1}{1+\gamma_E}\right)^{|\mathcal{R}_1|} = m \cdot \left(\frac{1}{1+\gamma_E}\right)^{|\mathcal{R}_1|}$$

Employing the same method, we can get

$$P\left(O_{R_{j^*} \to D}^{(S)}\right) \leq m \cdot \left(\frac{1}{1+\gamma_E}\right)^{|\mathcal{R}_2|}$$

Since the expected number of noise-generation nodes is given by $|\mathcal{R}_1| = |\mathcal{R}_2| = (n-1) \cdot (1 - e^{-\tau})$, thus, we can get

$$P_{out}^{(S)} \leq 2m \cdot \left(\frac{1}{1+\gamma_E}\right)^{(n-1)\cdot(1-e^{-\tau})} - \left[m \cdot \left(\frac{1}{1+\gamma_E}\right)^{(n-1)\cdot(1-e^{-\tau})}\right]^2$$

□

Appendix B. Proof of Lemma 2

Proof. The parameter τ should be set properly to satisfy both reliability and secrecy requirements.

• **Reliability Guarantee**

To ensure the reliability requirement $P_{out}^{(T)} \leq \varepsilon_t$, we know from the Lemma 1 that we just need

$$2\left[1 - e^{-2\gamma_R(n-1)(1-e^{-\tau})\tau}\right]^n - \left[1 - e^{-2\gamma_R(n-1)(1-e^{-\tau})\tau}\right]^{2n} \leq \varepsilon_t$$

Thus,

$$\left[1 - e^{-2\gamma_R(n-1)(1-e^{-\tau})\tau}\right]^n \leq 1 - \sqrt{1-\varepsilon_t}$$

That is,

$$-2\gamma_R(n-1)(1-e^{-\tau})\tau \geq \log\left[1 - \left(1 - \sqrt{1-\varepsilon_t}\right)^{\frac{1}{n}}\right]$$

By using Taylor formula, we have

$$\tau \leq \sqrt{\frac{-\log\left[1 - \left(1 - \sqrt{1-\varepsilon_t}\right)^{\frac{1}{n}}\right]}{2\gamma_R(n-1)}}$$

The above result indicates the maximum value the parameter τ we can take to ensure the reliability requirement.

• **Secrecy Guarantee**

To ensure the secrecy requirement $P_{out}^{(S)} \leq \varepsilon_s$, we know from the Lemma 1 that we just need

$$2m \cdot \left(\frac{1}{1+\gamma_E}\right)^{(n-1)(1-e^{-\tau})} - \left[m \cdot \left(\frac{1}{1+\gamma_E}\right)^{(n-1)(1-e^{-\tau})}\right]^2 \leq \varepsilon_s$$

Thus,

$$m \cdot \left(\frac{1}{1+\gamma_E}\right)^{(n-1)(1-e^{-\tau})} \leq 1 - \sqrt{1-\varepsilon_s}$$

That is,

$$\tau \geq -\log\left[1 + \frac{\log\left(\frac{1-\sqrt{1-\varepsilon_s}}{m}\right)}{(n-1)\log(1+\gamma_E)}\right]$$

The above result implies the minimum value parameter τ we can take to guarantee the secrecy requirement.

□

Appendix C. Proof of Theorem 1

Proof. From Lemma 2, we know that to ensure the reliability requirement, we have

$$\tau \leq \sqrt{\frac{-\log\left[1 - \left(1 - \sqrt{1-\varepsilon_t}\right)^{\frac{1}{n}}\right]}{2\gamma_R(n-1)}}$$

and

$$(n-1)(1-e^{-\tau}) \leq \frac{-\log\left[1 - \left(1 - \sqrt{1-\varepsilon_t}\right)^{\frac{1}{n}}\right]}{2\gamma_R\tau}$$

To ensure the secrecy requirement, we need

$$\left(\frac{1}{1+\gamma_E}\right)^{(n-1)(1-e^{-\tau})} \leq \frac{1 - \sqrt{1-\varepsilon_s}}{m}$$

Thus,

$$m \leq \frac{1 - \sqrt{1-\varepsilon_s}}{\left(\frac{1}{1+\gamma_E}\right)^{(n-1)(1-e^{-\tau})}} \leq \frac{1 - \sqrt{1-\varepsilon_s}}{\left(\frac{1}{1+\gamma_E}\right)^{\frac{-\log\left[1-\left(1-\sqrt{1-\varepsilon_t}\right)^{\frac{1}{n}}\right]}{2\gamma_R\tau}}}$$

By letting τ to take its maximum value for maximum interference at eavesdroppers, we get the following bound

$$m \leq \left(1 - \sqrt{1-\varepsilon_s}\right) \cdot (1+\gamma_E)^{\sqrt{\frac{-(n-1)\log\left[1-\left(1-\sqrt{1-\varepsilon_t}\right)^{\frac{1}{n}}\right]}{2\gamma_R}}}$$

□

Appendix D. Proof of Lemma 3

Proof. Similar to the proof of Lemma 1, we notice that $P_{out}^{(T)}$ is determined as

$$P_{out}^{(T)} = P\left(O_{S \to R_{j*}}^{(T)}\right) + P\left(O_{R_{j*} \to D}^{(T)}\right) - P\left(O_{S \to R_{j*}}^{(T)}\right) \cdot P\left(O_{R_{j*} \to D}^{(T)}\right)$$

Based on the definition of transmission outage probability, we have

$$P\left(O_{S \to R_{j*}}^{(T)}\right) = P\left(C_{S,R_{j*}} \leq \gamma_R\right)$$
$$\leq P\left(\frac{|h_{S,R_{j*}}|^2}{|\mathcal{R}_1|\tau} \leq \gamma_R\right)$$
$$= P\left(|h_{S,R_{j*}}|^2 \leq \gamma_R|\mathcal{R}_1|\tau\right)$$
$$= 1 - e^{-\gamma_R|\mathcal{R}_1|\tau}$$

Here $\mathcal{R}_1 = \left\{ j \neq j^* : |h_{R_j,R_{j^*}}|^2 < \tau \right\}$. Since the expected number of noise-generation nodes is given by $|\mathcal{R}_1| = (n-1) \cdot (1 - e^{-\tau})$. Then we have

$$P\left(O_{S \to R_{j^*}}^{(T)}\right) \leq 1 - e^{-\gamma_R(n-1)(1-e^{-\tau})\tau}$$

Employing the same method, we can get

$$P\left(O_{R_{j^*} \to D}^{(T)}\right) \leq 1 - e^{-\gamma_R(n-1)(1-e^{-\tau})\tau}$$

Thus, we have

$$P_{out}^{(T)} \leq 2\left[1 - e^{-\gamma_R(n-1)(1-e^{-\tau})\tau}\right] - \left[1 - e^{-\gamma_R(n-1)(1-e^{-\tau})\tau}\right]^2$$

Notice that the eavesdropper model of Protocol 1 is the same as that of Protocol 2, the method for ensuring secrecy is identical to that of in Lemma 1. Thus, we can see that the secrecy outage probability of Protocol 1 and Protocol 2 is the same, that is,

$$P_{out}^{(S)} \leq 2m \cdot \left(\frac{1}{1+\gamma_E}\right)^{(n-1)\cdot(1-e^{-\tau})} - \left[m \cdot \left(\frac{1}{1+\gamma_E}\right)^{(n-1)\cdot(1-e^{-\tau})}\right]^2$$

$\square$

Appendix E. Proof of Lemma 4

Proof. The parameter τ should be set properly to satisfy both reliability and secrecy requirements.

• **Reliability Guarantee**

To ensure the reliability requirement $P_{out}^{(T)} \leq \varepsilon_t$, we know from Lemma 4 that we just need

$$2\left[1 - e^{-\gamma_R(n-1)(1-e^{-\tau})\tau}\right] - \left[1 - e^{-\gamma_R(n-1)(1-e^{-\tau})\tau}\right]^2 \leq \varepsilon_t$$

That is,

$$1 - e^{-\gamma_R(n-1)(1-e^{-\tau})\tau} \leq 1 - \sqrt{1-\varepsilon_t}$$

By using Taylor formula, we have

$$\tau \leq \sqrt{\frac{-\log(1-\varepsilon_t)}{2\gamma_R(n-1)}}$$

• **Secrecy Guarantee**

Notice that the secrecy outage probability of Protocol 1 and Protocol 2 is same. Thus, to ensure the secrecy requirement, we need

$$\left(\frac{1}{1+\gamma_E}\right)^{(n-1)(1-e^{-\tau})} \leq \frac{1 - \sqrt{1-\varepsilon_s}}{m}$$

Thus,

$$\tau \geq -\log\left[1 + \frac{\log\left(\frac{1-\sqrt{1-\varepsilon_s}}{m}\right)}{(n-1)\log(1+\gamma_E)}\right]$$

The above result implies the minimum value parameter τ can take to guarantee the secrecy requirement.

$\square$

Appendix F. Proof of Theorem 2

Proof. From Lemma 4, we know that to ensure the reliability requirement, we have

$$\tau \leq \sqrt{\frac{-\log(1-\varepsilon_t)}{2\gamma_R(n-1)}}$$

and

$$(n-1)(1-e^{-\tau}) \leq \frac{-\log(1-\varepsilon_t)}{2\gamma_R\tau}$$

To ensure the secrecy requirement, we need

$$\left(\frac{1}{1+\gamma_E}\right)^{(n-1)(1-e^{-\tau})} \leq \frac{1 - \sqrt{1-\varepsilon_s}}{m}$$

Thus,

$$m \leq \frac{1 - \sqrt{1-\varepsilon_s}}{\left(\frac{1}{1+\gamma_E}\right)^{(n-1)(1-e^{-\tau})}} \leq \frac{1 - \sqrt{1-\varepsilon_s}}{\left(\frac{1}{1+\gamma_E}\right)^{\frac{-\log(1-\varepsilon_t)}{2\gamma_R\tau}}}$$

By letting τ to take its maximum value for maximum interference at eavesdroppers, we get the following bound

$$m \leq \left(1 - \sqrt{1-\varepsilon_s}\right) \cdot (1+\gamma_E)^{\sqrt{\frac{-(n-1)\log(1-\varepsilon_t)}{2\gamma_R}}}$$

$\square$

Appendix G. Proof of Lemma 5

Proof. Notice that two ways leading to transmission outage are: 1) there are no candidate relays in the relay selection region; 2) the SINR at the selected relay or the destination is less than γ_R. Let A_1 be the event that there is at least one relay in the relay selection region, and A_2 be the event that there are no relays in the relay selection region. We have

$$P_{out}^{(T)} = P_{out|A_1}^{(T)} P(A_1) + P_{out|A_2}^{(T)} P(A_2)$$

Since the relay is uniformly distributed, the number of candidate relays is a binomial distribution $\left(n, (1-2a)(1-2b)\right)$. We have

$$P(A_1) = 1 - \vartheta$$

and

$$P(A_2) = \vartheta$$

where $\vartheta = \left[1 - (1-2a)(1-2b)\right]^n$. When event A_2 happens, no relay is available. Then

$$P_{out|A_2}^{(T)} = 1$$

Thus, we have

$$P_{out}^{(T)} = P_{out|A_1}^{(T)} (1 - \vartheta) + 1 \cdot \vartheta$$

Notice that $P_{out|A_1}^{(T)}$ is determined as

$$P_{out|A_1}^{(T)} = P\left(O_{S \to R_{j^*}}^{(T)} \middle| A_1\right) + P\left(O_{R_{j^*} \to D}^{(T)} \middle| A_1\right)$$
$$- P\left(O_{S \to R_{j^*}}^{(T)} \middle| A_1\right) \cdot P\left(O_{R_{j^*} \to D}^{(T)} \middle| A_1\right)$$

Based on the definition of transmission outage probability, we have

$$P\left(O_{S \to R_{j^*}}^{(T)} \middle| A_1\right) = P\left(C_{S,R_{j^*}} \le \gamma_R \middle| A_1\right)$$
$$= P\left(\frac{E_s \cdot \frac{|h_{S,R_{j^*}}|^2}{d_{S,R_{j^*}}^\alpha}}{\sum_{R_j \in \mathcal{R}_1} E_s \cdot \frac{|h_{R_j,R_{j^*}}|^2}{d_{R_j,R_{j^*}}^\alpha} + \frac{N_0}{2}} \le \gamma_R \middle| A_1\right)$$
$$\doteq P\left(\frac{\frac{|h_{S,R_{j^*}}|^2}{d_{S,R_{j^*}}^\alpha}}{\sum_{R_j \in \mathcal{R}_1} \frac{|h_{R_j,R_{j^*}}|^2}{d_{R_j,R_{j^*}}^\alpha}} \le \gamma_R \middle| A_1\right)$$

Compared to the noise generated by multiple system nodes, the environment noise is negligible and thus is omitted here to simply the analysis. Notice that $\mathcal{R}_1 = \left\{j \neq j^* : |h_{R_j,R_{j^*}}|^2 < \tau\right\}$, then

$$P\left(O_{S \to R_{j^*}}^{(T)} \middle| A_1\right) \le P\left(\frac{|h_{S,R_{j^*}}|^2 d_{S,R_{j^*}}^{-\alpha}}{\sum_{R_j \in \mathcal{R}_1} \tau d_{R_j,R_{j^*}}^{-\alpha}} \le \gamma_R \middle| A_1\right)$$

As shown in Fig.2 that by assuming the coordinate of R_j as (x, y), we can see that the number of noise generating nodes in square $[x, x + dx] \times [y, y + dy]$ will be $(n-1)(1-e^{-\tau})dxdy$. Then, we have

$$\sum_{R_j \in \mathcal{R}_1} \frac{\tau}{d_{R_j,R_{j^*}}^\alpha} = \int_0^1 \int_0^1 \frac{\tau(n-1)(1-e^{-\tau})}{\left[\left(x - x_{R_{j^*}}\right)^2 + \left(y - y_{R_{j^*}}\right)^2\right]^{\frac{\alpha}{2}}} dxdy$$

where $\left(x_{R_{j^*}}, y_{R_{j^*}}\right)$ is the coordinate of the selected relay R_{j^*}, $x_{R_{j^*}} \in [a, 1-a], y_{R_{j^*}} \in [b, 1-b]$ and $a \in [0, 0.5], b \in [0, 0.5]$.

Notice that within the network area, where relays are uniformly distributed, the worst case location for the selected relay R_{j^*} is the point $(0.5, 0.5)$, at which the interference from the noise generating nodes is the largest; whereas, the best case location for the selected relay R_{j^*} is the four corner points $(a, b), (a, 1-b), (1-a, b)$ and $(1-a, 1-b)$ of the relay selection, where the interference from the noise generating nodes is the smallest. By considering the worst case location for the selected relay R_{j^*}, we have

$$P\left(O_{S \to R_{j^*}}^{(T)} \middle| A_1\right) \le P\left(\frac{|h_{S,R_{j^*}}|^2 d_{S,R_{j^*}}^{-\alpha}}{\tau(n-1)(1-e^{-\tau})\varphi_1} \le \gamma_R \middle| A_1\right)$$

Here

$$\varphi_1 = \int_0^1 \int_0^1 \frac{1}{\left[(x-0.5)^2 + (y-0.5)^2\right]^{\frac{\alpha}{2}}} dxdy$$

Due to $a \le d_{S,R_{j^*}} \le \sqrt{(1-a)^2 + (0.5-b)^2}$, we consider the worst case and let $\phi = \sqrt{(1-a)^2 + (0.5-b)^2}$, then

$$P\left(O_{S \to R_{j^*}}^{(T)} \middle| A_1\right) \le P\left(\frac{|h_{S,R_{j^*}}|^2 \phi^{-\alpha}}{\tau(n-1)(1-e^{-\tau})\varphi_1} \le \gamma_R \middle| A_1\right)$$
$$= P\left(|h_{S,R_{j^*}}|^2 \le \frac{\gamma_R \tau(n-1)(1-e^{-\tau})\varphi_1}{\phi^{-\alpha}} \middle| A_1\right)$$
$$= 1 - e^{-\frac{\gamma_R \tau(n-1)(1-e^{-\tau})\varphi_1}{\phi^{-\alpha}}}$$

Employing the same method, we can get

$$P\left(O_{R_{j^*}\to D}^{(T)}\middle|A_1\right) \le 1 - e^{-\frac{\gamma_R \tau (n-1)\left(1-e^{-\tau}\right)\varphi_2}{\phi - \alpha}}$$

here,

$$\varphi_2 = \int_0^1 \int_0^1 \frac{1}{\left[(x-1)^2 + (y-0.5)^2\right]^{\frac{\alpha}{2}}} dx dy$$

Then, we have

$$P_{out|A_1}^{(T)} \le 1 - e^{-\frac{\gamma_R \tau (n-1)\left(1-e^{-\tau}\right)}{\phi - \alpha}(\varphi_1 + \varphi_2)}$$

Thus, we have

$$P_{out}^{(T)} \le \left[1 - e^{-\frac{\gamma_R \tau (n-1)\left(1-e^{-\tau}\right)}{\phi - \alpha}(\varphi_1 + \varphi_2)}\right](1-\vartheta) + 1 \cdot \vartheta$$

Notice that $P_{out}^{(S)}$ is given by

$$P_{out}^{(S)} = P\left(O_{S\to R_{j^*}}^{(S)}\right) + P\left(O_{R_{j^*}\to D}^{(S)}\right) - P\left(O_{S\to R_{j^*}}^{(S)}\right) \cdot P\left(O_{R_{j^*}\to D}^{(S)}\right)$$

According to the definition of secrecy outage probability, we know that

$$P\left(O_{S\to R_{j^*}}^{(S)}\right) = P\left(\bigcup_{i=1}^m \{C_{S,E_i} \ge \gamma_E\}\right)$$

Thus, we have

$$P\left(O_{S\to R_{j^*}}^{(S)}\right) \le \sum_{i=1}^m P\left(C_{S,E_i} \ge \gamma_E\right)$$

Based on the definition of r_0, we denote by $G_1^{(i)}$ the event that the distance between E_i and the source is less than r_0, and denote by $G_2^{(i)}$ the event that distance between E_i and the source is lager than or equal to r_0. We have

$$P\left(C_{S,E_i} \ge \gamma_E\right)$$
$$= P\left(C_{S,E_i} \ge \gamma_E\middle|G_1^{(i)}\right)P\left(G_1^{(i)}\right) + P\left(C_{S,E_i} \ge \gamma_E\middle|G_2^{(i)}\right)P\left(G_2^{(i)}\right)$$
$$\le 1 \cdot \frac{1}{2}\pi r_0^2 + P\left(C_{S,E_i} \ge \gamma_E\middle|G_2^{(i)}\right)\left(1 - \frac{1}{2}\pi r_0^2\right)$$

of which

$$P\left(C_{S,E_i} \ge \gamma_E\middle|G_2^{(i)}\right) \le$$

$$P\left(\frac{|h_{S,E_i}|^2 r_0^{-\alpha}}{\Gamma \int_0^1 \int_0^1 \frac{1}{\left[(x-x_{E_i})^2 + (y-y_{E_i})^2\right]^{\frac{\alpha}{2}}} dx dy} \ge \gamma_E\middle|G_2^{(i)}\right)$$

where $\left(x_{E_i}, y_{E_i}\right)$ is the coordinate of the eavesdropper E_i. Γ is the sum of $(n-1)\left(1-e^{-\tau}\right)$ independent exponential random variables.

From Fig.2 we know that the largest interference at eavesdropper E_i happens when E_i is located at the point $(0.5, 0.5)$, while the smallest interference at E_i happens it is located at the four corners of the network region. By considering the smallest interference at eavesdroppers, we then have

$$P\left(C_{S,E_i} \ge \gamma_E\middle|G_2^{(i)}\right) \le P\left(\frac{|h_{S,E_i}|^2 r_0^{-\alpha}}{\Gamma \psi} \ge \gamma_E\right)$$
$$= P\left(|h_{S,E_i}|^2 \ge \Gamma \gamma_E \cdot \psi \cdot r_0^\alpha\right)$$

here

$$\psi = \int_0^1 \int_0^1 \frac{1}{\left(x^2 + y^2\right)^{\frac{\alpha}{2}}} dx dy$$

Based on the Markov inequality,

$$P\left(C_{S,E_i} \ge \gamma_E\middle|G_2^{(i)}\right) \le E_\Gamma\left[e^{-\Gamma \gamma_E \psi r_0^\alpha}\right]$$
$$= \left(\frac{1}{1 + \gamma_E \psi r_0^\alpha}\right)^{(n-1)\left(1-e^{-\tau}\right)}$$

Then, we have

$$P\left(C_{S,E_i} \ge \gamma_E\right) \le$$
$$\frac{1}{2}\pi r_0^2 + \left(\frac{1}{1 + \gamma_E \psi r_0^\alpha}\right)^{(n-1)\left(1-e^{-\tau}\right)}\left(1 - \frac{1}{2}\pi r_0^2\right)$$

Employee the same method, we have

$$P\left(C_{R_{j^*},E_i} \ge \gamma_E\right) \le$$
$$\pi r_0^2 + \left(\frac{1}{1 + \gamma_E \psi r_0^\alpha}\right)^{(n-1)\left(1-e^{-\tau}\right)}\left(1 - \pi r_0^2\right)$$

Notice that

$$\frac{1}{2}\pi r_0{}^2 + \left(\frac{1}{1+\gamma_E \psi r_0{}^\alpha}\right)^{(n-1)(1-e^{-\tau})}\left(1 - \frac{1}{2}\pi r_0{}^2\right)$$

$$= \pi r_0{}^2 + \left(\frac{1}{1+\gamma_E \psi r_0{}^\alpha}\right)^{(n-1)(1-e^{-\tau})}\left(1 - \pi r_0{}^2\right)$$

$$-\frac{1}{2}\pi r_0{}^2\left[1 - \left(\frac{1}{1+\gamma_E \psi r_0{}^\alpha}\right)^{(n-1)(1-e^{-\tau})}\right]$$

$$\leq \pi r_0{}^2 + \left(\frac{1}{1+\gamma_E \psi r_0{}^\alpha}\right)^{(n-1)(1-e^{-\tau})}\left(1 - \pi r_0{}^2\right)$$

Therefore

$$P\left(O^{(S)}_{S \to R_{j^*}}\right) \leq P\left(O^{(S)}_{R_{j^*} \to D}\right)$$

$$\leq m\left[\pi r_0{}^2 + \left(\frac{1}{1+\gamma_E \psi r_0{}^\alpha}\right)^{(n-1)(1-e^{-\tau})}\left(1 - \pi r_0{}^2\right)\right]$$

Then, we have

$$P^{(S)}_{out} \leq 2m\left[\pi r_0{}^2 + \left(\frac{1}{1+\gamma_E \psi r_0{}^\alpha}\right)^{(n-1)(1-e^{-\tau})}\left(1 - \pi r_0{}^2\right)\right]$$

$$-\left[m\left(\pi r_0{}^2 + \left(\frac{1}{1+\gamma_E \psi r_0{}^\alpha}\right)^{(n-1)(1-e^{-\tau})}\left(1 - \pi r_0{}^2\right)\right)\right]^2$$

$\square$

Appendix H. Proof of Lemma 6

Proof. The parameter τ should be set properly to satisfy both reliability and secrecy requirements.

• Reliability Guarantee

To ensure the reliability requirement $P^{(T)}_{out} \leq \varepsilon_t$, we know from Lemma 5 that we just need

$$\left[1 - e^{-\frac{\gamma_R \tau(n-1)(1-e^{-\tau})}{\phi^{-\alpha}}(\varphi_1+\varphi_2)}\right](1-\vartheta) + 1\cdot\vartheta \leq \varepsilon_t$$

that is,

$$-\frac{\gamma_R \tau (n-1)(1-e^{-\tau})}{\phi^{-\alpha}}(\varphi_1+\varphi_2) \geq \log\left(\frac{1-\varepsilon_t}{1-\vartheta}\right)$$

By using Taylor formula, we have

$$\tau \leq \sqrt{\frac{-\log\left(\frac{1-\varepsilon_t}{1-\vartheta}\right)\phi^{-\alpha}}{\gamma_R(n-1)(\varphi_1+\varphi_2)}}$$

• Secrecy Guarantee

To ensure the secrecy requirement $P^{(S)}_{out} \leq \varepsilon_s$, we know from Lemma 5 that we just need

$$2m\left[\pi r_0{}^2 + \left(\frac{1}{1+\gamma_E \psi r_0{}^\alpha}\right)^{(n-1)(1-e^{-\tau})}\left(1 - \pi r_0{}^2\right)\right] -$$

$$\left[m\left(\pi r_0{}^2 + \left(\frac{1}{1+\gamma_E \psi r_0{}^\alpha}\right)^{(n-1)(1-e^{-\tau})}\left(1 - \pi r_0{}^2\right)\right)\right]^2 \leq \varepsilon_s$$

Thus,

$$m\cdot\left[\pi r_0{}^2 + \left(\frac{1}{1+\gamma_E \psi r_0{}^\alpha}\right)^{(n-1)(1-e^{-\tau})}\left(1 - \pi r_0{}^2\right)\right] \leq$$

$$1 - \sqrt{1-\varepsilon_s}$$

That is,

$$\tau \geq -\log\left[1 + \frac{\log\left(\frac{\frac{1-\sqrt{1-\varepsilon_s}}{m}-\pi r_0{}^2}{1-\pi r_0{}^2}\right)}{(n-1)\log\left(1+\gamma_E \psi r_0{}^\alpha\right)}\right]$$

$\square$

Appendix I. Proof of Theorem 3

Proof. From Lemma 6, we know that to ensure the reliability requirement, we have

$$\tau \leq \sqrt{\frac{-\log\left(\frac{1-\varepsilon_t}{1-\vartheta}\right)\phi^{-\alpha}}{\gamma_R(n-1)(\varphi_1+\varphi_2)}}$$

and

$$(n-1)(1-e^{-\tau}) \leq \frac{-\log\left(\frac{1-\varepsilon_t}{1-\vartheta}\right)}{\gamma_R \tau \phi^\alpha(\varphi_1+\varphi_2)}$$

To ensure the secrecy requirement, we need

$$m\cdot\left[\pi r_0{}^2 + \left(\frac{1}{1+\gamma_E \psi r_0{}^\alpha}\right)^{(n-1)(1-e^{-\tau})}\left(1 - \pi r_0{}^2\right)\right] \leq$$

$$1 - \sqrt{1-\varepsilon_s}$$

Thus,

$$m \leq \frac{1-\sqrt{1-\varepsilon_s}}{\pi r_0{}^2 + \left(\frac{1}{1+\gamma_E \psi r_0{}^\alpha}\right)^{(n-1)(1-e^{-\tau})}\left(1 - \pi r_0{}^2\right)}$$

$$\leq \frac{1-\sqrt{1-\varepsilon_s}}{\pi r_0{}^2 + \left(\frac{1}{1+\gamma_E \psi r_0{}^\alpha}\right)^{\frac{-\log\left(\frac{1-\varepsilon_t}{1-\vartheta}\right)}{\gamma_R \tau \phi^\alpha(\varphi_1+\varphi_2)}}\left(1 - \pi r_0{}^2\right)}$$

By letting τ to take its maximum value for maximum interference at eavesdroppers, we get the following bound

$$m \leq \frac{1 - \sqrt{1 - \varepsilon_s}}{\pi r_0^2 + (1 - \pi r_0^2)\,\omega}$$

Here,

$$\omega = \left(1 + \gamma_E \psi r_0^\alpha\right)^{-\sqrt{\frac{-(n-1)\log\left(\frac{1-\varepsilon_t}{1-\delta}\right)}{\gamma_R(\varphi_1+\varphi_2)\phi^\alpha}}}$$

□

Acknowledgement. This work was Supported by The National Natural science foundation of china(61100153, 61373173, U1135002), the Fundamental Research Funds for the Central Universities(JY10000903005, JY10000903001).

References

[1] N. Sathya, Two-hop forwarding in wireless networks, Dissertation for the degree of Doctor of philosophy, Polytechnic University, 2006.

[2] H. Wang, Y. Zhang, J. Cao, Effective collaboration with information sharing in virtual universities, IEEE Transactions on Knowledge and Data Engineering, Vol. 21, No. 6, pp.840-853, 2009.

[3] Xiaoxun Sun, Hua Wang, Jiuyong Li, and Yanchun Zhang,Satisfying Privacy Requirements Before Data Anonymization. Comput. J. Vol.55, No.4, pp.422-437, 2012.

[4] Min Li, Xiaoxun Sun, Hua Wang, Yanchun Zhang, and Ji Zhang, Privacy-aware access control with trust management in web service, World Wide Web, Vol.14, No.4, pp.407-430, 2011.

[5] J. Talbot and D. Welsh, Complexity and Crytography : An Introduction,, Cambridge University Press, 2006.

[6] J. Zhang and M. C. Gursoy, Collaborative relay beam forming for secrecy, in Proceeding of 2010 IEEE International Conference on Communications (ICC), pp.1-5, 2010.

[7] I. Krikidis, J. S. Thompson and S. McLaughlin, Relay selection for secure cooperative networks with jamming, IEEE Transactions on Wireless Communications, vol. 8, no.10, pp.5003-5011, 2009.

[8] J. Li, A. Petropulu and S. Weber, On Cooperative Relaying Schemes for Wireless Physical Layer Security, IEEE Transactions on Signal Processing, vol. 59, no. 10, pp. 4985-4997, 2011.

[9] L. Dong, H. Yousefizadeh and H. Jafarkhani, Cooperative Jamming and Power Allocation for Wireless Relay Networks in Presence of Eavesdropper IEEE International Conference on Communications (ICC 2011), pp.1-5, 2011.

[10] J. Huang and A.L. Swindlehurst, Secure Communications via Cooperative Jamming in Two-hop Relay Systems, In 2010 IEEE Global Telecommunications Conference (GLOBECOM 2010), pp.1-5, 2010.

[11] L. Dong, Z. Han, A.P. Petropulu, and H.V. Poor, Secure wireless communications via cooperation, in Proc. 46th Annual Allerton Conference on Communication, Control, and Computing, pp. 1132-1138, 2008.

[12] L. Dong, Z. Han, A.P. Petropulu, and H.V. Poor, Improving wireless physical layer security via cooperating relays, IEEE Transactions on Signal Processing, vol. 58, no. 3, pp.1875-1888, 2010.

[13] S. Luo, J. Li, and A. Petropulu, Physical Layer Security with Uncoordinated Helpers Implementing Cooperative Jamming, in Proceeding of the seventh IEEE Sensor Array and Multichannel Signal Processing Workshop, pp.97-100, 2012.

[14] G. Zheng, L.C. Choo, and K. K. Wong, Optimal cooperative jamming to enhance physical layer security using relays, IEEE Transactions on Signal Processing, vol. 59, no. 3, pp. 1317-1322, 2011.

[15] K. Morrison, and D. Goeckel, Power allocation to noise-generating nodes for cooperative secrecy in the wireless environment, In the Forty Fifth Asilomar Conference on Signals, Systems and Computers (ASILOMAR), pp.275-279, 2011.

[16] S. Goel, and R. Negi, Guaranteeing secrecy using artificial noise, IEEE transactions on wireless communications, vol.7, no.6, pp.2180-2189, 2008.

[17] L. Lai and H.E. Gamal, The relay-eavesdropper channel: Cooperation for secrecy, IEEE Trans. Inf. Theory, vol. 54, no. 9, pp. 4005 - 4019, 2008.

[18] R. Negi and S. Goelm, Secret communication using artificial noise, In 62nd IEEE conference on Vehicular Technology Conference (VTC 2005), pp.1906-1910, 2005.

[19] R. Zhang, L. Song, Z. Han and B. Jiao. Physical Layer Security for Two-Way Untrusted Relaying With Friendly Jammers, IEEE Transactions on vehicular technology, vol. 61, no. 8, pp. 3693-3704, 2012.

[20] X. He and A. Yener, Two-hop secure communication using an untrusted relay: A case for cooperative jamming. in Proceeding of 2008 IEEE Global Telecommunications Conference, pp.1-5, 2008.

[21] Q. Guan, F. R. Yu, S. Jiang, and C. M. Leung. Joint Topology Control and Authentication Design in Mobile Ad Hoc Networks With Cooperative Communications, IEEE Transactions on vehicular technology, vol.61, no.6, pp.2674-2685, 2012

[22] Z. Ding, M. Xu, J. Lu, and F. Liu, Improving Wireless Security for Bidirectional Communication Scenarios, IEEE Transactions on vehicular technology, vol.61, no.6, pp.2842-2848, 2012.

[23] D. Goeckel, S. Vasudevan, D. Towsley, S. Adams, Z. Ding and K. Leung, Everlasting Secrecy in Two-Hop Wireless Networks Using Artificial Noise Generation from Relays, In proceeding of International Technology Alliance Collaboration System (ACITA 2011), 2011.

[24] D. Goeckel, S. Vasudevan, D. Towsley, S. Adams, Z. Ding and K. Leung, Artificial noise generation from cooperative relays for everlasting secrecy in two-hop wireless networks, IEEE Journal on Selected Areas in Communications, vol.29, no.10 pp.2067-2076, 2011.

[25] S. Vasudevan, S. Adams, D. Goeckel, Z. Ding, D. Towsley and K. Leung, Multi-User Diversity for Secrecy in Wireless Networks, In proceeding of Information Theory

and Applications Workshop (ITA 2010), pp.1-9, 2010.

[26] C. Capar, D. Goeckel, B. Liu and D. Towsley, Secret Communication in Large Wireless Networks without Eavesdropper Location Information, In Proceeding of IEEE INFOCOM 2012, pp.1152-1160, 2012.

[27] Z. Ding, K. Leung, D. Goeckel and D. Towsley, Opportunistic Relaying for Secrecy Communications: Cooperative Jamming vs Relay Chatting, IEEE Transactions on Wireless Communications, vol.10, no.6, pp.1725-1729, 2011.

[28] M. Dehghan, D. Goeckel, M. Ghaderi and Z. Ding, Energy Efficiency of Cooperative Jamming Strategies in Secure Wireless Networks, IEEE Transactions on Wireless Communications, vol.11, no.9, pp.3025-3029, 2012.

[29] A. Sheikholeslami, D. Goeckel, H. Pishro-Nik and D. Towsley, Physical Layer Security from Inter-Session Interference in Large Wireless Networks, In Proceeding of IEEE INFOCOM 2012, pp.1179-1187, 2012.

[30] C. Leow, C. Capar, D. Goeckel, and K. Leung, A Two-Way Secrecy Scheme for the Scalar Broadcast Channel with Internal Eavesdroppers, In the Forty Fifth Asilomar Conference on Signals, Systems and Computers (ASILOMAR 2011), pp.1840-1844, 2011.

[31] C. Capar and D. Goeckel, Network Coding for Facilitating Secrecy in Large Wireless Networks, In 46th Annual Conference on Information Sciences and Systems (CISS 2012), pp.1-6, 2012.

[32] H.David, Order Statistics, Wiley, New York, 1980.

17

Personalised Information Gathering and Recommender Systems: Techniques and Trends

Xiaohui Tao[†*], Xujuan Zhou[‡], Cher Han Lau[‡], Yuefeng Li[‡]

[†]Faculty of Sciences, University of Southern Queensland, Australia
[‡]Science and Engineering Faculty, Queensland University of Technology, Australia

Abstract

With the explosive growth of resources available through the Internet, information mismatching and overload have become a severe concern to users. Web users are commonly overwhelmed by huge volume of information and are faced with the challenge of finding the most relevant and reliable information in a timely manner. Personalised information gathering and recommender systems represent state-of-the-art tools for efficient selection of the most relevant and reliable information resources, and the interest in such systems has increased dramatically over the last few years. However, web personalization has not yet been well-exploited; difficulties arise while selecting resources through recommender systems from a technological and social perspective. Aiming to promote high quality research in order to overcome these challenges, this paper provides a comprehensive survey on the recent work and achievements in the areas of personalised web information gathering and recommender systems. The report covers concept-based techniques exploited in personalised information gathering and recommender systems.

Keywords: Personalisation, Information Gathering, Recommender Systems

1. Introduction

Over the last decade, the rapid growth and adoption of World Wide Web have further exacerbated user needs for efficient mechanisms for information and knowledge location, selection and retrieval. Web information covers a wide range of topics and serves a broad spectrum of communities [2]. Web users create Web information and new sources of knowledge at a rapid rate with various Web 2.0 applications such as blogs, social and professional networks, wikis, and many other types of social media. The abundance of information created by users explicitly and pro-actively contains rich semantic meaning and provides a huge potential to obtain deep knowledge about users. However, the massive User-Generated Content (UGC) in Web 2.0 era has made it increasingly difficult for users to effectively find exactly what they need. How to gather useful and meaningful information from the Web becomes a challenge to all users. This challenging issue is referred by many researchers as Web information gathering [27, 51].

Web information gathering aims to acquire useful and meaningful information for users from the Web. Web information gathering tasks are usually completed by the systems using keyword-based techniques. The keyword-based mechanism searches the Web by finding the documents with the specific terms matched. This mechanism is used by many existing Web search systems, for example, Google and Yahoo!, for their Web information gathering. Han and Chang [37] pointed out that by using keyword-based search, web information gathering systems can access the information quickly; however, the gathered information may possibly contain much useless and meaningless information. This is particularly referred as the fundamental issue in Web information gathering: information mismatching and information overloading [53–55, 57, 136].

Web-based recommender systems are the most notable application of the web personalization. With today's increasing information overload problem on the Web, the area of recommender systems research becomes more challenging than ever before. There remain difficulties that limit the full exploitation of personalization and resource selection through recommendation from both technology perspective and human and social perspective. Recommender systems have also been made by researchers as an important response to information overloading problems, for its ability to provide personalized and meaningful information recommendations by taking into account idiosyncratic user interests and information needs [90]. For example, while standard search engines are very likely to generate the same results to different users entering identical search queries, recommender systems are able to generate results to each user that are personalized and more relevant because they take into account each user's personal interests. The recommender technology has been successfully employed in many applications such as films, music, books. The richness of the online information challenges the current personalization techniques and also provides new possibilities for accurately users profiling. Thus, how to incorporate the new features and practices of Web 2.0 into personalized recommender applications becomes an important and urgent research topic.

Recommendation techniques can be divided into two major classes: content-based filtering (CBF) and collaborative filtering (CF). CBF focuses on the analysis of item content and user profiles are used to filter available objects. Collaborative filtering (CF) focuses on identification of other users with similar tastes, and utilisetheir opinions to recommend items. The user profiles are used to recommend to a user the information that satisfied previous users with a similar profile. In movie recommender application, for instance, a CBF system will typically rely on information such as genre, actors, director, producer etc. and match this against the learned preferences of the user in order to select a set of promising movie recommendations. CBF recommender systems need a technique to represent the features of the items. Feature representation can be created automatically for machine readable items (such as news or papers). However, for some items such as jokes, it is almost impossible to define the right set of describing features and to "objectively" classify them [73]. Collaborative filtering (CF) collects information about a user by asking them to rate items and makes recommendations based on highly rated items by users with similar taste. CF approaches make recommendations based on the ratings of items by a set of users (neighbours) whose rating profiles are most similar to that of the target user [7]. CF algorithms generally compute the overall similarity or correlation between users, and use that as weight when making recommendations. In book recommendation application, for example, the first step

of the CF system is to find the "neighbours" of the target user. The "neighbours" refer to other users who have similar tastes in books (rate the same books similarly). In the second step, only the books that are highly rated by the "neighbours" would be recommended.

In contrast with the content-base approaches, CF techniques rely on the availability of user profiles that capture past ratings histories of users [7] and do not require any human intervention for tagging content because item knowledge is not required. Therefore, the CF techniques can be applied to virtually any kind of items: papers, news, web sites, movies, songs, books, jokes, locations of holidays, stocks and promise to scale well to large item bases [73]. Collaborative filtering is the most widely used approach to build online recommender systems. It has been successfully employed in many applications, such as recommending books, CDs, and other products at *Amazon.com*, Movies by *MovieLens* [1]. Some methods combine both content and collaborative filtering approaches to make recommendations [96].

In the past decade, many researchers have aimed at gathering Web information and make recommendations for users with consideration of their personalised interest and preferences. In these works, Web user profiles are widely used for user modelling and personalization [49], because they reflect the interest and preferences of users [102]. User profiles are defined by Li and Zhong [57] as the interesting topics underlying user information needs. They are used in Web information gathering to describe user background knowledge, to capture user information needs, and to gather personalized Web information for users [31, 37, 57, 113]. This survey paper attempts to review the development of the concept-based, personalized Web information gathering techniques. The review notes the issues in Web personalization, focusing on Web user profiles and user information needs in personalized Web information gathering. The reviewed scholar reports that the concept-based models utilizing user background knowledge are capable of gathering useful and meaningful information for Web users. However, the representation and acquisition of user profiles need to be improved for the effectiveness of Web information gathering. This survey contributes to better understanding of existing Web information gathering systems.

The paper is organized as follows. Section 2 reviews the concept-based techniques employed by Web information gathering and recommender systems; Section 3 discusses the personalisation issues in the context of Web information gathering and recommendation, focusing on user profile representation and acquisition. Information gathering and recommender systems in social networks are discussed in Section 4. Section 5 makes the final remarks for the survey.

2. Exploiting Concepts for Web Information Gathering and Recommender Systems

Concept-based techniques that are used in web information gathering are also widely exploited in recommender systems as well. Recommender systems rely on concept-based techniques to access concepts of products. Concept-based information gathering techniques use semantic concepts extracted from documents and queries. Instead of matching keyword features representing the documents and queries, concept-based techniques attempt to compare the semantic concepts of documents to those of given queries. Similarity of documents to queries is determined by the matching level of their semantic concepts. The semantic concept representation and extraction are two typical issues in the concept-based techniques and are discussed in the following sections.

2.1. Semantic Concept Representation

Semantic concepts have various representations. In some models, concepts are represented by controlled lexicons defined in terminological ontologies, thesauruses, or dictionaries. In some models, they are represented by subjects in domain ontologies, library classification systems, or categorizations. Some models uses data mining techniques for concept extraction, semantic concepts are represented by patterns. The three representations given different strengths and weaknesses.

lexicon-based representation defines the concepts in terms and lexicons that are easily understood by users. WordNet [28] and its variations [9, 48] are typical models employing this kind of concept representation. In these models, semantic concepts are represented by the controlled vocabularies defined in terminological ontologies, thesauruses, or dictionaries. Because these are being controlled, they are also easily utilized by the computational systems. However, when extracting terms to represent concepts for information gathering, some noisy terms may also be extracted because of term ambiguity. As a result, information overloading problem may occur in gathering. Moreover, the lexicon-based representation relies largely on the quality of terminological ontologies, thesaurus, or dictionaries for definitions. However, the manual development of controlled lexicons or vocabularies (like WordNet) is usually costly. The automatic development is efficient, however, in sacrificing the quality of definitions and semantic relation specifications. Consequently, the lexicon-based representation of semantic concepts was reported to be able to improve the information gathering performance in some works [48, 68], but to be degrading the performance in other works [116].

Many Web systems rely upon subject-based representation of semantic concepts for information gathering. In this kind of representation, semantic concepts are represented by subjects defined in knowledge bases or taxonomies, including domain ontologies, digital library systems, and online categorizations. Typical information gathering systems exploiting domain ontologies for concept representation include those developed by Lim et al. [65], by Navigli [82], and by Velardi et al. [115]. Domain ontologies contain expert knowledge: the concepts described and specified in the ontologies are of high quality. However, expert knowledge acquisition is usually costly in ycapitalization and computation. Moreover, as discussed previously, the semantic concepts specified in many domain ontologies are structured only in the subsumption manner of *super-class* and *sub-class*, rather than the more specific *is-a*, *part-of*, and *related-to*, the ones developed by [31, 46] and [136]. Some attempted to describe more specified relations, like [13, 103] for *is-a*, [33, 92] for *part-of*, and [41] for *related-to* relations only. Tao et al. [107, 108] made a further progress from these works and portrayed the basic *is-a*, *part-of*, and *related-to* semantic relations in one single computational model for concept representation.

Also used for subject-based concept representation are the library systems, like Dewey Decimal Classification (DDC) used by [46, 118], Library of Congress Classification (LCC) and Library of Congress Subject Headings (LCSH) [107, 108], and the variants of these systems, such as the "China Library Classification Standard" used by [132] and Alexandria Digital Library (ADL) [117]. These library systems represent the natural growth and distribution of human intellectual work that covers the comprehensive and exhaustive topics of world knowledge [15]. In these systems, concepts are represented by subjects that are defined by librarians and linguists manually under a well-controlled process [15]. Concepts are constructed in taxonomic structure, originally designed for information retrieval from libraries. These are beneficial to the information gathering systems. The concepts are linked by semantic relations, such as subsumption like *super-class* and *sub-class* in the DDC and LCC, and *broader*, *used-for*, and *related-to* in the LCSH. However, information gathering systems using library systems for concept representation largely rely on knowledge bases. The limitations of the library systems, for example, focus on the United States more than on other regions by the LCC and LCSH, would be incorporated by the information gathering systems that use them for concept representation.

The online categorizations are also widely relied on by many information gathering systems for concept representation. Typical online categorization used for concept representation include the Yahoo! categorization used by [31] and *Open Directory Project*[1] used by [16, 86]. In these categorizations, concepts are represented by categorization subjects and organized in a taxonomical structure. However, the nature of categorizations is in the subsumption manner of one containing another (*super-class* and *sub-class*), but not the semantic *is-a*, *part-of*, and *related-to* relations. Thus, the semantic relations associated with the concepts in such representations are not in adequate details and specific levels. These problems weaken the quality of concept representation and thus the reducing performance of information gathering systems.

Another semantic concept representation in Web information gathering systems is pattern-based representation that uses multiple terms (e.g. phrases) to represent a single semantic concept. Phrases contain more content than any one of their containing terms. Research representing concepts by patterns include Li and Zhong [53–55, 57–59], Wu *et al.* [122, 124, 125], Zhou *et al.* [138–140], Dou *et al.* [22], and Ruiz-Casado *et al.* [93]. However, pattern-based semantic concept representation poses some drawbacks. The concepts represented by patterns can have only subsumption specified for relations. Usually, the relations exist between patterns are specified by investigation of their containing terms, like [57, 125, 138]. If more terms are added into a phrase, to make the phrase more specific, the phrase becomes a sub-class concept of any concepts represented by the sub-phrases in it. Consequently, no specific semantic concepts like *is-a* and *part-of* can be specified and thus some semantic information may be missing in pattern-based concept representations. Another problem of pattern-based concept representation is caused by the length of patterns. The concepts can be adequately specific for discriminating one from others only if the patterns representing the concepts are long enough. However, if the patterns are too long, the patterns extracted from Web documents would be of low frequency and thus, cannot support the concept-based information gathering systems substantially [125]. Although the pattern-based concept representation poses such drawbacks, it is still one of the major concept representations in information gathering systems.

The semantic content of text documents has different representations, such as controlled lexicons, categories, or patterns. The lexicon-based representation of documents is easy to be understood by users or computational systems. With such a representation, text documents are represented by a set of terms chosen from controlled vocabularies defined in terminological ontologies, thesauruses, or dictionaries. However, when extracting lexical descriptors, some noisy terms are also extracted along with meaningful, representative terms, due to term ambiguity problem. The development of terminological ontologies, thesauruses, or dictionaries is also costly in finance, time, and usually requires a large amount of human power involvement. As a result, lexicon-based semantic content representation is ineffective and costly.

Categorizations are also widely used to represent document contents [40, 86, 88, 109]. In such a representation, concepts are represented by categories and organized in a tree or graphic structure. The relationships existing between concept nodes in the structure are explored in order to measure the capacity of a concept describing or representing the semantic content of a document. However, the natural relationship in categorizations is subsumption of one containing another (super-class and sub-class), but not the detailed, specific semantic relations (like is-a, part-of, and related-to). Thus, the concept specification needs to improve toward a more detailed and specific level.

Another representation is pattern-based that uses multiple phrases to represent document contents [23, 57, 61, 140]. However, pattern-based semantic annotation suffers from a problem by the length of patterns. Concepts are specific and discriminating only if patterns are substantially long. However, if a pattern is too long, its frequency would be very low in documents. Consequently, such pattern becomes useless because of poor applicability [60]. In addition, because of using text mining for pattern extraction, the quality of patterns is difficult to control. As a result, noisy patterns are extracted as well as useful patterns.

2.2. Semantic Concept Extraction

Text classification is the process of classifying an incoming stream of documents into categories by using the classifiers learned from the training samples [66]. In generally, text classification problem can be a "binary" classification problem If there are exactly two classes or a "multi-class" problem if there are more than two classes and each document falls into exactly one class or a "multi-label categorization" problem if a document may have more than one associated category in a classification scheme. Multi-label and multi-class tasks are often handled by reducing them to k binary classification tasks, one for each category [128]. The works conducted by Tao's and Yang et al. [129] are about multi-label text classification. The former worked on categorizing library catalogue items into multiple subjects and the latter adopted active learning algorithms for multi-label classification.

---

[1] http://www.dmoz.org

There are different types of text classifier. Fung et al. [29] categorized them into two types: *kernel-based classifiers* and *instance-based classifiers*. Typical kernel-based classifier learning approaches include the *Support Vector Machines* (SVMs) [43] and regression models [98]. These approaches may incorrectly classify many negative samples from an unlabelled set into a positive set, thus causing the problem of information overloading in Web information gathering. Typical instance-based classification approaches include the *K*-Nearest Neighbour (*K*-NN) [19] and its variants, which do not rely upon the statistical distribution of training samples. However, instance-based approaches are not capable of extracting highly accurate positive samples from the unlabeled set. Other research works, such as [31, 88], have a different way of categorizing the classifier learning techniques: *document representations based classifiers*, including SVMs and *K*-NN; and *word probabilities based classifiers*, including Naive Bayesian, decision trees [43] and neural networks used by [133]. These classifier learning techniques have different strengths and weaknesses, and should be chosen based upon the problems they are attempting to solve.

Machine learning for text classification can be categorised into three groups: supervised, semi-supervised, and unsupervised. When there is a set of pre-classified documents available to train classifiers, the process is referred to as supervised classification. Sometimes, samples may be inadequate or insufficient, though available. Such a problem is referred to as semi-supervised text classification. Nguyen and Caruana [83] proposed a semi-supervised approach to address the problem and learned classifiers from only partial label samples (the training documents are pre-classified into a set of possible classes with only one correct class). Fung et al. [29] introduced an approach to learn classifiers from only positive and unlabelled samples, without negative ones. The approach first extracts negative samples from unlabelled set and builds classifiers as usual. Supervised and semi-supervised text classification techniques more or less rely on pre-classified samples to learn classifiers. Yang et al. [130] proposed to build classification model for a target class without associated training samples, by analysing the correlating auxiliary classes.

Text classification techniques are widely used in concept-based Web information gathering systems. Gauch et al. [31] described how text classification techniques are used for concept-based Web information gathering. Web users submit a topic associated with some specified concepts. The gathering agents then search for the Web documents that are referred to by the concepts. Sebastiani [98] outlined a list of tasks in Web information gathering to which text classification techniques may contribute: automatic indexing for Boolean information retrieval systems, document organization

(particularly in personal organization or structuring of a corporate document base), text filtering, word sense disambiguation, and hierarchical categorization of web pages. Also, as specified by Meretakis et al.[75], the Web information gathering areas contributed to by text classification may include sorting emails, filtering junk emails, cataloguing news articles, providing relevance feedback, and reorganizing large document collections. Text classification techniques have been utilized by to classify Web documents into the best matching interest categories, based on their referring semantic concepts [69].

Some limitations and weaknesses of these text classification techniques employed in concept-based Web information gathering exist. Glover et al. [34] pointed out that Web information gathering performance substantially relies on the accuracy of predefined categories. If the arbitration of a given category is wrong, the performance is degraded. Another challenging problem, referred to as "cold start", occurs when there is an inadequate number of training samples available to learning classifiers. Also, as pointed out by Han and Chang [37], concept-based Web information gathering systems rely on an assumption that the content of Web documents is adequate to make descriptions for classification. When the assumption fails, using text classification techniques alone becomes unreliable for Web information gathering systems. The solution to this problem is to use high quality semantic concepts, as argued by Han and Chang [37], and to integrate both text classification and Web mining techniques.

Ontologies have been studied and exploited by many works to facilitate text classification. Gabrilovich and Markovitch [30] enhanced text classification by generating features using domain-specific and common-sense knowledge in large ontologies with hundreds of thousands of concepts. Camous et al. [11] also introduced domain-independent method that uses the Medical Subject Headings (MeSH) ontology. The method investigates the inter-concept relationships and represents documents by MeSH subjects. Similarly, Camous' work considers the semantic relations exist in concepts. However, their work focuses only on the medical domain. Whereas, Wang and Domeniconi [119] and Hu et al. [40] derived background knowledge from Wikipedia to represent documents and attempted to deal with the sparsity and high dimensionality problems in text classification. Instead of Wikipedia with free-contributed entries, the approach proposed in [] uses the superior LCSH, which is a world knowledge ontology and has been under continuous development for a hundred years by knowledge engineers.

Many works exploited pattern mining techniques to help build classification models. Malik and Kender [71] proposed the "Democratic Classifier", which is a pattern-based classification algorithm using short patterns. Their democratic classifier relies on the quality of training samples and cannot deal with the "no training set available" problem. Bekkerman and Matan [3] argued that most of the information on documents can be captured in phrases and proposed a text classification method that employs lazy learning from labelled phrases. The phrases in their work are in fact a special form of sequential patterns.

Web content mining is an emerging field of applying knowledge discovery technology to Web data. Web content mining discovers knowledge from the content of Web documents, and attempts to understand the semantics of Web data [49, 57]. Based on various Web data types, Web content mining can be categorised into Web text mining, Web multimedia data mining (e.g. image, audio, video), and Web structure mining [49]. In this paper, Web information is particularly referred to text documents existing on the Web. Thus, the term "Web content mining" here refers to "Web text content mining", the knowledge discovery from the content of Web text documents. Kosala and Blockeel [49] categorized Web content mining techniques into database views and information retrieval views. From the database view, the goal of Web content mining is to model the Web data so that Web information gathering may be performed based on concepts rather than on keywords. From the information retrieval view, the goal is to improve Web information gathering based on either inferred or solicited Web user profiles. Web content mining contributes significantly to Web information gathering in either view.

Many techniques are applied in Web content mining, including pattern mining, association rules mining, text classification and clustering, and data generalization and summarisation [53, 55]. Li and Zhong [53–55, 57], Wu *et al.* [124], and Zhou *et al.* [138–140] represented semantic concepts by maximal patterns, sequential patterns, and closed sequential patterns, and attempted to discover these patterns for semantic concepts extracted from Web documents. Their experiments reported substantial improvements achieved by their proposed models, in comparison with the traditional *Rocchio*, *Dempster-Shafer*, and probabilistic models. Association rules mining extracts meaningful content from Web documents and discovers their underlying knowledge. Existing models using association rules mining include Li and Zhong [52], Li *et al.* [56], and Yang *et al.* [127], who used the granule techniques to discover association rules; Xu and Li [126] and Shaw *et al.* [100], who attempted to discover concise association rules; and Wu *et al.* [123], who discovered positive and negative association rules. Some works, such as Dou *et al.* [22], attempted to integrate multiple Web content mining techniques for concept extraction. These works were claimed capable of extracting concepts from Web

documents and improving the performance of Web information gathering. However, as pointed out by Li and Zhong [54, 55], the existing Web content mining techniques have some limitations. The main problem is that these techniques are incapable of specifying the specific semantic relations (e.g. *is-a* and *part-of*) that exist in the concepts. Their concept extraction needs to be improved for more specific semantic relation specification, considering the fact that the current Web is nowadays moving toward the Semantic Web [4].

3. Personalisation in Web Information Gathering and Recommender Systems

Web user profiles are widely used by Web information systems for user modelling and personalization [49]. User profiles reflect the interests of users [102]. In terms of Web information gathering, user profiles are defined by Li and Zhong [57] as the interesting topics underlying user information needs . Hence, user profiles are used in Web information gathering to capture user information needs from the user submitted queries, in order to gather personalized Web information for users [31, 37, 57, 113].

Web user profiles are categorized by Li and Zhong [57] into two types: the *data diagram* and *information diagram* profiles (also called *behaviour-based profiles* and *knowledge-based profiles* by [76]). The data diagram profiles are usually acquired by analyzing a database or a set of transactions [31, 57, 76, 104, 105]. These kinds of user profiles aim to discover interesting registration data and user profile portfolios. The information diagram profiles are usually acquired by using manual techniques; such as questionnaires and interviews [76, 113], or by using information retrieval and machine-learning techniques [31]. They aim to discover interesting topics for Web user information needs.

Personalized recommender systems have ability to provide meaningful information recommendations [90] by taking into account idiosyncratic user interests and information needs. The representation of user information needs is variously referred to as ï£¡user profilesï£¡, or ï£¡topic profilesï£¡. Recommender systems can be divided into two major classes: content-based filtering [78] and collaborative filtering recommender [47]. Content-based filtering focuses on the analysis of item content. The user profiles are used to filter available objects. Collaborative filtering focuses on identification of other users with similar tastes, and the use of their opinions to recommend items. The user profiles are used to recommend to a user information that satisfied previous users with a similar profile. The recommender systems success depend on large extent on the ability of the learned profiles to represent the users actual interests. Learning a personalized user profile is one of the most challenging tasks in developing the next generation of information filtering and recommender systems [62, 140].

User profiles are widely used in not only Web information gathering [57, 107, 108], but also personalized Web services [37], personalized recommendations [76], and marketing research [137]. User profile representation and construction are very important issues within these research fields. In this section, the methods and techniques for profiles representation and construction will be discussed.

3.1. User Profile Representation

User profiles have various representations. As defined by [102], user profiles are represented by a previously prepared collection of data reflecting user interests. In many approaches including conventional information gathering and information recommendation, this "collection of data" refers to a set of terms (or vector space of terms) that can be directly used to expand the queries submitted by users [18, 76, 113]. For instance, traditional content-based information filtering uses single-vector or multi-vector models that produce one term-weight or more than one term-weight vectors [98] to represent the relevant information of the topic of likely interest for a user. Such profiles are called term-based user profiles.

These term-based user profiles, however, may cause poor interpretation of user interests to the users, as pointed out by [55, 57]. Also, term-based user profiles suffer from the problems introduced by the keyword-match techniques because many terms are usually ambiguous. Attempting to solve this problem, Li and Zhong [57] represented user profiles by patterns. However, pattern-based user profiles also suffer from problems of inadequate semantic relations specification and the dilemma of pattern length and pattern frequency, as discussed previously in Section 2 for pattern-based concept representation. Recently, the two-stage information filtering (i.e., recommender system) and decision making support system have been developed by Zhou et.al. [140] and Li et.al. [62] using both the term-based and pattern-based profiles.

User profiles can also be represented by personalized ontologies. Tao *et al.* [107, 108], Gauch *et al.* [31], Trajkova and Gauch [113], and Sieg *et al.* [104] represented user profiles by a sub-taxonomy of a predefined hierarchy of concepts. The concepts exist in the concepts are associated with weights indicating the user-perceived interests in these concepts. This kind of user profiles describes user interests explicitly. The concepts specified in user profiles have clear definitions and extents. They are thus excellent for inferences performed to capture user information needs. However, clearly specifying user interests in ontologies is a

difficult task, especially for their semantic relations, such as *is-a* and *part-of*. In these aforementioned works, only Tao *et al.* [107, 108] could emphasis these semantic relations in user interest specification.

User profiles can also be represented by a training set of documents, as the user profiles in TREC-11 Filtering Track [89] and the model proposed by Tao *et al.* [106] for acquiring user profiles from the Web. User profiles (the training sets) consist of positive documents that contain user interest topics, and negative documents that contain ambiguous or paradoxical topics. This kind of user profiles describes user interests implicitly, and thus have great flexibility to be used with any concept extraction techniques. The drawback is that noise may be extracted from user profiles as well as meaningful and useful concepts. This may cause an information overloading problem in Web information gathering.

3.2. User Profile Construction

When acquiring user profiles, the content, life cycle, and applications need to be considered [97]. Although user interests are approximate and explicit, it was argued by [31, 57, 107, 108] that they can be specified by using ontologies. The life cycle of user profiles refers to the period that the user profiles are valuable for Web information gathering. User profiles can be long-term or short-term. For instance, persistent and ephemeral user profiles were built by Sugiyama *et al.* [105], based on the long term and short term observation of user behaviour. Applications are also important factors requiring consideration in user profile acquisition. These factors considered in user profile acquisition also define the utilization of user profiles for their contributing areas and period.

User profile acquisition techniques can be categorized into three groups: *interviewing, non-interviewing*, and *semi-interviewing* techniques. Interviewing user profiles are entirely acquired using manual techniques; such as questionnaires, interviews, and user classified training sets. Trajkova and Gauch [113] argued that user profiles can be acquired explicitly by asking users questions. One typical model using user-interview profiles acquisition techniques is the TREC-11 Filtering Track model [89]. User profiles are represented by training sets in this model, and acquired by users manually. Users read training documents and assign positive or negative judgements to the documents against given topics. Based upon the assumption that users know their interests and preferences exactly, these training documents perfectly reflect users' interests. However, this kind of user profile acquisition mechanism is costly. Web users have to invest a great deal of effort in reading the documents and providing their opinions and judgements. However, it is unlikely that Web users wish to burden themselves with answering questions or reading many training documents in order to elicit profiles [55, 57].

The non-interviewing techniques do not involve users directly but ascertain user interests instead. Such user profiles are usually acquired by observing and mining knowledge from user activity and behaviour [57, 101, 105, 113]. Typical model is the personalized, ontological user profiles acquired by [108] using a world knowledge base and user local instance repositories. Some other works, like [31, 113] and [104], acquire non-interviewing ontological user profiles by using global categorizations such as Yahoo! categorization and Online Directory Project. The machine-learning techniques are utilized to analyse the user-browsed Web documents, and classification techniques are used to classify the documents into the concepts specified in the global categorization. As a result, user profiles in these models are a sub-taxonomy of the global categorizations. However, because the categorizations used are not well-constructed ontologies, the user profiles acquired in these models cannot describe the specific semantic relations. Instead of classifying interesting documents into the supervised categorizations, Li and Zhong [55, 57] used unsupervised methods to discover interesting patterns from the user-browsed Web documents, and illustrated the patterns to represent user profiles in ontologies. The model developed by [67] acquired user profiles adaptively, based on the content study of user queries and online browsing history. In order to acquire user profiles, Chirita *et al.* [17] and Teevan *et al.* [111] extracted user interests from the collection of user desktop information such as text documents, emails, and cached Web pages. Makris *et al.* [70] comprised user profiles by a ranked local set of categories and then utilized Web pages to personalize search results for a user. These non-interviewing techniques, however, have a common limitation of ineffectiveness. Their user profiles usually contain much noise and uncertainties because of the use of automatic acquiring techniques.

With the aim of reducing user involvement and improving effectiveness, semi-interviewing user profiles are acquired by semi-automated techniques. This kind of user profiles may be deemed as that acquired by the hybrid mechanism of interviewing and non-interviewing techniques. Rather than providing users with documents to read, some approaches annotate the documents first and attempt to seek user feedback for just the annotated concepts. Because annotating documents may generate noisy concepts, global knowledge bases are used by some user profile acquisition approaches. They extract potentially interesting concepts from the knowledge bases and then explicitly ask users for feedback, like the model proposed by [107]. Also, by using a so-called Quickstep topic ontology, Middleton *et al.* [76] acquired user profiles

from unobtrusively monitored behaviour and explicit relevance feedback. The limitation of semi-interviewing techniques is that they largely rely upon knowledge bases for user background knowledge specification.

In recommender systems, the construction of accurate profiles is a key task since accurate profiles enable both content-based filtering (to insure recommendations are appropriate) and collaborative filtering (to insure users with similar profiles are indeed similar) [72]. Current existing user profiling for the recommender systems is mainly using user rating data and selected items' content. Usually, hundreds of thousands of users and items are involved in a recommender system, but only a few items are viewed, selected or rated by users. As Sarwar et.al reported in [95], the density of the available ratings in commercial recommender systems is often less than 1%. Moreover, as for new users, they will start with a blank profile without selecting or rating any items at all. These situations are commonly referred to as the data **sparseness** and **cold start** problem [96]. With the increasing use of recommender systems in e-commerce and social networks, maliciously or unfairly influences to the outcomes of recommender systems by creating false user rating data are also intensified. For example: a simple but effective attack to recommender system is to deliberately create a bunch of fake users with pseudo ratings favour or disfavour to some particular products. With the fake information, user profile data becomes unreal and not reliable hence recommender algorithms are impeded by the sparsity, cold start and malicious data problems.

The user profile information can be input explicitly by users or implicitly gathered by software agents that monitor user activity [32]. Explicit acquisition techniques usually require information such as how users rate or select items; whereas implicit acquisition techniques passively observe user behaviours to discover user interests by inferring from user-system interactions. Currently the user profile information for online recommendation is mainly obtained by analysing usage log data such as users' click streams and navigation patterns. Both the explicit and implicit methods have their respective strengths and weaknesses. Explicit techniques are capable of constructing accurate user profile, because information comes directly from the users (e.g., a user rates the relevance of a set of items). However, they may place an increased cognitive burden on users [79]. Implicit acquisition techniques place little or no burden on the users. However, inferences drawn from user interaction are not always valid, as the indicators of user interests are often erratic [45].

In the Web 2.0 era, people engage in a growing variety and number of Web activities on social websites, from buying on commercial sites, to blogging, to tagging, to online dating, to twittering, to post personal pictures. These interactions can serve as a valuable source of the usersï£¡ implicit feedback. For example, the tags are pieces of light weight textual information but contain very rich and explicit topic information because users proactively provided these tags. They are independent of the content of the items, which makes it possible to achieve content filtering for any items such as video or music files [140]. For example, in recent years, Liang et.al. [63, 64] developed the personalized item recommendation systems by using tag and item information .

4. Information Gathering and Recommender Systems in Social Web

Although the term "social networking" is being used in new ways since the availability of the digital medium, the concepts behind it have been studied for quite a long time. The modern digital medium technology makes sharing contents, collaborating with others, and connecting with each other to create a community faster, easier and more accessible to a wider population than ever before. Social networking is a type of virtual community that has grown tremendously in popularity over the past few years. The network of users is the platform; the community drives the content. Users actively participate in social networks, upload their personal photos, share their bookmarks, write blogs, and annotate and comment on the information provided by others. They create information, build content, and establish online communities. Nowadays, massive quantities of User Generated Content (UGC) on social networks are available.

Unlike the user rating data which is numeric, the UGC comprises various forms of media and creative works such as text, audio, visual, and combined created by users explicitly and pro-actively. Therefore, it contains rich semantic information and provides a huge potential to obtain deep knowledge about users, items, the various relationships among users and items. The UGC has become an important information resource in addition to traditional website materials. From the UGC information, it is possible to acquire users' opinions, perspectives, or tastes towards items or other users. The growing and readily available user-generated content is rising the new opportunity to construct user profiles accurately compared with the existing personalized recommender techniques, as well as to mitigate the cold start and malicious rating problems considerably.

There has been a tremendous increase in user-generated content (UGC) in the past a few years via the technologies of Web 2.0. It is now well recognized that the user-generated content (e.g., product reviews, tags, forum discussions and blogs) contains valuable user opinions that can be exploited for many applications. By exploiting the UGC more effectively via the use of the latest collaborative filtering, data mining

techniques, more accurate and sophisticate user profiles can be built. Based on the enhanced user profiles, high quality and reliable recommendations can be generated. Many significant researches have been done to investigate new strategies available in Web 2.0 framework. In this section, we review some new strategies for social recommender systems.

4.1. Using Tag Information for Recommendation

Like other UGC information, tag is becoming an important information source to profile user's topic interests as well as to describe the content or classification of items. A tag is a keyword that is added to a digital object (e.g. a website, picture or video clip) to describe it, but not as part of a formal classification system. Tags are freely chosen keywords and they are a simple yet powerful tool for organizing, searching and exploring the resources. Compared with other traditional implicit user information such as click stream and web log, the tag information has some distinctive advantages. One important advantage is that tags are pieces of light weighted textual information but contain very rich and explicit topic information since they are given by users explicitly and pro-actively. Another important advantage is that it is independent with the content of the items, which makes it possible to do content filtering for any items such as videos, music files etc. Moreover, tagging behaviour forms a three dimensional relationship among users, items and tags such as the additional implied item-tag, user-tag besides the typical implicit user-item relationship.

However, since there is no restriction or boundary on selecting words for tagging items, the tags used by users are free-style and contain a lot of noise such as semantic ambiguity which means that the same tag name has different meaning for different users, tag synonym which means that different tags actually have the same meaning. Another serious situation of tags is that nearly 60% tags are personal tags that are only used by one user [99]. All these disadvantages of tags bring challenges to make use of tags to profile users' topic preferences accurately or describe the topics of the items correctly. Thus, how to solve these problems caused by the free-style vocabulary of tags is a key issue to improve the accuracy of recommendation systems based on tag information.

The work of Tso-shuter *et al.* [114] extended the user-item matrix to user-item-tag matrix to make collaborative filtering item recommendation. However, this work didn't consider the noise of tags. More recently, the noise of tags has become an important research question. In the recent work of [99], a special tag rating function was used to find user's preferences for tags. Along with the tag preferences, the click streams, tag search history of each user were used to get user's preferences for items through the inferred tags preferences. However, Sen' work needs various kinds of extra information or special function, which makes the work incomparable and gives restrictions to the application of the work. Moreover, it is difficult to determine the influence of tag information when the click streams, search queries were combined together.

Different from Sen's work the approach proposed by Liang *et al.* [63] makes use of the standard item taxonomy or ontology given by experts to represent each user's tag individually to remove the noise of tags. Item taxonomy is a set of controlled vocabulary terms or topics designed to describe or classify items, which is available for various domains. Because item taxonomy is usually designed and developed by experts, reflecting the common views to the description and classification of items, providing not only a standard vocabulary but also a hierarchical structure to represent the relationships among concepts or categories, it can be used to eliminate the inaccuracy caused by the users' free-style vocabulary in social tags.

4.2. Blogs as a Valuable Information Resource

Social media such as *blog*, *Flickr*, *Youtube*, *Facebook*, and *Twitter* has arisen as the major user-generated media platform in recent years. A blog is a simple web page consisting of brief paragraphs of opinion, information, personal diary entries, or links. People express their opinions, ideas, experiences, thoughts, wishes through these free-form writings. A typical blog post can combine text, images, and links to other blogs, web pages, and other media related to its topic. The individual users show their interest in online opinions about products or services. They share their brand experiences and opinions, positive or negative, regarding any product or service. The vendors of these items are increasingly coming to realize that these consumer voices can potentially wield enormous influence in shaping the opinions of other consumers and they are paying more and more attention to these issues [38]. Currently, many sentiment analysis works are focus on product reviews or movie review [121], [134], [85], [8] on blogs, customer review sites, and Web Pages. The opinion mining and sentiment analysis such as customer opinion summarisation [141] and sentiment analysis of user reviews [21] are possibly as augmentations to recommendation systems [110], since it might behoove such a system not to recommend items that receive a lot of negative feedback. The researchers Joshi and Belsare [44] developed a blog mining program called *BlogHarvest*, which searches for, and extracts, a blogger's interests in order to recommend blogs with similar topics. The program uses classification, links, topic similarity clustering and tagging based on opinion

mining to provide these features. The program design is based on the knowledge that blogging communities are not formed randomly, but as a result of shared interests. It is also designed to provide a useful search facility to bloggers while generating large amounts of revenue for advertising services and providers.

4.3. Microblogs as Real Time Information Resource

Twitter is one of the micro-blog service providers that achieves great success. Twitter is a micro-blogging service where users send messages (a.k.a., tweets) to a network of associates from a variety of devices. A tweet is a text-based post and only has 140 characters, which is approximately the length of a typical newspaper headline and subhead [77]. The short messages are very easy and convenience to both sender and reader to share things of interest and communicate their thoughts anywhere and anytime in the world. Today, Twitter has gained popularity with over 200 million users and averaging 1600 tweets sent per second. Twitter consists users from different fields including celebrities (@ladygaga, @justinbieber), national leaders (@barackobama, @kevinrudd), news publishers (@cnn, @ap) to general public. Twitter's user base has grown rapidly and the volumes of messages produced by Twitter everyday is vast. According to [131], in April 2010, Twitter had 106 million registered users, 180 million unique visitors everyday. Users often perform search task in microblog (for example, Twitter) to answer their information need [25]. Searching in micro-blog can be different as compared to traditional web search in the following aspects [112]:

1. Users search twitter for information about people and temporal related information.

2. Twitter search is less varying and can be used to monitor content while web search is used to gain knowledge about a topic.

3. Twitter provides more social content and event-related information while web results are more factual and navigational.

4. Language used in Twitter and Web result is very different.

As more and more users post reviews about products and services they use, or express their political and religious views on Twitter, tweets become a valuable source of peoples opinions and sentiments. Tweets data can be efficiently used to infer people's opinions for marketing or social studies. Given its popularity, Twitter is seen as a potential new form of eWOM (electronic word-of-mouth) marketing by the businesses and organizations concerned with reputation management [42]. Twitter has also been witnessed as the major online platform where news of significant events were broken such as presidential election debate [20], earthquake [94] and the death of Michael Jackson [35]. Twitter is also used as the primary tool in critical situations when communication channel is limited such as the Iranian election 2009 [10] and Mumbai terrorist attack [84]. Micro-blog, *Twitter* and *Facebook* in particular, has become an important tool for users to share various information from personal updates, question answering [26], news [50] to general babbling [6].

One unique characteristic of Twitter is its rapid response to the change of the Twitter sphere. While it cannot be considered as a reliable information source as compared to authoritative media outlets, its ability to gather emerging topics is impressive. This can be achieved by performing trend analysis and topic detection. Naaman *et al.*[80] analyse the characteristics of emerging trends on Twitter and identify two types of trends: exogenous (broadcast events, global news, important days, physical events) and endogenous (memes, retweets, fan activities). The study also presents five key features: content, interaction, participation, time and social, to collect content aggregation statistics for trend analysis. [12] proposed a 5-steps process to model life cycle of term using a novel aging theory based on user authority, calculated using PageRank algorithm. The emerging term selection is based on nutrition (term quality) and energy (term burstiness). Twitter trends can also be mined with data mining technique such as Kohonen's Self-Organising Map (SOM) to visualise users demographic of trending topics, to reveal underlying pattern and characteristics for decision making [14].

An entry point for micro-blog study is to understand the characteristics of Twitter. [50] conducted a study based on 41.7 million users and 106 million tweets to answer whether Twitter is "social network, or a news media?" They extracted 1.47 billion relationship tuples and revealed that users who topped the chart with over 1 million followers are mostly celebrities (e.g. @oprah, @kimkardashian) or mass media (e.g. @cnn, @nytimes). About 77.9% of the relationships are one-way connection with only 22.1% of reciprocal relationship exists. Moreover, 67.6% of users do not even follow any of their followers at all, which shows a very weak social relationship. Interestingly, another study by Weng *et al.*[120] which based on top-1000 Singapore based twitters listed in *Twitterholic.com* shows the opposite from Kwak et al.'s finding. Study performed by Weng *et al.* revealed that 72.4% of users follow more than 80% of their followers and 80.5% of the users have 80% of their friends follow them back. Both studies agree that Twitter is an excellent news

alternatives but the social relationship varies among different user groups.

Apart from trending topic identification, topic modelling can also be used to understand tweet content. Latent Dirichlet allocation (LDA) is one of the popular techniques for its performance and flexibility [5]. LDA is a generative topic model that presents underlying topics as a set of infinite mixture. Each document is considered as a probability distribution of topics and their probabilities can then estimated through sampling methods. Alternatively, Ramage *et al.*[87] uses an variation of LDA, namely Labelled LDA (L-LDA) to model tweets for post ranking and user recommendation task. L-LDA allows mixing labelled topics together with latent topics discovered by original LDA. While it is questionable of the performance of bag-of-words (BOW) model such as TF-IDF in micro-blogs retrieval task, this study shows that performance of TF-IDF and L-LDA are almost identical in ranking task and TF-IDF actually outperforms L-LDA in user recommendation task by about 30%. A combination of L-LDA and TF-IDF improves the result of tweets ranking by a slight 3% but significantly boost the performance in user recommendation task by 66%. This shows that TF-IDF can still be an important feature for micro-blogs task.

Conversely, various studies suggested that LDA may not work on Twitter due to the short length of tweets [39, 120]. One way to overcome such problem is to group tweets together to provide more context. Tweets can be grouped by content and terms [39], or by topics [120] or based on users, as an application of author-topic (AT) model [91]. However, studies show that direct application of AT model does not yield significant improvement as compared to simple term-based approach [39, 135]. A Twitter-LDA model proposed by [135], which considers "a single tweet is usually about a single topic" also shows that content aggregation in Twitter performs better compared to author based aggregation. This might be due to less variation in content-based aggregation than author-based.

While information search in micro-blog is important, its research is still budding. Preliminary works includes the understanding the search type [36] , investigation of hash-tag based retrieval [24], Researchers begin to identify ways to deal with the short length of micro-blogs [81] and query expansion to capture more context [74]. TREC 2011 has also created a dedicated micro-blog track to tackle various ad hoc micro-blogs retrieval problems.

Opinion mining in Twitter is different from the opinion mining from the blogs, review sites or other Web pages. Reviews tend to be longer and more verbose than tweets which may only be a few words long and often contain significant spelling errors. Reviews usually focus on a specific product or entity and contains little irrelevant information. However, tweets tend to be much more diverse in terms of topics with issues ranging from politics and recent news to religion. As the largest, most well-known, and most popular of the micro-blogging sites, Twitter is an ideal sources for spotting the information about societal interest and general people's opinions. However, there has been little prior opinion mining work in the micro-blogging area since Twitter is relative new technology.

5. Conclusions

This paper has discussed the challenges existing in the current Web information gathering and recommender systems, as well as state-of-the-art techniques employed by them to deal with the challenges. In the recent years, much effort has been dedicated by many research groups on effectively accessing web information. Their results have demonstrated that the key to deal with the challenges is moving the current systems towards concept-based and personalised. Moving on this trend, the researchers have made many great achievements, particularly in personalised information gathering and recommender systems. However, many challenging problems still exist in these areas, for example, how to make breakthrough on the current information gathering and recommender systems on social networks.

Some recent study on micro-blog has also been covered in this paper, including the micro-blog background, application, trend analysis, topic detection, opinion mining, and information gathering. While many of the studies performed on micro-blog related problems are still in its early stage, the preliminary results are promising. Many of the problems are studied extensively in traditional information retrieval field, but the technical difficulty and its applicability when applied in micro-blog are still uncertain. This is due to the short and dynamic nature of micro-blog and the challenge in gathering context. Substantial amount of noise always exist in micro-blogs, but the nature of topic-specific for non-spam messages are definitely indicative and expressive for various micro-blog related task.

The future research direction of information gathering and recommender systems on social networking environment is exciting and vibrant. Network analysis and trend detection models that exploit various twitter characteristics, will be beneficial from the large volume of tweets. Data mining techniques (e.g. pattern mining, association rules) can be applied to further improve the retrieval performance. Lastly, high-level semantic feature such as sentiment analysis and entity detection can then be applied in different real-world applications,

which will fully unleash the power of social networks as a valuable, wealthy source of information.

References

[1] G. Adomavicius and A. Tuzhilin. Toward the next generation of recommender systems: A survey of the state-of-the-art and possible extensions. *IEEE Transactions on Knowledge and Data Engineering*, 17(6):734–749, 2005.

[2] G. Antoniou and F. van Harmelen. *A Semantic Web Primer*. The MIT Press, 2004.
—— BBB ——

[3] R. Bekkerman and M. Gavish. High-precision phrase-based document classification on a modern scale. In *Proceedings of the 17th ACM SIGKDD international conference on Knowledge discovery and data mining*, KDD '11, pages 231–239, New York, NY, USA, 2011. ACM.

[4] T. Berners-Lee, J. Hendler, and O. Lassila. The semantic Web. *Scientific American*, 5:29–37, 2001.

[5] D. Blei, A. Ng, and M. Jordan. Latent dirichlet allocation. *The Journal of Machine Learning Research*, 3:993–1022, 2003.

[6] D. Boyd, S. Golder, and G. Lotan. Tweet, tweet, retweet: Conversational aspects of retweeting on twitter. In *Proceedings of the 2010 43rd Hawaii International Conference on System Sciences*, HICSS '10, pages 1–10, Washington, DC, USA, 2010. IEEE Computer Society.

[7] J. S. Breese, D. Heckerman, and C. Kadie. Empirical analysis of predictive algorithms for collaborative filtering. pages 43–52, 1998.

[8] S. Brody and N. Elhadad. An unsupervised aspect-sentiment model for online reviews. In *HLT '10: Human Language Technologies: The 2010 Annual Conference of the North American Chapter of the Association for Computational Linguistics*, pages 804–812, Morristown, NJ, USA, 2010. Association for Computational Linguistics.

[9] A. Budanitsky and G. Hirst. Evaluating WordNet-based measures of lexical semantic relatedness. *Computational Linguistics*, 32(1):13–47, 2006.

[10] A. Burns and B. Eltham. Twitter free iran: an evaluation of twitterâĂŹs role in public diplomacy and information operations in iranâĂŹs 2009 election crisis. In *Record of the Communications Policy & Research Forum*, pages 298–310, 2009.

[11] F. Camous, S. Blott, and A. F. Smeaton. Ontology-based medline document classification. In *Proceedings of the 1st international conference on Bioinformatics research and development*, BIRD'07, pages 439–452, Berlin, Heidelberg, 2007. Springer-Verlag.

[12] M. Cataldi, L. Di Caro, and C. Schifanella. Emerging topic detection on twitter based on temporal and social terms evaluation. In *Proceedings of the Tenth International Workshop on Multimedia Data Mining*, MDMKDD '10, pages 4:1–4:10, New York, NY, USA, 2010. ACM.

[13] S. Cederberg and D. Widdows. Using lsa and noun coordination information to improve the precision and recall of automatic hyponymy extraction. In *Proceedings of the seventh conference on Natural language learning at HLT-NAACL 2003*, pages 111–118, Morristown, NJ, USA, 2003. Association for Computational Linguistics.

[14] M. Cheong and V. Lee. Integrating web-based intelligence retrieval and decision-making from the twitter trends knowledge base. In *SWSM '09: Proceeding of the 2nd ACM workshop on Social web search and mining*, pages 1–8, New York, NY, USA, 2009. ACM.

[15] L. M. Chan. *Library of Congress Subject Headings: Principle and Application*. Libraries Unlimited, 2005.

[16] P. A. Chirita, W. Nejdl, R. Paiu, and C. Kohlschütter. Using ODP metadata to personalize search. In *Proceedings of the 28th annual international ACM SIGIR conference on Research and development in information retrieval*, pages 178–185. ACM Press, 2005.

[17] P. A. Chirita, C. S. Firan, and W. Nejdl. Personalized query expansion for the Web. In *Proceedings of the 30th annual international ACM SIGIR conference on Research and development in information retrieval*, pages 7–14, 2007.

[18] H. Cui, J.-R. Wen, J.-Y. Nie, and W.-Y. Ma. Probabilistic query expansion using query logs. In *Proceedings of the 11th international conference on World Wide Web*, pages 325–332. ACM Press, Honolulu, Hawaii, USA, 2002.

[19] B. V. Dasarathy, editor. *Nearest Neighbor (NN) Norms: NN Pattern Classification Techniques*. Los Alamitos: IEEE Computer Society Press, 1990.

[20] N. Diakopoulos and D. Shamma. Characterizing debate performance via aggregated twitter sentiment. In *Proceedings of the 28th international conference on Human factors in computing systems*, pages 1195–1198. ACM, 2010.

[21] X. Ding, B. Liu, and P. S. Yu. A holistic lexicon-based approach to opinion mining. In *Proceedings of the Conference on Web Search and Web Data Mining (WSDM)*, 2008.

[22] D. Dou, G. Frishkoff, J. Rong, R. Frank, A. Malony, and D. Tucker. Development of neuroelectromagnetic ontologies(NEMO): a framework for mining brainwave ontologies. In *Proceedings of the 13th ACM SIGKDD international conference on Knowledge discovery and data mining*, pages 270–279, 2007.

[23] Z. Dou, R. Song, and J.-R. Wen. A large-scale evaluation and analysis of personalized search strategies. In *WWW '07: Proceedings of the 16th international conference on World Wide Web*, pages 581–590, New York, NY, USA, 2007. ACM Press.

[24] M. Efron. Hashtag retrieval in a microblogging environment. In *Proceeding of the 33rd international ACM SIGIR conference on Research and development in information retrieval*, SIGIR '10, pages 787–788, New York, NY, USA, 2010. ACM.

[25] M. Efron. Information search and retrieval in microblogs. *Journal of the American Society for Information Science and Technology*, 62(6):996–1008, 2011.

[26] M. Efron and M. Winget. Questions are content: A taxonomy of questions in a microblogging environment. *Proceedings of the American Society for Information Science and Technology*, 47(1):1–10, 2010.

[27] B. Espinasse, S. Fournier, and F. Freitas. Agent and ontology based information gathering on restricted web domains with AGATHE. In *Proceedings of the 2008 ACM symposium on Applied computing*, pages 2381–2386, Brazil, 2008.

[28] C. Fellbaum, editor. *WordNet: An Electronic Lexical Database*. ISBN: 0-262-06197-X. MIT Press, Cambridge, MA, 1998.

[29] G. P. C. Fung, J. X. Yu, H. Lu, and P. S. Yu. Text classification without negative examples revisit. *IEEE Transactions on Knowledge and Data Engineering*, 18(1):6–20, January 2006.

[30] E. Gabrilovich and S. Markovitch. Feature generation for text categorization using world knowledge. In *Proceedings of The Nineteenth International Joint Conference for Artificial Intelligence*, pages 1048–1053, Edinburgh, Scotland, 2005.

[31] S. Gauch, J. Chaffee, and A. Pretschner. Ontology-based personalized search and browsing. *Web Intelligence and Agent Systems*, 1(3-4):219–234, 2003.

[32] S. Gauch, M. Speretta, A. Chandramouli, and A. Micarelli. User profiles for personalized information access. *The Adaptive Web*, Volume 4321/2007, pp54-89, 2007.

[33] R. Girju, A. Badulescu, and D. Moldovan. Automatic discovery of part-whole relations. *Comput. Linguist.*, 32(1):83–135, 2006.

[34] E. J. Glover, K. Tsioutsiouliklis, S. Lawrence, D. M. Pennock, and G. W. Flake. Using Web structure for classifying and describing Web pages. In *WWW '02: Proceedings of the 11th international conference on World Wide Web*, pages 562–569, New York, NY, USA, 2002. ACM Press.

[35] D. Goh and C. Lee. An analysis of tweets in response to the death of michael jackson. In *Aslib Proceedings*, volume 63, pages 432–444. Emerald Group Publishing Limited, 2011.

[36] G. Golovchinsky and M. Efron. Making sense of twitter search. 2010.

[37] J. Han and K.-C. Chang. Data mining for Web intelligence. *Computer*, 35(11):64–70, 2002.

[38] T. Hoffman. Online reputation management is hot – but is it ethical? ComputerWorld, 2 2008.

[39] L. Hong and B. D. Davison. Empirical study of topic modeling in twitter. In *Proceedings of the First Workshop on Social Media Analytics*, SOMA '10, pages 80–88, New York, NY, USA, 2010. ACM.

[40] X. Hu, X. Zhang, C. Lu, E. K. Park, and X. Zhou. Exploiting wikipedia as external knowledge for document clustering. In *KDD '09: Proceedings of the 15th ACM SIGKDD international conference on Knowledge discovery and data mining*, pages 389–396, New York, NY, USA, 2009. ACM.

[41] D. Inkpen and G. Hirst. Building and using a lexical knowledge base of near-synonym differences. *Computational Linguistics*, 32(2):223–262, 2006.

[42] B. J. Jansen, M. Zhang, K. Sobel, and A. Chowdury. Twitter power: Tweets as electronic word of mouth. *J. Am. Soc. Inf. Sci.*, 60(11):2169–2188, 2009.

[43] T. Joachims. Text categorization with Support Vector Machines: learning with many relevant features. In *Proceedings of the 10th European conference on machine learning*, number 1398, pages 137–142, Chemnitz, DE, 1998. Springer Verlag, Heidelberg, DE.

[44] M. Joshi and N. Belsare. Blogharvest: Blog mining and search framework. In *International Conference on*

Management of Data, Delhi, India, , 2006, December 14-16 2006. Computer Society of India.

[45] D. Kelly and J. Teevan. Implicit feedback for inferring user preference: a bibliography. *SIGIR Forum*, 37(2):18–28, 2003.

[46] J. D. King, Y. Li, X. Tao, and R. Nayak. Mining World Knowledge for Analysis of Search Engine Content. *Web Intelligence and Agent Systems*, 5(3):233–253, 2007.

[47] J. Konstan et al. Grouplens:Applying collaborative filtering to usenet news. *Communications of the ACM*, 40(3) (1997) 77ï£¡87.

[48] H. Kornilakis, M. Grigoriadou, K. Papanikolaou, and E. Gouli. Using WordNet to support interactive concept map construction. In *Proceedings. IEEE International Conference on Advanced Learning Technologies, 2004.*, pages 600–604, 2004.

[49] R. Kosala and H. Blockeel. Web mining research: A survey. *ACM SIGKDD Explorations Newsletter*, 2(1):1–15, 2000.

[50] H. Kwak, C. Lee, H. Park, and S. Moon. What is twitter, a social network or a news media? In *Proceedings of the 19th international conference on World wide web*, pages 591–600. ACM, 2010.

[51] Y. Li. Information fusion for intelligent agent-based information gathering. In *WI '01: Proceedings of the First Asia-Pacific Conference on Web Intelligence: Research and Development*, pages 433–437, London, UK, 2001. Springer-Verlag.

[52] Y. Li and N. Zhong. Interpretations of association rules by granular computing. In *Proceedings of IEEE International Conference on Data Mining, Melbourne, Florida, USA*, pages 593–596, 2003.

[53] Y. Li and N. Zhong. Ontology-based Web mining model. In *Proceedings of the IEEE/WIC International Conference on Web Intelligence, Canada*, pages 96–103, 2003.

[54] Y. Li and N. Zhong. Capturing evolving patterns for ontology-based web mining. In *Proceedings of the 2004 IEEE/WIC/ACM International Conference on Web Intelligence*, pages 256–263, Washington, DC, USA, 2004. IEEE Computer Society.

[55] Y. Li and N. Zhong. Web Mining Model and its Applications for Information Gathering. *Knowledge-Based Systems*, 17:207–217, 2004.

[56] Y. Li, W. Yang, and Y. Xu. Multi-tier granule mining for representations of multidimensional association rules. In *Proceedings of the Sixth IEEE International Conference on Data Mining*, pages 953–958, 2006.

[57] Y. Li and N. Zhong. Mining Ontology for Automatically Acquiring Web User Information Needs. *IEEE Transactions on Knowledge and Data Engineering*, 18(4):554–568, 2006.

[58] Y. Li, X. Zhou, P. Bruza, Y. Xu, and R. Y. Lau. A two-stage text mining model for information filtering. In *CIKM '08: Proceeding of the 17th ACM conference on Information and knowledge management*, pages 1023–1032, New York, NY, USA, 2008. ACM.

[59] Y. Li, S.-T. Wu, and X. Tao. Effective pattern taxonomy mining in text documents. In *CIKM '08: Proceeding of the 17th ACM conference on Information and knowledge management*, pages 1509–1510, New York, NY, USA, 2008. ACM.

[60] Y. Li, A. Algarni, S.-T. Wu, and Y. Xu. Mining negative relevance feedback for information filtering. In *Proceedings of the IEEE/WIC/ACM international conference on Web Intelligence*, pages 606–613, 2009.

[61] Y. Li, A. Algarni, and N. Zhong. Mining positive and negative patterns for relevance feature discovery. In *Proceedings of 16th ACM SIGKDD Conference on Knowledge Discovery and Data Mining*, pages 753–762, 2010.

[62] Y. Li, X. Zhou, P. Bruza, Y. Xu, and R. Y. Lau. A Two-stage Decision Model for Information Filtering. In *Decision Support System*, 52(2012), p706-716.

[63] H. Liang, Y. Xu, Y. Li, R. Nayak, and L. Weng. Personalized recommender systems integrating social tags and item taxonomy. In *Proc. of WI 09*, 2009.

[64] H. Liang, Y. Xu, Y. Li, R. Nayak, and X. Tao Connecting Users and Items with Weighted Tags for Personalized Item Recommendations. In *Proceedings of the 21st ACM conference on Hypertext and hypermedia*, 2010.

[65] S.-Y. Lim, M.-H. Song, K.-J. Son, and S.-J. Lee. Domain ontology construction based on semantic relation information of terminology. In *30th Annual Conference of the IEEE Industrial Electronics Society*, volume 3, pages 2213–2217 Vol. 3, 2004.

[66] B. Liu, Y. Dai, X. Li, W. Lee, and P. Yu. Building text classifiers using positive and unlabeled examples. In *Proceedings of the Third IEEE International Conference on Data Mining, ICDM2003*, pages 179–186, 2003.

[67] F. Liu, C. Yu, and W. Meng. Personalized web search for improving retrieval effectiveness. *IEEE Transactions on Knowledge and Data Engineering*, 16(1):28–40, 2004.

[68] S. Liu, F. Liu, C. Yu, and W. Meng. An effective approach to document retrieval via utilizing WordNet and recognizing phrases. In *SIGIR '04: Proceedings of the 27th annual international ACM SIGIR conference on Research and development in information retrieval*, pages 266–272, New York, NY, USA, 2004. ACM Press.

[69] Z. Ma, G. Pant, and O. R. L. Sheng. Interest-based personalized search. *ACM Transactions on Information Systems (TOIS)*, 25(1):5, 2007.

[70] C. Makris, Y. Panagis, E. Sakkopoulos, and A. Tsakalidis. Category ranking for personalized search. *Data & Knowledge Engineering*, 60(1):109–125, Jan. 2007.

[71] H. H. Malik and J. R. Kender. Classifying high-dimensional text and web data using very short patterns. In *Proceedings of the 2008 Eighth IEEE International Conference on Data Mining*, pages 923–928, Washington, DC, USA, 2008. IEEE Computer Society.

[72] B. Marko and S. Yoav. Fab: Content-Based, Collaborative Recommendation. *Communication of the ACM*, 40(3), pp66-72, 1997

[73] P. Massa and B. Bhattacharjee. Using trust in recommender systems: An experimental analysis. In *In Proceedings of iTrust2004 International Conference*, pages 221–235, 2004.

[74] K. Massoudi, M. Tsagkias, M. de Rijke, and W. Weerkamp. Incorporating query expansion and quality indicators in searching microblog posts. *Advances in Information Retrieval*, pages 362–367, 2011.

[75] D. Meretakis, D. Fragoudis, H. Lu, and S. Likothanassis. Scalable association-based text classification. In *CIKM '00: Proceedings of the ninth international conference on Information and knowledge management*, pages 5–11, New York, NY, USA, 2000. ACM Press.

[76] S. E. Middleton, N. R. Shadbolt, and D. C. D. Roure. Ontological user profiling in recommender systems. *ACM Transactions on Information Systems (TOIS)*, 22(1):54–88, 2004.

[77] S. Milstein, A. Chowdhury, G. Hochmuth, B. Lorica, and R. Magoulas. Twitter and the micro-messaging revolution: Communication, connections, and immediacyÂřx140 characters at a time. An O'Reilly Radar Report . 54 pages, November 2008.

[78] R. Mooney and L. Roy. Information filtering based on user behavior analysis and best match text retrieval. *Proceedings of 5th ACM Conference on Digital Libraries*, pages 195ï£¡204, 2002.

[79] M. Morita and Y. Shinoda. Content-based book recommending using learning for text categorization. *Proceedings of SIGIR '94 ACM*, pages 272–281, 1994.

[80] M. Naaman, H. Becker, and L. Gravano. Hip and trendy: Characterizing emerging trends on twitter. *Journal of the American Society for Information Science and Technology*, 2011.

[81] N. Naveed, T. Gottron, J. Kunegis, and A. C. Alhadi. Searching microblogs: coping with sparsity and document quality. In *Proceedings of the 20th ACM international conference on Information and knowledge management*, CIKM '11, pages 183–188, New York, NY, USA, 2011. ACM.

[82] R. Navigli, P. Velardi, and A. Gangemi. Ontology learning and its application to automated terminology translation. *Intelligent Systems, IEEE*, 18:22–31, 2003.

[83] N. Nguyen and R. Caruana. Classification with partial labels. In *KDD '08: Proceeding of the 14th ACM SIGKDD international conference on Knowledge discovery and data mining*, pages 551–559, New York, NY, USA, 2008. ACM.

[84] O. Oh, M. Agrawal, and H. Rao. Information control and terrorism: Tracking the mumbai terrorist attack through twitter. *Information Systems Frontiers*, 13:33–43, 2011. 10.1007/s10796-010-9275-8.

[85] A.-M. Popescu and O. Etzioni. Extracting product features and opinions from reviews. In *HLT '05: Proceedings of the conference on Human Language Technology and Empirical Methods in Natural Language Processing*, pages 339–346, Morristown, NJ, USA, 2005. Association for Computational Linguistics.

[86] G. Qiu, K. Liu, J. Bu, C. Chen, and Z. Kang. Quantify query ambiguity using odp metadata. In *SIGIR '07: Proceedings of the 30th annual international ACM SIGIR conference on Research and development in information retrieval*, pages 697–698, New York, NY, USA, 2007. ACM Press.

[87] D. Ramage, S. Dumais, and D. Liebling. Characterizing microblogs with topic models. In *International AAAI Conference on Weblogs and Social Media*. The AAAI Press, 2010.

[88] D. Ravindran and S. Gauch. Exploiting hierarchical relationships in conceptual search. In *Proceedings of the 13th ACM international conference on Information and Knowledge Management*, pages 238–239, New York, USA, 2004. ACM Press.

[89] S. E. Robertson and I. Soboroff. The TREC 2002 filtering track report. In *Text REtrieval Conference*, 2002.

[90] P. Resnick, N. Iacovou, M. Suchak, P. Bergstrom, and J. Riedl. Grouplens: An open architecture for collaborative filtering of netnews. In *CSCW*, pages 175–186, 1994.

[91] M. Rosen-Zvi, T. Griffiths, M. Steyvers, and P. Smyth. The author-topic model for authors and documents. In *Proceedings of the 20th conference on Uncertainty in artificial intelligence*, pages 487–494. AUAI Press, 2004.

[92] D. A. Ross and R. S. Zemel. Learning parts-based representations of data. *The Journal of Machine Learning Research*, 7:2369–2397, 2006.

[93] M. Ruiz-Casado, E. Alfonseca, and P. Castells. Automatising the learning of lexical patterns: An application to the enrichment of WordNet by extracting semantic relationships from Wikipedia. *Data & Knowledge Engineering*, 61(3):484–499, June 2007.

[94] T. Sakaki, M. Okazaki, and Y. Matsuo. Earthquake shakes twitter users: real-time event detection by social sensors. In *WWW '10: Proceedings of the 19th international conference on World wide web*, pages 851–860, New York, NY, USA, 2010. ACM. p851-sakaki.pdf.

[95] B. Sarwar, G. Karypis, J. Konstan, and J. Reidl. Item-based collaborative filtering recommendation algorithms. In *Proceedings of the 10th international conference on World Wide Web*, WWW '01, pages 285–295, New York, NY, USA, 2001. ACM.

[96] A. I. Schein, A. Popescul, L. H. Ungar, and D. M. Pennock. Methods and metrics for cold-start recommendations. In *SIGIR '02: Proceedings of the 25th annual international ACM SIGIR conference on Research and development in information retrieval*, pages 253–260, New York, NY, USA, 2002. ACM.

[97] J. Schuurmans, B. de Ruyter, and H. van Vliet. User profiling. In *CHI '04: CHI '04 extended abstracts on Human factors in computing systems*, pages 1739–1740, New York, NY, USA, 2004. ACM Press.

[98] F. Sebastiani. Machine learning in automated text categorization. *ACM Computing Surveys (CSUR)*, 34(1):1–47, 2002.

[99] S. Sen, J. Vig, and J. Riedl. Tagommenders: Connecting users to items through tags. In *Proc. of WWW' 09*, pages 671–680, 2009.

[100] G. Shaw, Y. Xu, and S. Geva. Deriving non-redundant approximate association rules from hierarchical datasets. In *CIKM '08: Proceeding of the 17th ACM conference on Information and knowledge management*, pages 1451–1452, New York, NY, USA, 2008. ACM.

[101] X. Shen, B. Tan, and C. Zhai. Implicit user modeling for personalized search. In *CIKM '05: Proceedings of the 14th ACM international conference on Information and knowledge management*, pages 824–831, New York, NY, USA, 2005. ACM Press.

[102] M. A. Shepherd, A. Lo, and W. J. Phillips. A study of the relationship between user profiles and user queries. In *Proceedings of the 8th annual international ACM SIGIR conference on Research and development in information retrieval*, pages 274–281, 1985.

[103] K. Shinzato and K. Torisawa. Extracting hyponyms of prespecified hypernyms from itemizations and headings in web documents. In *COLING '04: Proceedings of the 20th international conference on Computational Linguistics*, page 938, Morristown, NJ, USA, 2004. Association for Computational Linguistics.

[104] A. Sieg, B. Mobasher, and R. Burke. Web search personalization with ontological user profiles. In *Proceedings of the sixteenth ACM conference on Conference on information and knowledge management*, pages 525–534, New York, NY, USA, 2007. ACM.

[105] K. Sugiyama, K. Hatano, and M. Yoshikawa. Adaptive web search based on user profile constructed without any effort from users. In *Proceedings of the 13th international conference on World Wide Web*, pages 675–684, 2004.

[106] X. Tao, Y. Li, N. Zhong, and R. Nayak. Automatic Acquiring Training Sets for Web Information Gathering. In *Proceedings of the 2006 IEEE/WIC/ACM International Conference on Web Intelligence*, pages 532–535, HK, China, 2006.

[107] X. Tao, Y. Li, N. Zhong, and R. Nayak. Ontology mining for personalized web information gathering. In *Proceedings of the 2007 IEEE/WIC/ACM International Conference on Web Intelligence*, pages 351–358, 2007.

[108] X. Tao, Y. Li, N. Zhong, and R. Nayak. An ontology-based framework for knowledge retrieval. In *Proceedings of the 2008 IEEE/WIC/ACM International Conference on Web Intelligence*, pages 510–517, 2008.

[109] X. Tao, Y. Li, and N. Zhong. A personalized ontology model for web information gathering. *IEEE Transactions on Knowledge and Data Engineering, IEEE computer Society Digital Library. IEEE Computer Society*, 23(4):496–511, 2011.

[110] J. Tatemura. Virtual reviewers for collaborative exploration of movie reviews. In *Proceedings of Intelligent User Interfaces (IUI)*, pages 272–275, 2000.

[111] J. Teevan, S. T. Dumais, and E. Horvitz. Personalizing search via automated analysis of interests and activities. In *Proceedings of the 28th annual international ACM SIGIR conference on Research and development in information retrieval*, pages 449–456, 2005.

[112] J. Teevan, D. Ramage, and M. Morris. #twittersearch: a comparison of microblog search and web search. In *Proceedings of the fourth ACM international conference on Web search and data mining*, pages 35–44. ACM, 2011.

[113] J. Trajkova and S. Gauch. Improving ontology-based user profiles. In *Proceedings of RIAO 2004*, pages 380–389, 2004.

[114] Tso-Sutter, K.H.L., L. Marinho, and L. Schmidt-Thieme. Tag-aware recommender systems by fusion of collaborative filtering algorithms. In *Proc. of Applied Computing*, pages 1995–1999, 2008.

[115] P. Velardi, P. Fabriani, and M. Missikoff. Using text processing techniques to automatically enrich a domain ontology. In *FOIS '01: Proceedings of the international conference on Formal Ontology in Information Systems*, pages 270–284, New York, NY, USA, 2001. ACM Press.

[116] G. Varelas, E. Voutsakis, P. Raftopoulou, E. G. Petrakis, and E. E. Milios. Semantic similarity methods in WordNet and their application to information retrieval on the Web. In *WIDM '05: Proceedings of the 7th annual ACM international workshop on Web information and data management*, pages 10–16, New York, NY, USA, 2005. ACM Press.

[117] J. Wang and N. Ge. Automatic feature thesaurus enrichment: extracting generic terms from digital gazetteer. In *JCDL '06: Proceedings of the 6th ACM/IEEE-CS joint conference on Digital libraries*, pages 326–333, New York, NY, USA, 2006. ACM.

[118] J. Wang and M. C. Lee. Reconstructing DDC for interactive classification. In *Proceedings of the sixteenth ACM conference on Conference on information and knowledge management*, pages 137–146, New York, NY, USA, 2007. ACM.

[119] P. Wang and C. Domeniconi. Building semantic kernels for text classification using wikipedia. In *KDD '08: Proceeding of the 14th ACM SIGKDD international conference on Knowledge discovery and data mining*, pages 713–721, New York, NY, USA, 2008. ACM.

[120] J. Weng, E. Lim, J. Jiang, and Q. He. Twitterrank: finding topic-sensitive influential twitterers. In *Proceedings of the third ACM international conference on Web search and data mining*, pages 261–270. ACM, 2010.

[121] D. T. Wijaya and S. Bressan. A random walk on the red carpet: rating movies with user reviews and pagerank. In *CIKM '08: Proceeding of the 17th ACM conference on Information and knowledge management*, pages 951–960. ACM, 2008.

[122] S.-T. Wu, Y. Li, Y. Xu, B. Pham, and C. P. Automatic pattern taxonomy exatraction for web mining. In *Proceedings of IEEE/WIC/ACM International Conference on Web Intelligence*, pages 242–248, Beijing, China, 2004.

[123] X. Wu, C. Zhang, and S. Zhang. Efficient mining of both positive and negative association rules. *ACM Transactions on Information Systems (TOIS)*, 22(3):381–405, 2004.

[124] S.-T. Wu, Y. Li, and Y. Xu. Deploying approaches for pattern refinement in text mining. In *Proceedings of the Sixth International Conference on Data Mining*, pages 1157–1161, 2006.

[125] S.-T. Wu. *Knowledge Discovery Using Pattern Taxonomy Model in Text Mining*. PhD thesis, Faculty of Information Technology, Queensland University of Technology, 2007.

[126] Y. Xu and Y. Li. Generating concise association rules. In *CIKM '07: Proceedings of the sixteenth ACM conference on Conference on information and knowledge management*, pages 781–790, New York, NY, USA, 2007. ACM.

[127] W. Yang, Y. Li, J. Wu, and Y. Xu. Granule mining oriented data warehousing model for representations of multidimensional association rules. *International Journal of Intelligent Information and Database Systems*, 2(1):125–145, 2008.

[128] Y. Yang and T. Joachims. Text Categorization. *Scholarpedia*, 3(5):4242, 2008.

[129] B. Yang, J.-T. Sun, T. Wang, and Z. Chen. Effective multi-label active learning for text classification. In *KDD '09: Proceedings of the 15th ACM SIGKDD international conference on Knowledge discovery and data mining*, pages 917–926, New York, NY, USA, 2009. ACM.

[130] T. Yang, R. Jin, A. K. Jain, Y. Zhou, and W. Tong. Unsupervised transfer classification: application to text

categorization. In *Proceedings of the 16th ACM SIGKDD international conference on Knowledge discovery and data mining*, KDD '10, pages 1159–1168, New York, NY, USA, 2010. ACM.

[131] J.Yarow.   Twitter   finally   reveals   all   its secret   stats.   BusinessInsider   Weblog   Article, http://www.businessinsider.com/twitter-stats-2010-4/, 04 2010.

[132] Z. Yu, Z. Zheng, S. Gao, and J. Guo.   Personalized information recommendation in digital library domain based on ontology. In *IEEE International Symposium on Communications and Information Technology, 2005. ISCIT 2005.*, volume 2, pages 1249–1252, 2005.

[133] L. Yu, S. Wang, and K. K. Lai.   An integrated data preparation scheme for neural network data analysis. *IEEE Transactions on Knowledge and Data Engineering*, 18(2):217–230, 2006.

[134] W. Zhang, C. Yu, and W. Meng. Opinion retrieval from blogs.   In *Proceedings of the sixteenth ACM conference on Conference on information and knowledge management*, CIKM '07, pages 831–840. ACM, 2007.

[135] W. Zhao, J. Jiang, J. Weng, J. He, E. Lim, H. Yan, and X. Li.   Comparing twitter and traditional media using topic models. *Advances in Information Retrieval*, pages 338–349, 2011.

[136] N. Zhong.   Representation   and   construction   of ontologies for Web intelligence. *International Journal of Foundation of Computer Science*, 13(4):555–570, 2002.

[137] N. Zhong.   Toward Web Intelligence.   In *Proceedings of 1st International Atlantic Web Intelligence Conference*, pages 1–14, 2003.

[138] X. Zhou, Y. Li, P. Bruza, Y. Xu, and R. Y. Lau. Pattern taxonomy mining for information filtering. In *AI '08: Proceedings of the 21st Australasian Joint Conference on Artificial Intelligence*, pages 416–422, Berlin, Heidelberg, 2008. Springer-Verlag.

[139] X. Zhou, Y. Li, P. Bruza, Y. Xu, and R. Y. K. Lau. Two-stage model for information filtering. In *WI-IAT '08: Proceedings of the 2008 IEEE/WIC/ACM International Conference on Web Intelligence and Intelligent Agent Technology*, pages 685–689, Sydney, Australia, 2008. IEEE Computer Society.

[140] X. Zhou, Y. Li, P. Bruza, Y. Xu, and R. Y. K. Lau. Pattern Mining for a Two-stage Information Filtering System. In *Proceedings of the 15th Pacific-Asia Conference on Knowledge Discovery and Data Mining (PAKDD2011)*, pages p363-374, Shenzhen, China, 2011.

[141] L. Zhuang, F. Jing, X. Zhu, and L. Zhang.   Movie review mining and summarization.   In *Proceedings of the ACM SIGIR Conference on Information and Knowledge Management (CIKM)*, 2006.

Permissions

List of Contributors

Kai Chen
Institute of Information Engineering, Chinese Academy of Sciences

Jun Shao
Zhejiang Gongshang University

Maryam Najafian Razavi and Denis Gillet
Ecole Polytechnique Fédérale de Lausanne (EPFL), 1015 Lausanne, Switzerland

Rafiqul Haque and Nenad B. Krdzavac
Department of Accounting, Finance, and Information System, College of Business and Law, University College Cork, Cork City, Cork, Ireland

Nayan fang
School of Communication and Information Engineering, University of Electronic Science and Technology of China, Chengdu, China
Tsinghua National Laboratory for Information Science and Technology, Tsinghua University, Beijing, China

Jie Zeng and Xin Su
Tsinghua National Laboratory for Information Science and Technology, Tsinghua University, Beijing, China

Yujun Kuang
School of Communication and Information Engineering, University of Electronic Science and Technology of China, Chengdu, China

Xin Su and Jie Zeng
Tsinghua National Laboratory for Information Science and Technology, Tsinghua University, Beijing, China

Shichao Yu
Tsinghua National Laboratory for Information Science and Technology, Tsinghua University, Beijing, China
University of Electronic Science and Technology of China (UESTC), Chengdu, China

Yujun Kuang
University of Electronic Science and Technology of China (UESTC), Chengdu, China

Raoul Strackx, Pieter Agten, Niels Avonds and Frank Piessens
iMinds-DistriNet - KU Leuven, Celestijnenlaan 200A, 3001 Heverlee, Belgium

Heng Xu
College of Information Sciences and Technology, Pennsylvania State University, University Park, PA 16802, USA

Tamara Dinev
Barry Kaye College of Business, Florida Atlantic University, Boca Raton, FL 33431, USA

Han Li
School of Business, Minnesota State University, Moorhead, MN 56563, USA

Ahmad Y. Javaid and Mansoor Alam
2801 W. Bancroft St., EECS Department, College of Engineering, The University of Toledo, Toledo, Ohio, USA

Weiqing Sun
2801 W. Bancroft St., ET Department, College of Engineering, The University of Toledo, Toledo, Ohio, USA

Ji Zhang
Department of Mathematics and Computing University of Southern Queensland, Australia

GuoYan Meng and QingShan Zhao
Department of Compute Science, Xinzhou teacher University, Xinzhou 034000, Shanxi Province, P. R. China

ChuanLong Wang and XiHong Yan
Department of mathematics, Taiyuan Normal University, Taiyuan 030012, Shanxi Province, P. R. China

Fahad Alarfi and Maribel Fernández
King's College London, Department of Informatics, Strand, London WC2R 2LS, UK

Xiaoxun Sun
Australian Council for Educational Research, Australia

Lili Sun
Department of Mathematics & Computing, University of Southern Queensland, Australia

Daniel Ohene-Kwofie and E. J. Otoo
School Of Electrical and Information Engineering, University Of the Witwatersrand, South Africa

Gideon Nimako
School of Public Health, University Of the Witwatersrand, South Africa

Daouda Ahmat and Damien Magoni
LaBRI, University of Bordeaux, France

Tegawendé F. Bissyandé
SnT, University of Luxembourg, Luxembourg

Sara Motahari, Julia Mayer and Quentin Jones
New Jersey Institute of Technology, Newark, New Jersey, NJ 07103-3513, USA

Yulong Shen
School of Computer Science and Technology, Xidian University, China
State Key Lab. of Integrated Service Network, Xi'an, Shaanxi, China

Yuanyu Zhang
School of Computer Science and Technology, Xidian University, China
School of Systems Information Science, Future University Hakodate, Japan

Xiaohui Tao
Faculty of Sciences, University of Southern Queensland, Australia

Xujuan Zhou, Cher Han Lau and Yuefeng Li
Science and Engineering Faculty, Queensland University of Technology, Australia

Index

A

A Cryptographic Signature, 51

Adaptive Selection, 41-43, 45

Anonymity, 3, 8, 63, 72, 128, 134, 136, 138-140, 143, 146-147, 184-185, 188-193

B

Backtracking, 166-168

Buffer Overflow, 1-2, 48

Business Process, 24-25, 29, 32-33

C

Computer-mediated Communication,187, 189, 191

Cooperative Relay, 194-195

Cryptographic Hardware, 120

D

Data Mining, 86, 108-111, 138, 146-147, 182, 185, 192, 215, 218, 221, 223-226, 228-229

Data Publishing, 136, 138

Density-based Methods, 96, 98

Description Graph, 24-26

Diffie-hellman Algorithm, 166

Direct Anonymous Attestation Protocol, 127

Distance-based Methods, 91

E

Emerging Security Problems, 1

F

Fourier-based Perturbation, 41, 43, 45

G

Grounded Theory, 5-8, 11, 13-14, 22-23

H

High-dimensional Datasets, 86, 100, 103, 106

Histograms, 89-90

I

In-memory Databases, 148, 158

Inference Problem, 181-182, 184-185, 192

Inflo Method, 96-97

Information Gathering, 213-219, 224-226, 228

Information Mismatching, 213

Institutional Trust, 62, 65-66, 70

Intelligent Activity, 24-26, 28-29, 32

J

Jamming Attack, 76, 79, 84

K

Kerdock Codebook, 37, 41-43, 45

Kernel Functions, 90-91

Knn-distance Methods, 92, 95

Knowledge-based Trust, 62, 65

M

Massive Mimo, 35-37, 39, 41-43, 45

Microblogs, 222, 225, 227

Modularization, 46

Multipath Routing, 166, 168-169

Multisplitting, 112, 119

N

Nescient Activity, 24-25, 28-29, 31-32

Network & Distributed Systems Security, 120

Non-stationary, 112, 114, 119

O

Online Communities, 62-63, 70, 221

Opportunistic Behavior, 62-64, 67

Outlier Detection, 86-87, 89-92, 96, 98, 101-104, 106-111

P

P2p Networks, 62-63, 65, 70, 166-169, 179-180

Partitioning Clustering Methods, 98

Peer-to-peer (p2p) File Sharing, 62

People Tagging, 5, 18

Performance Evaluation, 56, 75, 77, 85, 158-159

Personalisation, 213-214

Pessimistic Concurrency, 148-149

Physical Layer Security, 194-195, 211-212

Principle Of Least Privilege, 46, 60

Privacy Certification Authority, 127

Privacy Models, 139

Privacy Protection, 120, 147, 186, 192

Privacy-preserving, 1, 3-4, 120-121, 123-124, 126-127, 133, 136, 146, 192

Privacy-preserving Technologies, 120

Privilege Separation, 46, 61

Q

Quantized Precoding, 35

R

Real Time Information Resource, 222

Recommender Systems, 213-215, 218, 220-221, 224, 226, 228

Regression Models, 89

Relationship Control, 15-16, 18

Risk Prediction Framework, 181, 185-187

S

Satisfying Privacy Requirements, 17, 211

Secrecy Outage, 194-195, 197, 204

Secure Communication, 51, 173, 211

Security And Safety, 1

Self-adaptive Weighting Matrices, 112, 114, 116

Situational Normality, 64-66, 68-71

Smart Cards And Privacy, 120

Social Computing, 62-63, 65-66, 70, 72, 181, 183-184, 186-187, 190-193

Social Inference Protection Systems, 181-182, 189-190, 192

Social Software, 5-9, 13-15, 18-19, 22-23, 62-63

T

Traffic Modification, 166

Transmission Outage, 194, 197, 204

Two-hop Wireless Networks, 194-195

U

Uav Cyber-security, 75

Ubiquitous Computing, 22-23, 76, 181, 192

User Selection, 35-37, 39-40

User Selection Algorithm, 35-37, 40

V

Valuable Information Resource, 222

Volatile Social Relationships, 5

W

Web Metering, 3-4, 120-129, 131-134

Web Metering Operations, 124, 133

Printed in the USA
CPSIA information can be obtained
at www.ICGtesting.com
JSHW051432221024
72173JS00006B/1450